AF362056

Phylogenomic, Biogeographic, and Evolutionary Research Trends in Arachnology

Phylogenomic, Biogeographic, and Evolutionary Research Trends in Arachnology

Editor

Matjaž Kuntner

MDPI • Basel • Beijing • Wuhan • Barcelona • Belgrade • Manchester • Tokyo • Cluj • Tianjin

Editor
Matjaž Kuntner
National Institute of Biology
Slovenia

Editorial Office
MDPI
St. Alban-Anlage 66
4052 Basel, Switzerland

This is a reprint of articles from the Special Issue published online in the open access journal *Diversity* (ISSN 1424-2818) (available at: https://www.mdpi.com/journal/diversity/special_issues/Evo_Arachnology).

For citation purposes, cite each article independently as indicated on the article page online and as indicated below:

LastName, A.A.; LastName, B.B.; LastName, C.C. Article Title. *Journal Name* **Year**, *Volume Number*, Page Range.

ISBN 978-3-0365-4165-5 (Hbk)
ISBN 978-3-0365-4166-2 (PDF)

Cover image courtesy of Matjaž Kuntner

Contents

About the Editor

Matjaž Kuntner

Matjaž Kuntner, PhD, studies evolution, phylogenetics, biogeography and conservation, focusing on spider biology and covering topics from biodiversity and taxonomic discovery to morphological adaptation, sexual conflict, genomics and silk biology. A graduate from the George Washington University, USA, Kuntner is employed at the National Institute of Biology and the ZRC SAZU, both in Slovenia. He works globally, with adjunct academic titles at the Smithsonian, USA, at the University of Ljubljana, Slovenia, and at Hubei University, China

Preface to "Phylogenomic, Biogeographic, and Evolutionary Research Trends in Arachnology"

You are about to read a book dedicated to arachnids and arachnology, focusing on arachnid systematics, biogeography, and evolution. Arachnids represent a hyperdiverse, yet understudied, group of invertebrate animals. The diversity of spiders, scorpions, and harvestmen may be relatively well documented, while additional orders remain enigmatic. Arachnids range from sub-millimeter to dinner plate sizes and exhibit an astonishing array of morphologies and sexual biologies, making them excellent models in the study of the interplay of natural and sexual selection. Only a handful of arachnid genomes have so far been annotated, but genomic data are beginning to be utilized in phylogenetic analyses at species and higher taxonomic levels. In fact, systematics focusing on arachnids has been at the forefront of this discipline, with recent contributions uncovering the utility of transcriptomic and genomic data in deciphering the tree of life. Into this wealth of phylogenomic data, arachnologists routinely weave phenotypic and ecological variables for truly integrative evolutionary studies. Arachnids show diverse, and oftentimes clade-predictable, patterns in dispersal biology. As a consequence, some clades have become textbook examples of vicariant biogeography, while others, e.g., those traveling aerially by silken sails, maintain lively patterns of gene flow over continents. Together, arachnids can help us understand the Earth's biogeographic history as well as the evolution of complex biodiversity hotspots. Considering arachnid age and deep phylogenetic splits, their evolutionary landscape is uniquely diverse and calls for new original and synthetic research.

Matjaž Kuntner
Editor

 *diversity*

Editorial

Phylogenomic, Biogeographic, and Evolutionary Research Trends in Arachnology

Matjaž Kuntner [1,2,3,4]

[1] Department of Organisms and Ecosystems Research, National Institute of Biology, 1000 Ljubljana, Slovenia; matjaz.kuntner@nib.si
[2] Jovan Hadži Institute of Biology, Research Centre of the Slovenian Academy of the Sciences and Arts, 1000 Ljubljana, Slovenia
[3] Department of Entomology, National Museum of Natural History, Smithsonian Institution, Washington, DC 20560, USA
[4] Centre for Behavioural Ecology and Evolution, College of Life Sciences, Hubei University, Wuhan 430011, China

1. Introduction

Textbook knowledge tells us that arachnids are a hyper diverse clade of chelicerates that have taken on terrestrial lifestyles. Original papers published in prestigious venues routinely reconstruct details of this purported single terrestrialization event that would have been followed by arachnid diversification on land. However, we are beginning to understand that arachnids are very likely paraphyletic; as such, Arachnida can only circumscribe an assemblage of chelicerates that live terrestrially. If so, arachnid terrestrialization may have taken several independent routes at different historic times. While the diversity and phylogeny of spiders, scorpions and harvestmen may be relatively well documented and understood, additional groups that we deem to be arachnids remain enigmatic and will likely continue to be more or less neglected after this Special Issue. We have here assembled examples of contemporary studies that include both original research as well as reviews focusing on "arachnids" and cover loosely defined biological subdisciplines of phylogenomics, biogeography, and evolution. The latter includes systematics, taxonomy, DNA barcoding, and trait evolution. In this editorial, I introduce the authors of these papers and their featured research, and through this narrative, I pose two questions. The first one is what is arachnology given that arachnids may not be monophyletic? The second question is where should our field be headed toward in the near future?

2. What Is Arachnology?

In a paper titled "What Is an "Arachnid"? Consensus, Consilience, and Confirmation Bias in the Phylogenetics of Chelicerata" [1], Prashant Sharma and colleagues review the systematics of the group we refer to as arachnids. They focus on the evidence for arachnid monophyly; it seems to be weak at best and seems to have been repeatedly confirmed through biased interpretations of hypotheses and the evidence in their support. By showing the fragility of phylogenies and the research bias of works that confirm rather than challenge classification hypotheses, as well as the paucity and deficiency of classical morphological characters, these authors question the standards and trends in the field.

Arachnologists such as myself have rarely doubted the validity of the classical arachnid orders, such as spiders, scorpions, harvestmen, and mites, and classical systematic literature would additionally suggest that these major groups share common ancestry with other terrestrial chelicerates that are known as arachnids. However, if the time has arrived to reassess our understanding on what an arachnid really is, as questioned by Sharma and colleagues, then by extension we need to ask ourselves this: What is arachnology and who is an arachnologist? To scientists who have considered themselves as arachnologists

Citation: Kuntner, M. Phylogenomic, Biogeographic, and Evolutionary Research Trends in Arachnology. *Diversity* **2022**, *14*, 347. https://doi.org/10.3390/d14050347

Received: 26 April 2022
Accepted: 27 April 2022
Published: 28 April 2022

Publisher's Note: MDPI stays neutral with regard to jurisdictional claims in published maps and institutional affiliations.

throughout their careers, this is a tough question indeed. A handy explanation, considering all evidence from the above review, is this. Having a basis in morphological and ecological definitions and perhaps defying a solid phylogenetic definition, arachnology refers to any biological investigation of the terrestrial (and secondarily aquatic) lineages of chelicerates, both extinct as well as extant. Arachnologists, by extension of this logic, study these organisms.

Even if arachnology unites students of a paraphyletic assemblage of evolutionary lineages, arachnologists will continue our quest in getting to know our organisms and their role in ecosystems. In this respect, arachnology resembles other thriving biological disciplines that study non-monophyletic groups, for example, ichthyology, herpetology, or microbiology. Even if these fields are defined as research of para- or polyphyletic groups of organisms, they nonetheless continue to unite practical societies and produce relevan science.

3. Phylogenomic and Evolutionary Research Trends in Arachnology

Taxonomy has been and remains the fundamental biological discipline that provides the token of biological communication as it defines and describes species and classifies them in the tree of life. Its importance notwithstanding, taxonomy has stagnated recently despite the availability of modern tools that the discipline should utilize. In the paper titled "Improving taxonomic practices and enhancing its extensibility—an example from araneology" [2], Jason Bond and colleagues review a decade of publications on spider taxonomy. They evaluate the types of data used to delineate species, whether data were made freely available, whether an explicit species hypothesis was stated, what types of media were used, the sample sizes, and the degree to which species constructs were integrative. The results they report are worrying, and they may be true for most invertebrate groups and not only spiders. Namely, the study concludes that taxonomy remains largely descriptive, not integrative, and provides no explicit conceptual framework. Bond et al. make four recommendations that would, if the taxonomic community implements them, enhance the rigor, repeatability, and scientific standards in taxonomy.

Systematics has seen tremendous leaps towards phylogenomic data capture as genomic sequencing in non-model organisms has become routine. However, given that whole genome or transcriptome capture is not always feasible and its costs are substantial, research groups have focused on developing protocols for reduced representation sequencing. Among these efforts, the most widely used approach is to focus on ultra-conserved elements (UCE). Indeed, arachnology has hopped on this train early on, and papers continue to demonstrate the effectiveness of modern phylogenomics using UCE. In a paper titled "In Silico Assessment of Probe-Capturing Strategies and Effectiveness in the Spider Sub-Lineage Araneoidea (Order: Araneae)" [3], Yi-Yen Li and colleagues report on development of a probe set specific for orb-weaving spiders, lineage Araneoidea. This research opens the doors for numerous studies that require araneoid UCE data at the species or higher levels.

Why are some clades hyper rich with species while other clades of similar ages are species poor? Many factors can be at play, including diversification and extinction tempos. In a paper titled "*Solenysa*, a Cretaceous Relict Spider Group in East Asia" [4], Jiahui Tian and colleagues explore the long evolutionary history of one of the major clades of linyphiids and the reasons for its relative poverty in species diversity compared with the other linyphiid lineages. The authors found that *Solenysa* diverged from other linyphiids in the Cretaceous and underwent diversification stasis well into Oligocene. They explained this stasis followed by modest diversification with the Cenozoic ecosystem transition triggered by global climate changes. Jiahui Tian and colleagues conclude that *Solenysa* is a Cretaceous relict that has survived mass extinction around the K-T boundary.

The next title refers to arachnid specimens, not to arachnologists, as one might incorrectly think. In a paper titled "Old Brains in Alcohol: The Usability of Legacy Collection Material to Study the Spider Neuroarchitecture" [5], Andres Rivera-Quiroz and Jeremy

Miller explore whether or not the central nervous system in spiders can be reconstructed with minimal invasion and from old museum specimens. It can. Using a minimally destructive method of specimen preparation for micro-CT investigation of ganglia on a range of specimens of varying ages, these authors found no significant differences in the brain shape nor brain relative volume. This is good news for students of soft internal anatomies who should go ahead and study important, rare, legacy specimens along with newly collected ones.

4. Biogeographic Research Trends in Arachnology

The chelicerate orders collectively referred to as arachnids show diverse, and oftentimes clade-predictable patterns in dispersal biology. As their consequence, some clades have become textbook examples of vicariant biogeography, e.g., trapdoor and liphistiid spiders, scorpions, and harvestmen. Other arachnids, such as spiders that balloon (that is, haphazardly travel aerially by silken sails), maintain lively patterns of gene flow over continents. In the best dispersing groups, such as long-jawed spiders (*Tetragnatha*) and the giant golden orb web spiders (*Nephila* and *Trichonephila*), uninterrupted gene flow can easily span intercontinentally and over thousands of kilometers. Biogeography of arachnids can help us better understand the history of the Earth's biotas and the evolution of complex biodiversity hotspots.

Indeed, biogeography and the history of hotspot formation feature prominently in this Special Issue. In a paper titled "Incorporating Topological and Age Uncertainty into Event-Based Biogeography of Sand Spiders Supports Paleo-Islands in Galapagos and Ancient Connections among Neotropical Dry Forests" [6], Ivan Magalhaes and colleagues present an elegant biogeographic study of sand spiders (Sicariidae: *Sicarius*) from Neotropical xeric biomes. This research found that *Sicarius* must have dispersed to the Galapagos Islands when the archipelago consisted of paleo-islands that are now submerged; thus, this colonization must have occurred before the emergence of modern Galapagos Islands. This paper is likely to advance the analytical methods applied in historical biogeography as it presents an approach for evaluating competing hypotheses given phylogenetic topological instability and vagueness in time split estimation.

In another biogeographical paper titled "A Natural Colonisation of Asia: Phylogenomic and Biogeographic History of Coin Spiders (Araneae: Nephilidae: *Herennia*)" [7], Eva Turk and colleagues report on a reconstructed evolutionary history of the nephilid spider genus *Herennia*. Known as coin spiders for their undulating abdominal shape, *Herennia* features numerous species that are narrow endemics in Southeast Asia and Australasia, as well as one widespread and common species, *H. multipuncta*. Based on a phylogenomic scaffold and an ultrametric phylogeny, these authors tested and discarded the hypothesis of a human mediated colonization of *H. multipuncta* in favor of its alternative, paraphrasing this as a natural "coinquest". This study further used an innovative biogeographic approach with dispersal probabilities depending on continental and island tectonic histories in appropriate time slices in Earth's past, as well as the natural history of the organisms. First proposed for nephilid spiders globally, Turk et al. [7] modified this novel approach here.

No fewer than three original papers dissected the biogeography of the Caribbean archipelago, one of the global biodiversity hotspots. In a paper titled "Single-Island Endemism despite Repeated Dispersal in Caribbean *Micrathena* (Araneae: Araneidae): An Updated Phylogeographic Analysis" [8], Lily Shapiro and colleagues reconstruct the biogeographic history of spiny orb weavers. *Micrathena*, according to these authors, colonized the archipelago on five occasions, but despite such efficiency at crossing the ocean barrier, which might be seen as facilitating continuous gene flow, the patterns of diversification on islands resulted in a pronounced single-island endemism in *Micrathena*. This study and the next one both failed to find corroborative evidence for the existence of a land bridge that may have connected the Greater Antilles with the American mainland—the GAARlandia scenario.

Also focusing on the Caribbean biogeographic history is the paper by Klemen Čandek and colleagues titled "Biogeography of Long-Jawed Spiders Reveals Multiple Colonization of the Caribbean" [9]. Čandek and colleagues provided a phylogenetic context for originally collected representatives of *Tetragnatha* spanning the Caribbean islands by adding numerous global terminals. The resulting chronogram and the reconstructed ancestral areas both revealed a pattern that contrasts the one from a spiny orb weaver; *Tetragnatha* instead showed low levels of island endemism despite its high species richness on the archipelago. These authors also attempted to test the predictions from the Intermediate dispersal model of biogeography, something that would require an a priori definition of three categories of dispersers. However, long-jawed spiders did not fit one of these three categories as the genus uniquely comprises both excellently and poorly dispersing species. Čandek et al. concluded that *Tetragnatha* represents a 'dynamic disperser', i.e., a taxon that readily undergoes evolutionary changes in dispersal propensity.

These papers do not yet exhaust the studies on Caribbean biogeography as reported in this Special Issue of *Diversity*. In a paper titled "Island-to-Island Vicariance, Founder-Events and within-Area Speciation: The Biogeographic History of the Antillattus Clade (Salticidae: Euophryini)" [10], Franklyn Cala-Riquelme and colleagues study the Antillattus clade of jumping spiders (genera *Antillattus*, *Truncattus*, and *Petemethis*) of the arhipelago. This study particularly tested the GAARlandia land bridge scenario to explain spider diversity of the Greater Antilles. In contrast to the above studies, Franklyn Cala-Riquelme and colleagues found GAARlandia as a credible explanation of the biogeographic patterns, with an inferred historic dispersal from northern South America to Hispaniola. Subsequently to that inferred event, jumping spiders show imprints of vicariance, founder-events, within-island speciation, as well as multiple dispersal events in parts of the phylogeny.

The Baja peninsula in Mexico is among the biogeographically understudied yet diverse areas of the New World. In a paper titled "New Distributional Records of *Phidippus* (Araneae: Salticidae) for Baja California and Mexico: An Integrative Approach" [11], Luis Hernández Salgado and colleagues report on a survey of *Phidippus* jumping spiders of Baja using DNA barcoding combined with morphology. They augment the species list of Baja to now comprise 10 *Phidippus* species with evidence of an undescribed one.

Moving south to the Guayana region of South America, a paper titled "Beta Diversity along an Elevational Gradient at the Pico Da Neblina (Brazil): Is Spider (Arachnida-Araneae) Community Composition Congruent with the Guayana Region Elevational Zonation?" [12] authored by André Nogueira and colleagues reports on a thorough sampling of spiders from a Brazilian mountain along an elevation gradient. These authors detected high beta diversity among the sites, but they found several unexpected patterns related to species abundances and dominance. Samplings of arachnids as intensive as the one reported in this paper are rare indeed, but they are critical to begin to understand geographical variation in species diversity.

5. Arachnology's Direction

If arachnology is the study of terrestrial chelicerate lineages, where is our field headed? More and more arachnid genomes are being annotated on a yearly basis, and genomic data are beginning to be utilized in phylogenetic analyses at the species and higher taxonomic levels. In fact, systematics focusing on several arachnid lineages has been at the forefront of this discipline, with recent contributions uncovering the utility of transcriptomic and genomic data in deciphering the tree of life and in testing evolutionary and biogeographic hypotheses and scenarios. Into this wealth of phylogenomic data, arachnologists routinely weave phenotypic and ecological variables for truly integrative evolutionary studies.

Nature has selected the evolution of certain traits and animal products that arachnids are renowned for. Take spider silk, for example, which represents nature's toughest biomaterial. Only recently have we found that silk proteins are many times as diverse as we understood only a decade ago. Genomic and transcriptomic analyses are helping us discover new and new genes that code for various types of silk, and proteomics and functional ecology of silk are emerging fields that may potentially revolutionize biotechnological efforts towards utilizing these amazing materials. Spider and scorpion venoms are another wealth of animal products worthy of precise biochemical and genomic scrutiny and call for medical applications. Finally, morphology is not going to retire any time soon. Spider orb weaver lineages, such as the giant wood spider (*Nephila*), widow spiders (*Latrodectus*), jewel spiders (*Gasteracantha*), and others, have reached, evolutionarily speaking, nature's greatest differences in male and female shape and size, and students of sexual size dimorphism regularly make these their model organisms.

In closing, let me call for even higher outputs and standards in arachnological research. Considering arachnid age and deep phylogenetic splits, their evolutionary landscape is uniquely diverse, and this calls for continuous original and synthetic research. Our Special Issue should serve as an invitation to arachnology for the new generation of biologists. Come equipped with specialized skills, join the existing labs, and create new ones; then, help us transform arachnology into modern science.

Funding: This research was funded by the Slovenian Research Agency, grant numbers P1-0255 and J1-1703.

Acknowledgments: I thank the authors of the 12 papers that we had the privilege to assemble in this Special Issue, and who are helping us define new directions in arachnology. I have had tremendous support from the editorial staff of *Diversity*: Thank you all, and particularly Phyllis He, Emma Li, and Caitlin Sheng. I furthermore thank Mark Harvey, Luc Legal, Martín J. Ramírez, and Michael Wink for taking over editorial roles for some of the papers. Numerous reviewers have helped with their voluntary work for the good of the field. This editorial benefited from comments by Matjaž Gregorič and Prashant Sharma.

Conflicts of Interest: The author declares no conflict of interest.

References

1. Sharma, P.P.; Ballesteros, J.A.; Santibáñez-López, C.E. What Is an "Arachnid"? Consensus, Consilience, and Confirmation Bias in the Phylogenetics of Chelicerata. *Diversity* **2021**, *13*, 568. [CrossRef]
2. Bond, J.E.; Godwin, R.L.; Colby, J.D.; Newton, L.G.; Zahnle, X.J.; Agnarsson, I.; Hamilton, C.A.; Kuntner, M. Improving Taxonomic Practices and Enhancing Its Extensibility—An Example from Araneology. *Diversity* **2021**, *14*, 5. [CrossRef]
3. Li, Y.-Y.; Tsai, J.-M.; Wu, C.-Y.; Chiu, Y.-F.; Li, H.-Y.; Warrit, N.; Wan, Y.-C.; Lin, Y.-P.; Cheng, R.-C.; Su, Y.-C. In Silico Assessment of Probe-Capturing Strategies and Effectiveness in the Spider Sub-Lineage Araneoidea (Order: Araneae). *Diversity* **2022**, *14*, 184. [CrossRef]
4. Tian, J.; Zhan, Y.; Shi, C.; Ono, H.; Tu, L. *Solenysa*, a Cretaceous Relict Spider Group in East Asia. *Diversity* **2022**, *14*, 120. [CrossRef]
5. Rivera-Quiroz, F.A.; Miller, J.A. Old Brains in Alcohol: The Usability of Legacy Collection Material to Study the Spider Neuroarchitecture. *Diversity* **2021**, *13*, 601. [CrossRef]
6. Magalhaes, I.L.F.; Santos, A.J.; Ramírez, M.J. Incorporating Topological and Age Uncertainty into Event-Based Biogeography of Sand Spiders Supports Paleo-Islands in Galapagos and Ancient Connections among Neotropical Dry Forests. *Diversity* **2021**, *13*, 418. [CrossRef]
7. Turk, E.; Bond, J.E.; Cheng, R.C.; Čandek, K.; Hamilton, C.A.; Gregorič, M.; Kralj-Fišer, S.; Kuntner, M. A Natural Colonisation of Asia: Phylogenomic and Biogeographic History of Coin Spiders (Araneae: Nephilidae: *Herennia*). *Diversity* **2021**, *13*, 515. [CrossRef]
8. Shapiro, L.; Binford, G.J.; Agnarsson, I. Single-Island Endemism despite Repeated Dispersal in Caribbean *Micrathena* (Araneae: Araneidae): An Updated Phylogeographic Analysis. *Diversity* **2022**, *14*, 128. [CrossRef]
9. Čandek, K.; Agnarsson, I.; Binford, G.J.; Kuntner, M. Biogeography of Long-Jawed Spiders Reveals Multiple Colonization of the Caribbean. *Diversity* **2021**, *13*, 622. [CrossRef]
10. Cala-Riquelme, F.; Wiencek, P.; Florez-Daza, E.; Binford, G.J.; Agnarsson, I. Island-to-Island Vicariance, Founder-Events and within-Area Speciation: The Biogeographic History of the Antillattus Clade (Salticidae: Euophryini). *Diversity* **2022**, *14*, 224. [CrossRef]
11. Hernández Salgado, L.C.; Guerrero Fuentes, D.R.; Garduño Villaseñor, L.A.; Castañeda Betancur, L.; López Reyes, E.; Ceccarelli, F.S. New Distributional Records of *Phidippus* (Araneae: Salticidae) for Baja California and Mexico: An Integrative Approach. *Diversity* **2022**, *14*, 159. [CrossRef]
12. Nogueira, A.A.; Brescovit, A.D.; Perbiche-Neves, G.; Venticinque, E.M. Beta Diversity along an Elevational Gradient at the Pico Da Neblina (Brazil): Is Spider (Arachnida-Araneae) Community Composition Congruent with the Guayana Region Elevational Zonation? *Diversity* **2021**, *13*, 620. [CrossRef]

diversity

MDPI

Review

What Is an "Arachnid"? Consensus, Consilience, and Confirmation Bias in the Phylogenetics of Chelicerata

Prashant P. Sharma [1,*], Jesús A. Ballesteros [2] and Carlos E. Santibáñez-López [3]

[1] Department of Integrative Biology, University of Wisconsin–Madison, Madison, WI 53706, USA
[2] Department of Biology, Kean University, Union, NJ 07083, USA; jeballes@kean.edu
[3] Department of Biology, Western Connecticut State University, Danbury, CT 06810, USA; santibanezlopezc@wcsu.edu
* Correspondence: prashant.sharma@wisc.edu

Abstract: The basal phylogeny of Chelicerata is one of the opaquest parts of the animal Tree of Life, defying resolution despite application of thousands of loci and millions of sites. At the forefront of the debate over chelicerate relationships is the monophyly of Arachnida, which has been refuted by most analyses of molecular sequence data. A number of phylogenomic datasets have suggested that Xiphosura (horseshoe crabs) are derived arachnids, refuting the traditional understanding of arachnid monophyly. This result is regarded as controversial, not least by paleontologists and morphologists, due to the widespread perception that arachnid monophyly is unambiguously supported by morphological data. Moreover, some molecular datasets have been able to recover arachnid monophyly, galvanizing the belief that any result that challenges arachnid monophyly is artefactual. Here, we explore the problems of distinguishing phylogenetic signal from noise through a series of in silico experiments, focusing on datasets that have recently supported arachnid monophyly. We assess the claim that filtering by saturation rate is a valid criterion for recovering Arachnida. We demonstrate that neither saturation rate, nor the ability to assemble a molecular phylogenetic dataset supporting a given outcome with maximal nodal support, is a guarantor of phylogenetic accuracy. Separately, we review empirical morphological phylogenetic datasets to examine characters supporting Arachnida and the downstream implication of a single colonization of terrestrial habitats. We show that morphological support of arachnid monophyly is contingent upon a small number of ambiguous or incorrectly coded characters, most of these tautologically linked to adaptation to terrestrial habitats.

Keywords: Arthropoda; circular reasoning; investigator bias; paleontology; phylogenomics

Citation: Sharma, P.P.; Ballesteros, J.A.; Santibáñez-López, C.E. What Is an "Arachnid"? Consensus, Consilience, and Confirmation Bias in the Phylogenetics of Chelicerata. *Diversity* **2021**, *13*, 568. https://doi.org/10.3390/d13110568

Academic Editors: Matjaž Kuntner and Eric Buffetaut

Received: 3 October 2021
Accepted: 3 November 2021
Published: 6 November 2021

Publisher's Note: MDPI stays neutral with regard to jurisdictional claims in published maps and institutional affiliations.

1. A Lesson from the Toucans

Around the time that we were Ph.D. students, one of the most memorable parables in science was passed down to us from Scott Edwards, a renowned ornithologist at the Museum of Comparative Zoology at Harvard University. At the very dawn of molecular phylogenetics in 1985, Scott, an undergraduate and aspiring junior scientist, ran into a colleague of his in the hallway of the Museum. His colleague, another student, proudly relayed, "I have found a synapomorphy for toucans!". The synapomorphy he was referring to pertained to a morphological character that united this particular group of birds, as evidence of their evolutionary relationship. Scott, to this day one of the most affable and collegial of evolutionary biologists, of course congratulated his fellow student. But he then mused: Did you encounter this synapomorphy because toucans are a natural group, *or because you went looking for it?*

Scott's coda addressed a broader question for phylogeneticists and scientists at large: How do we as natural historians distinguish observations of natural phenomena from investigator bias? That story, relayed to us 14 years ago at the time of this writing, deeply influenced the way we approach phylogenetic inquiry. It underscored the importance of doubt in our outlook on the meaning of consilience in science, as well as how different phylogenetic data classes and competing topologies should be explored and evaluated.

That lesson bears heavily upon the phylogeny of Chelicerata, the subdivision of arthropods that includes groups like spiders, scorpions, and horseshoe crabs. In this review, we assess recent ideas and hypotheses pertaining to chelicerate relationships, with emphasis on the question of arachnid monophyly. We specifically scrutinize the practice of preferring only the topologies that are consistent with traditional, morphology-based relationships.

2. A Brief History of a Gordian Knot in Metazoan Phylogeny

The traditional understanding of chelicerate relationships is rooted in morphological data, which divide extant Chelicerata into three groups: Pycnogonida (sea spiders), Xiphosura (horseshoe crabs), and Arachnida (an assemblage of 12 terrestrial orders) [1–3]. Implicit in this topology is the hypothesis of a single colonization of land by the common ancestor of Arachnida. Within Euchelicerata (Xiphosura + Arachnida), extinct marine groups like Synziphosurina, Eurypterida, and Chasmataspidida (collectively thought to constitute a paraphyletic assemblage called "Merostomata") are thought to form a grade subtending Arachnida, reflecting a stepping-stone to the colonization of land by an aquatic ancestor [4–6]. Based upon an array of morphological characters, as well as their early appearance in Silurian and Devonian deposits, scorpions are generally reconstructed as the earliest (or one of the earliest) branching groups of arachnids by an array of historical and recent paleontological analyses [5,7–9] (Figure 1).

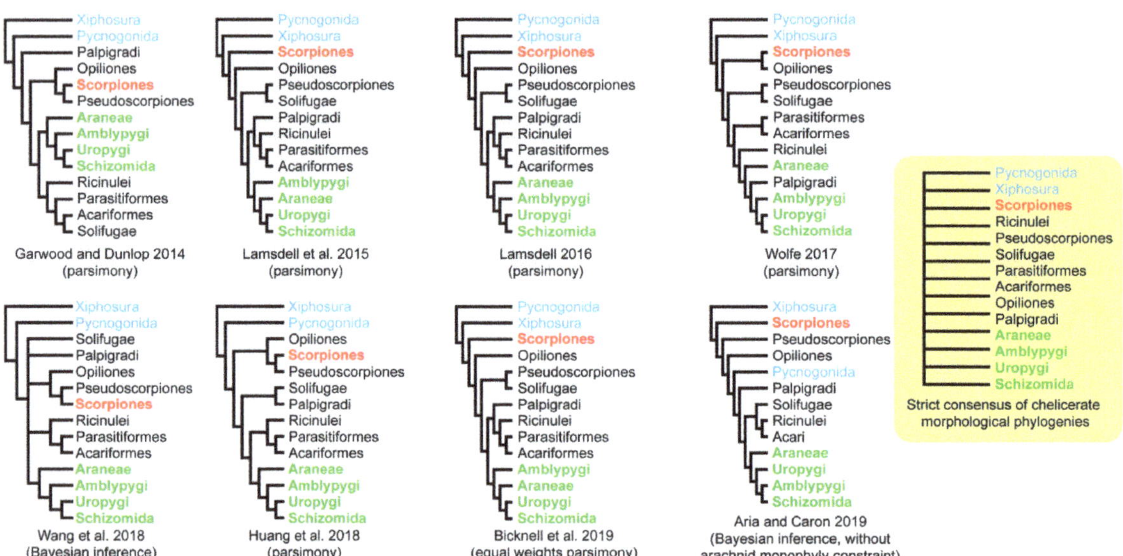

Figure 1. Phylogenetic relationships of chelicerate orders based on recent analyses of morphological datasets. Fossil taxa have been removed for clarity. Blue: marine orders; green: Tetrapulmonata; red: Scorpiones.

While the monophyly of Chelicerata and Euchelicerata (Xiphosura + Arachnida) has received broad support from analyses of morphological and molecular datasets, relationships within the arachnids remain unresolved, with the exception of a few key nodes (e.g., Arachnopulmonata sensu Ontano et al. [10]; Panscorpiones; Tetrapulmonata; Figure 1), which have largely been elucidated through phylogenomic datasets and rare genomic changes like whole genome duplication events. The fulcrum of the disagreement between morphological and molecular datasets comprises the monophyly of Arachnida itself. While morphologists have typically considered arachnids to constitute a strongly supported monophyletic group, molecular sequence data have consistently struggled to recover this relationship (in some cases, even upon the inclusion of morphological data), since the first molecular phylogenetic datasets for Chelicerata were generated [11,12]. Sanger-sequenced datasets and analyses of mitochondrial genomes in particular could not recover arachnid monophyly, despite various approaches to phylogenetic analysis and inference [13–15]. This result was often dismissed on grounds of a wide array of putative artifacts, such as the aberrant morphology of the sea spiders, long branch attraction, and model misspecification. Lending credence to this argument was the clear evidence for heterogeneity in branch lengths across some arachnid taxa, with Acariformes, the non-opilioacariform Parasitiformes, and Pseudoscorpiones in particular exhibiting accelerated evolutionary rates that incurred the threat of long branch attraction artifacts [15–17].

The advent of genome-scale datasets facilitated a more precise examination of the source of this discordance. A 62-gene analysis of Regier et al. [16] was able to recover arachnid monophyly in only two of four analyses, and with limited support (68% and 80% bootstrap support). The authors of the work interpreted this to mean that Arachnida was "strongly recovered", though other relationships of comparable depth were recovered more consistently and with higher support values (typically, maximal nodal support values across all four analyses). Subsequently, a 3644-gene analysis of Sharma et al. [17] discovered a faint signal supporting arachnid monophyly in slowly evolving genes. It was shown that the 500- and 600-slowest evolving genes in the matrix, when concatenated, could recover the monophyly of Arachnida with maximal nodal support, though this signal vanished upon addition of faster evolving genes.

This study was often mistakenly interpreted to mean that slowly evolving genes were more accurate than faster-evolving genes in the face of long branch attraction, reflecting the underlying assumption that arachnids must be monophyletic. However, this conclusion is contradicted by the existence of localized peaks of nodal support—comparable to that of Arachnida—for mutually exclusive hypotheses also supported by slowly-evolving genes (e.g., the sister group relationships of pseudoscorpions to either scorpions or Acari exhibit localized peaks of support in rate-subsampled matrices [17] (Figure 2a). Therefore, the ability to find local peaks of nodal support within an ordered subsampling of genes when long branch attraction is incident cannot be a guarantor of phylogenetic accuracy. Indeed, subsequent works based on phylogenetically informed orthology criteria for gene selection could not replicate this signal for Arachnida (Figure 2b). Sharma et al. [17] articulated the concern that searching for arachnid monophyly may represent an idiosyncratic goal and that the traditional interpretation of a single terrestrialization event was undermined by the robust and derived placement of scorpions as the sister group of the tetrapulmonates. This group was named Arachnopulmonata [17], a clade of the only extant arachnid orders that bear book lungs.

The significance of the scorpion placement is twofold. First, paleontologists had previously inferred several Paleozoic stem-group scorpion fossils to be marine or at least aquatic, based on the fossil assemblages where they were discovered and the existence of putative gills in the mesosoma of these extinct groups. Under this scenario, a derived placement of scorpions would imply multiple terrestrialization events, or a return to aquatic habitat in the branch subtending the modern scorpions (more recently, given the proliferation of support for Arachnopulmonata, paleontologists have subsequently revised this interpretation to suggest that all Paleozoic scorpions must have been terrestrial, the presence of gilled fossils like *Waeringoscorpio* notwithstanding; see Howard et al. [18]). Second, the derived placement of scorpions proximal to tetrapulmonates suggested a derived origin of the book lung—a respiratory organ thought to represent the internalized counterpart of the horseshoe crab book gill [19]. The robust recovery of Arachnopulmonata in phylogenomic datasets was further substantiated by a series of rare genomic changes stemming from the discovery of a shared whole genome duplication event in the common ancestor of the arachnopulmonates [20–24]. Paralleling the history of the waves of genome duplication in the vertebrates, genomes of Arachnopulmonata were shown to exhibit two Hox clusters [23], broad retention and transcriptional activity of duplicated paralogs of developmental patterning genes [22], enrichment of specific microRNA families [21], and arachnopulmonate-specific divergence of gene expression patterns of duplicated paralogs [25–27]. Additional datasets from recently established tetrapulmonate model systems (e.g., tarantula; whip spiders [28–31]), as well as non-arachnopulmonate outgroups (e.g., mite; tick; harvestman; [32–34]) further buttressed this inference, and additionally revealed that the fast-evolving pseudoscorpions definitively constitute a member of Arachnopulmonata [10]. Intriguingly, extant Xiphosura exhibit a shared two- or possibly three-fold genome duplication on the branch subtending the four living species, but these events are unrelated to the arachnopulmonate duplication [35–38].

Due to the systemic nature of whole genome duplication events, the weight of evidence was decidedly in favor of phylogenomic results and contrary to morphological analyses (Figure 2c). This is because the incidence of shared duplications results in specific and replicated gene tree topologies, as well as divergent gene expression patterns ensuing paralogs, across hundreds of retained duplicates; explaining these genomic phenomena as the result of independent events becomes implausibly non-parsimonious. As a result, in the span of a few years, decades-old hypotheses based on morphology—such as scorpions constituting the sister group of harvestmen [2,39] or the remaining arachnids [1,40], or pseudoscorpions the sister group of solifuges [1,2]—were refuted, now considered the likely consequences of morphological convergence and ensuing misinterpretations of similar feeding structures and shared arrangements of respiratory organs (e.g., two-segmented chelate chelicerae and tracheal arrangement in pseudoscorpions and solifuges; anatomy of the book lungs and book gills of scorpions and merostomates, respectively; preoral chambers formed from gnathobases in scorpions and harvestmen [2,19,39,40]).

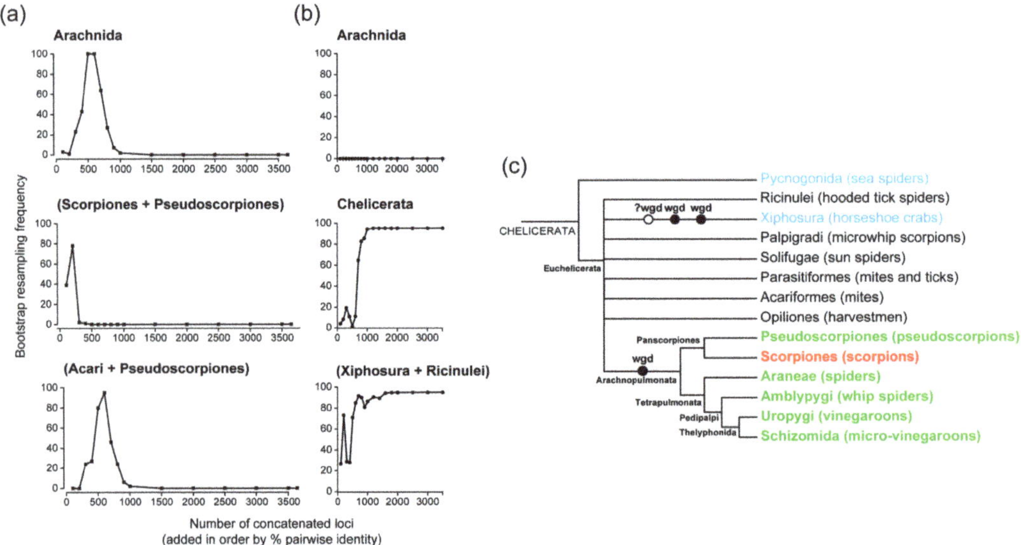

Figure 2. The ability to find localized peaks of support for nodes in slowly evolving genes is not a guarantor of phylogenetic accuracy. (**a**) Selected nodal support trajectories from Sharma et al. [17]. The bottom pair of nodes are mutually exclusive, and the bottom-most node (Acari + Pseudoscorpiones) has been falsified by rare genomic changes [10]. (**b**) Selected nodal support trajectories from Ballesteros and Sharma [41]; no support was recovered for Arachnida even among slowly evolving genes. The support trajectory for a nested placement of Xiphosura (bottom) is closely comparable to that of a robustly supported node (Chelicerata; middle). (**c**) Simplified phylogeny of Chelicerata showing inferred locations of whole genome duplications (WGD). Note that no morphological phylogeny has ever recovered this composition of Arachnopulmonata. Colors in (**c**) follow Figure 1.

3. The Debate over Arachnid Monophyly

A more definitive debate about arachnid monophyly in molecular datasets was addressed in the works of Ballesteros and Sharma [41] and Lozano-Fernández et al. [42]. The former work came to the conclusion that there was no support for arachnid monophyly. Ballesteros and Sharma [41] implemented analyses of gene-wise log-likelihood score (ΔGLS) to dissect phylogenetic signal; simulations of incomplete lineage sorting to assess explanatory vehicles for arachnid non-monophyly; and multiple approaches to species tree inference. Their orthology inference approach made use of phylogenetically unrooted topologies, which has been shown to outperform BLAST-based or distance-based calculations of orthologs [43]. Of the 106 analyses performed, the only analysis that could recover Arachnida was one where ΔGLS was used to identify and concatenate a minority of genes (33.6%) that supported this relationship. This minority of genes showed no evidence of being "better" or more accurate than the majority, with respect to an analysis of 70 metrics for systematic bias, such as saturation, evolutionary rate, or missing data (Figure 3 of Ballesteros and Sharma [41]). They were also able to show that the nested placement of Xiphosura was not attributable to long branch attraction; upon removing all long orders from the analysis in taxon deletion experiments, they still found Xiphosura as derived within the arachnids (Figure 5 of Ballesteros and Sharma [41]). Moreover, they were able to show that detection and isolation of genes supporting specific relationships with ΔGLS could be used to recover nonsensical relationships, such as a grouping of chelicerates with Pancrustacea, with maximal nodal support (Figure 8 of Ballesteros and Sharma [41]). This thought experiment served to reinforce that the ability to find matrices supporting specific relationships with maximal support is not synonymous with phylogenetic accuracy, as cherry-picking genes that support preconceived hypotheses is a circular exercise.

These lessons were ignored by the subsequent work of Lozano-Fernández et al. [42] who analyzed chelicerate phylogeny with a larger sampling of taxa. This work analyzed three matrices using maximum likelihood and Bayesian inference approaches. One of these (Matrix A) was composed using preselected genes, whose origins and basis for selection are not reproducible (ref. [44]). The second (Matrix B) was the result of a distance-based algorithmic approach to orthology inference (OMA; [45]). The third (Matrix C) was the result of uniting the first two matrices, after excluding duplicates. They found that one of their three matrices (Matrix A) was able to recover arachnid monophyly, though only after the removal of six taxa from the analysis, and only using the site heterogeneous CAT+GTR+Γ model implemented by PhyloBayes-mpi [46], a computationally intensive Bayesian inference approach. The same approach, when applied to their other matrices, failed to recover arachnid monophyly. To understand why, the authors examined saturation plots of each matrix (a metric for multiple substitutions at the same sites; a correlate of evolutionary rate) and observed that Matrix A had a slightly better value for saturation (a slope of 0.38) in comparison to the other two matrices (0.33 for both Matrices B and C). Lozano-Fernández et al. [42] reached the conclusion that concomitant application of denser taxonomic sampling, the use of site heterogeneous models, and filtering for unsaturated genes could recover Arachnida (as well as Acari), and therefore, phylogenetic accuracy. The ability to recover arachnid and acarine monophyly in one out of seven analyses was touted as an example of "consilience" in phylogenetics.

Notably, both of these previous studies [41,42] were missing a handful of key lineages—namely, the miniaturized arachnid orders Palpigradi and Schizomida, and the rare, slowly evolving parasitiform lineage Opilioacariformes [15,47]. A subsequent investigation of chelicerate phylogeny by Ballesteros et al. [44] was able to include these taxa for the first time in a combined phylogenomic framework. This work was similarly unable to recover arachnid monophyly, despite the application of site heterogeneous models developed for maximum likelihood frameworks, as well as through the use of PhyloBayes-mpi. To understand why, Ballesteros et al. [44] added these three taxa to Matrices A and B of Lozano-Fernández et al. [42] and reran analyses using both maximum likelihood and

PhyloBayes-mpi. The results were the same—Ballesteros et al. [44] could not recover arachnid monophyly using this approach either.

In the course of those analyses, Ballesteros et al. [44] discovered a series of grave analytical and bioinformatic errors in the works of Lozano-Fernández et al. [42], which were detailed in a supplementary document (Supplementary Text S2 of Ballesteros et al. [44]). These included errors in the filtering of loci by taxonomic completeness; incorrect measurement of per-locus saturation; incorrect calculation of linear regressions for reporting saturation due to a well-known error in a previous version of the software Microsoft Excel; inconsistency in the definition of saturation across studies by the same research team (compare the calculation of slope in [42] versus [48]); and lack of convergence of results generated using PhyloBayes-mpi. While the Lozano-Fernández et al. research team attempted to address these issues in a follow-up study using a derivation of Matrix B (by Howard et al. [18]), this work was similarly shown to suffer from additional bioinformatic errors, with some genes appearing more than once in the principal Howard et al. matrix (ref. reanalyses by Ontano et al. [10]). In addition, Howard et al. [18] preselected another pair of historical datasets (two that were known to yield arachnid monophyly under certain substitution models) and analyzed all datasets using the CAT-Poisson model in PhyloBayes-mpi. Upon recovering support for arachnid monophyly in all applications of CAT-Poisson, they reached the conclusion that these results substantiated the accuracy of Arachnida.

This result is outright contradicted in at least one case (the 500-slowest evolving genes matrix of Sharma et al. [17]), due to (1) the demonstrable analytical superiority of the CAT-GTR+Γ [49] model (explicitly deemed the best model and the one preferred across works by Lozano et al. [42,48]) over the CAT-Poisson model, and (2) a previous analysis of this same matrix using CAT-GTR+Γ by Sharma et al. [17], which recovered arachnid paraphyly with maximal nodal support (Figure 7 of Sharma et al. [17]). Howard et al. [18] also ignored as part of their reanalyses any matrices built using slowly evolving genes that did not recover arachnid monophyly, and for which previous analyses using CAT-GTR+Γ also refuted arachnid monophyly [41,44].

Putting aside the error-prone phylogenetic analyses, flexible criteria for dataset diagnosis, and willingness to dismiss contradictory evidence from the literature, there are two major concerns with the conclusions drawn by Lozano-Fernández et al. [42] and Howard et al. [18].

First, the addition of just two to three terminals to the matrices of both studies (which achieves complete sampling of all extant chelicerate orders), as well as a key parasitiform lineage (Palpigradi, Schizomida, and Opilioacariformes, respectively), consistently results in the disruption of both arachnid and acarine monophyly with support, even when these matrices were reanalyzed using identical algorithmic approaches and substitution models [10,44]. These reanalyses suggest that the monophyly of Arachnida and Acari requires the exclusion of certain lineages that undermine these traditional groupings.

Second, the matrices that were able to recover these groups are rife with paralogs; 29% (68/233) of loci in the Lozano et al. [42] Matrix A and 41% (82/200) of loci in the Howard et al. [18] matrix were detected as including clear paralogs, using an annotation approach based on the *Drosophila melanogaster* proteome [50]. This discovery suggests that the monophyly of Arachnida may be an artifact reflecting noise and bioinformatic error, rather than phylogenetic signal.

Despite the flaws of the Lozano-Fernández et al. [42] and Howard et al. [18] studies, these works, and the attendant conclusion of arachnid monophyly, continue to receive widespread support among morphologists and paleontologists. Adherents of arachnid monophyly seize upon the existence of molecular matrices that can recover arachnid monophyly with maximal support (as well as the claim that these partitions are less saturated or somehow more accurate), citing these as either: (1) outright evidence for the accuracy of arachnid monophyly [42]; or (2) evidence that molecular data are at least partly congruent with arachnid monophyly (or are ambiguous on the matter), which is argued should trigger a deference to the traditional morphological understanding of chelicerate relationships [18]. The implicit argument of the latter claim is steeped in the assumption that morphology is a reliable arbiter of deep phylogenetic relationships in Chelicerata.

We put both of these claims to the test.

4. How to Get the Tree You Want without Really Trying

Lisa: "By your logic I could claim that this rock keeps tigers away."

Homer: "Oh, how does it work?"

Lisa: "It doesn't work."

Homer: "Uh-huh."

Lisa: "It's just a stupid rock."

Homer: "Uh-huh."

Lisa: "But I don't see any tigers around, do you?"

Homer: "Lisa, I want to buy your rock."

—*The Simpsons*, 1996

For phylogenomic datasets, it is generally understood that nodal support in the form of resampling techniques and posterior probabilities is not a reliable measure of the underlying signal in concatenation-based phylogenetic approaches. Maximal support values are commonly obtained in phylogenomic datasets above a certain size, but these may be attributable to amplification of noise, rather than signal. Exploration of datasets and dissection of systematic biases at the level of genes and sites is critical to understanding which nodes exhibit inter-partition conflict or lack of phylogenetic signal. Rigorous investigations of phylogenetic signal within supermatrices are especially crucial when long branch attraction artifacts are incident, as these tend to drive high nodal support values in cases of model misspecification or undersampling of fast-evolving lineages.

As for saturation, minimizing bias resulting from saturated sites not a controversial goal, but approaches to mitigating saturation are mixed in efficacy for Chelicerata. A partial solution is to restrict analyses to slowly evolving genes (as saturation is correlated with evolutionary rate; but see Figure 2b). Solutions based on the recoding of datasets (e.g., Dayhoff recoding) have retrieved uninformative results, due to the abrogation of signal at the base of Euchelicerata [42,44]. Site heterogeneous models are similarly variable as potential solutions for recovering Arachnida (discussed above). For this review, we used the same contextual definition of saturation (unusually, measured for a supermatrix, rather than individual genes) used by Lozano-Fernández et al. [42].

To explore the reliability of nodal support and saturation as metrics for phylogenetic accuracy, we undertook a series of thought experiments originally proposed by Ballesteros and Sharma [41], using as our source data Matrices A and B of Lozano-Fernández et al. [42]. We specifically explored the possibility of discovering partitions that support alternative groupings, with comparison of saturation rates for these matrices versus the original values of Matrices A and B.

Using the approach of Shen et al. [51], we dissected support for four debunked hypotheses of arthropod relationships: Cormogonida (the sister group relationship of Pycnogonida to the remaining arthropods); Atelocerata (Myriapoda + Hexapoda); Schizoramia (Crustacea + Chelicerata); and Myriochelata (Myriapoda + Chelicerata). To discover partitions supporting each grouping, we compared ΔSLS distributions for unconstrained trees versus trees constrained to recover each of these older groupings, computing both sets of maximum likelihood trees under site heterogeneous (PMSF) models, following approaches previously detailed by us [41]. We concatenated partitions supporting each grouping and inferred the resulting maximum likelihood trees using a site heterogeneous model. Nodal support was inferred using ultrafast bootstrap resampling. As shown in Figure 3, we were able to construct matrices from both Matrix A or Matrix B that could recover each of these groups, with nodal support (>95% bootstrap) for all four groupings in one or both matrices.

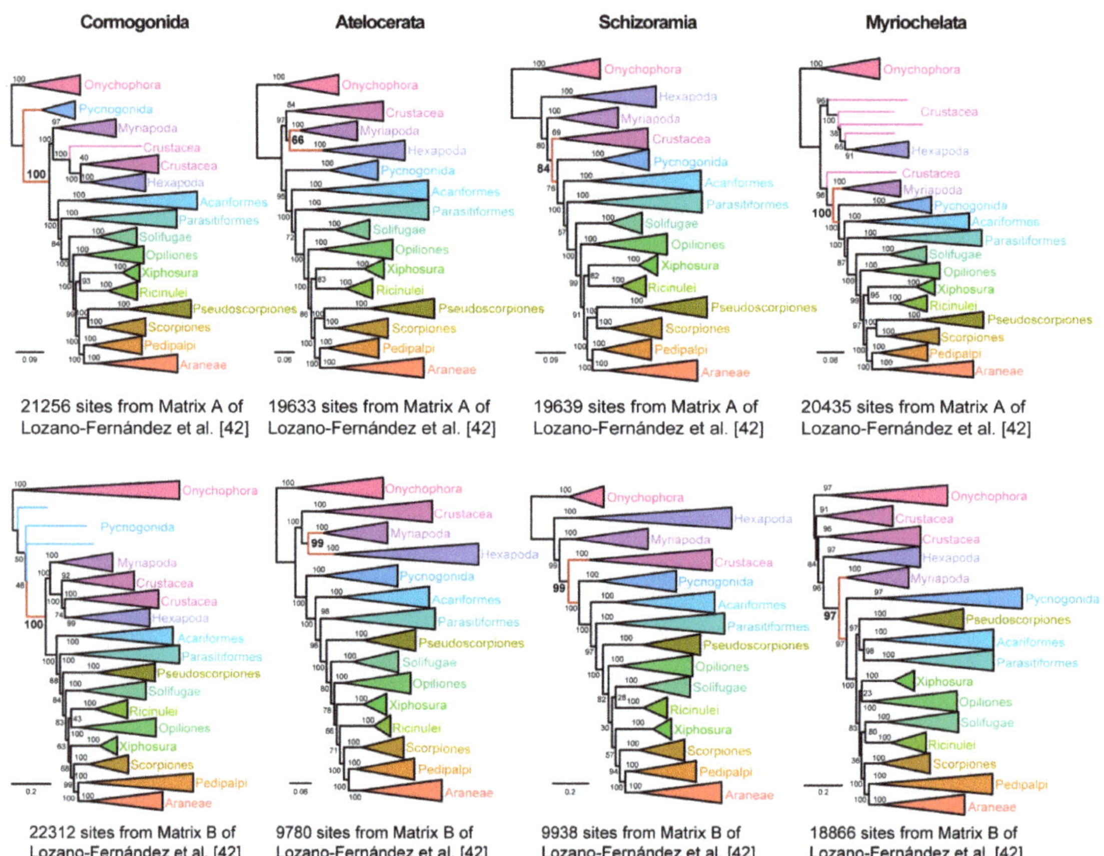

Figure 3. Cherry-picking molecular data to recover preconceived, traditional relationships can be used to justify debunked groupings. Top row: Topologies derived from Matrix A of Lozano-Fernández et al. [42]. Bottom row: Topologies derived from Matrix B of Lozano-Fernández et al. [42].

To those uninitiated in arthropod phylogeny, the ability to generate such data matrices may be construed as a lack of clear signal in basal arthropod relationships. Could such matrices imply a hidden signal for traditional groupings that were once supported by certain subsets of morphological characters? To dispel this notion, we proceeded to generate three additional matrices that supported completely nonsensical groupings: a clade of Xiphosura and Crustacea, to the exclusion of hexapods and other chelicerates (inspired by the whimsical notion of "making horseshoe crabs crabs again"); a clade of scorpions and spiders, to the exclusion of the other tetrapulmonates (the arachnids that most frighten people); and a clade of Pycnogonida + *Drosophila melanogaster*, to the exclusion of all other hexapods and chelicerates (taxa studied by Thomas Hunt Morgan; it is a little-known fact that the earliest works of the father of the first arthropod model organism addressed the development of sea spiders [52]). As shown in Figure 4, we were just as able to easily construct matrices from both Matrix A and Matrix B that could recover absurd groupings, with nodal support (>90% ultrafast bootstrap) in one or both analyses.

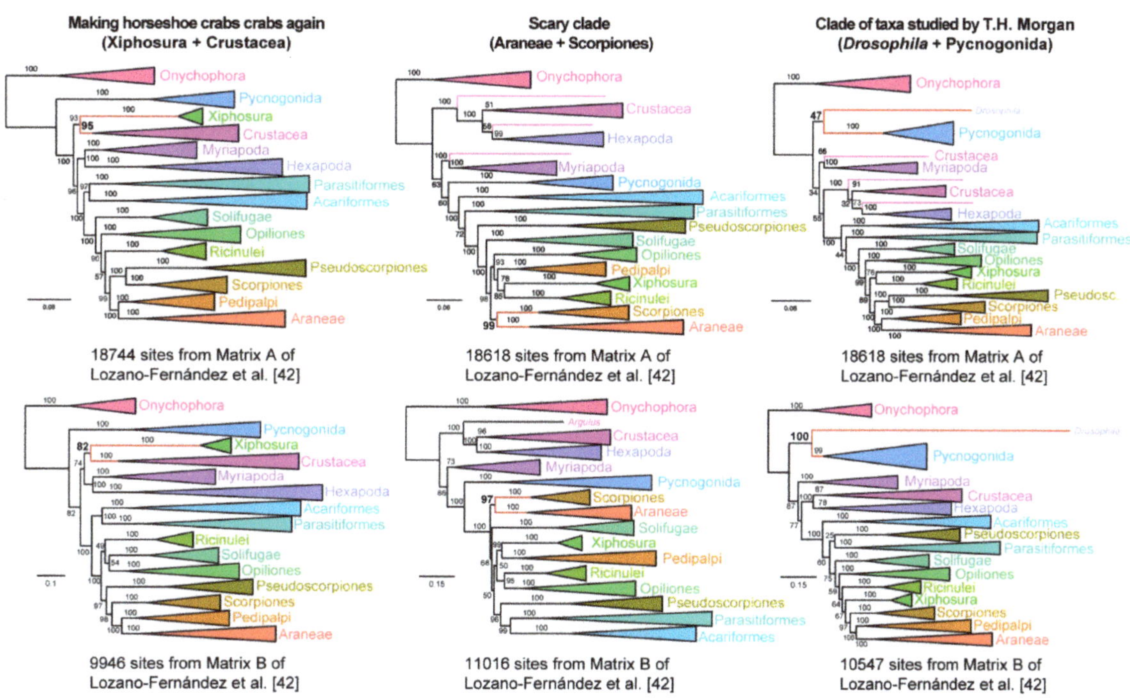

Figure 4. Taken to an extreme, cherry-picking molecular data can recover completely nonsensical relationships, often with support. Top row: Topologies derived from Matrix A of Lozano-Fernández et al. [42]. Bottom row: Topologies derived from Matrix B of Lozano-Fernández et al. [42].

Surely, one might think these matrices must compare poorly to actual phylogenetic matrices. We would expect such datasets to be smaller than unconstrained subsets or exhibit aberrant values for various measures of systematic bias, such as saturation. To test this, we generated values for saturation for every dataset generated from genes in Matrices A and B, to compare these to the original values reported in the study of Lozano-Fernández et al. [42]. Our approach to measuring saturation was identical to that implemented by Lozano-Fernández et al. [42], with the exception that we did not use an erroneous version of Excel to calculate coefficients of correlation (correctly measured, all R^{22} values exceeded 97%). As shown in Figure 5, every Matrix A-derived dataset recovering spurious relationships exhibited equal or better saturation values than Matrix A; and three Matrix B-derived datasets recovering absurd groupings outperformed Matrix B.

These analyses underscore that the ability to generate a matrix that can recover a preconceived result with maximal support does not equate with phylogenetic accuracy. They further reinforce the broadly understood principle that looking for a post hoc justification that validates a specific outcome in science (in the case of Lozano-Fernández et al., a metric for saturation, which they have defined inconsistently from one study to the next [42,48]) is the epitome of confirmation bias. In the specific case of saturation, it is well known that this metric is not a strong predictor of phylogenetic accuracy by itself, as elegantly demonstrated by Mongiardino Koch [53].

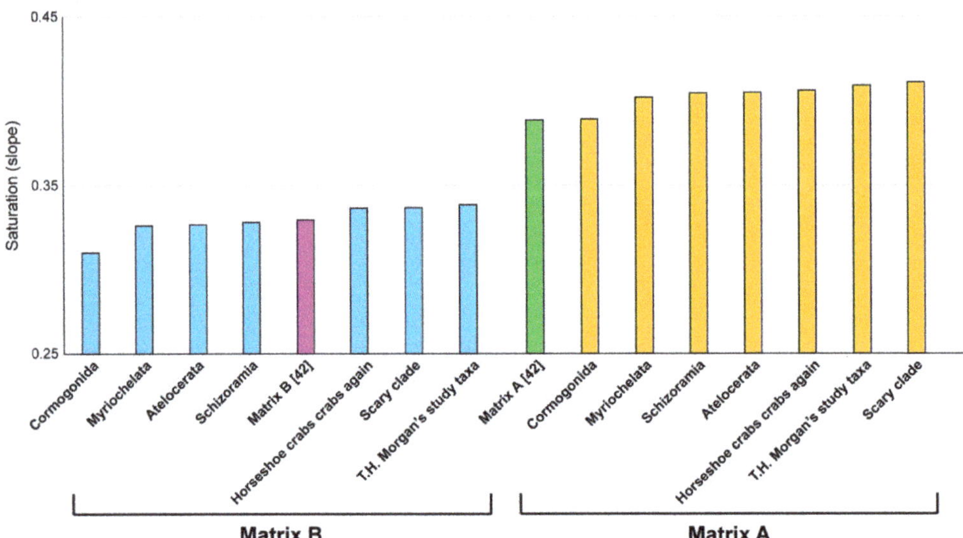

Figure 5. The saturation value of a supermatrix is not a valid proxy for phylogenetic accuracy (contra [42,48]). Slopes measured for matrices supporting debunked or absurd groupings can outperform Matrices A and B of Lozano-Fernández et al. [42], when drawn from the same underlying genes. (In other words, if you fell for this argument, as articulated by Lozano-Fernández et al. [42] in support of arachnid monophyly, you got fooled.).

This bears directly upon the epistemology of Lozano-Fernández et al. [42] and Howard et al. [18], whose interpretation of phylogenetic accuracy is predicated on the ability to recover a preconceived, traditional hypothesis (which is circular). Notably, Howard et al. [18] used the same approach as Sharma et al. [17] with respect to approach to orthology inference and the selection of slowly-evolving genes as the filtering criterion—they unwittingly evaluated the same set of orthogroups that had already been analyzed by the earlier work, but cast their 200-locus dataset as independent verification; in reality, the overlap between the gene sets is so extensive that the Howard et al. [18] 200-locus matrix cannot be treated as an independent analysis, as much as a recapitulation of a published result. Moreover, neither Lozano-Fernández et al. [42] nor Howard et al. [18] ever directly addressed the observation that other workers were unable to recover arachnid monophyly when restricting analyses to slowly-evolving or less saturated genes in other datasets, nor using site heterogeneous models (including CAT-GTR+Γ in PhyloBayes-mpi) on such filtered datasets [41,44]—the silver bullets they proposed for consistent recovery of arachnid monophyly across datasets do not work. It is particularly damning that expanding taxonomic sampling to sample all extant chelicerate orders in their datasets, and thereafter reanalyzing those datasets using identical methods, only serves to destabilize this node, with support [10,44]. This rejection of evidence that contradicts a preconceived or preferred hypothesis, as well as the unwillingness to test a broader range of datasets more rigorously and dispassionately, may reflect a calcified bias by adherents of "molecular paleobiology" in support of traditional, morphology-based relationships.

Parenthetically, "molecular paleobiology" refers to a loosely defined school of systematics that aims to reconcile paleontological and molecular phylogenies, ostensibly placing priority on the morphological and paleontological data partitions (as exemplified by the works in question [18,42]). Note that here we distinguish between molecular paleontology in the strict sense (e.g., the study of fossilized biomarkers; sequencing of ancient DNA) and "molecular paleobiology" in the sense of a specific phylogenetic research program (hence, we use quotation marks here to distinguish this approach to phylogenetics). "Molecular paleobiologists" champion the integration of morphological and molecular datasets as the best approach to phylogenetics. Most of the activities of this school are no different from what molecular phylogeneticists simply call molecular dating, which is a widely implemented technique. Some of the recurring controversies engendered by "molecular paleobiologists" have stemmed from overconfidence in the completeness of the fossil record, overconfidence in morphological phylogenies (or specific, arbitrarily selected morphological phylogenies, when competing hypotheses are available), overvaluing morphological and paleontological evidence in the assessment of competing molecular hypotheses, overinterpreting paleontological and morphological data in establishing calibrations in molecular dating; outright misconstruing or misrepresenting the morphological literature to accord with a preferred molecular topology; and consistently ignoring the repeated observation by other research groups that site heterogeneous models (as well as other proposed silver bullets to achieving morphological trees with molecular data) do not actually achieve the desired result across datasets [10,44,54–59]. However, given that morphological hypotheses of relationships predate molecular counterparts by decades or sometimes centuries, "molecular paleobiological" results are often unguardedly accepted by the broader community for their palatability, specifically by those who do not examine the underlying phylogenomic data and analyses or lack the expertise to do so.

Returning to the matter of chelicerate relationships, the most recent work addressing these issues comprehensively sampled all extant chelicerate orders with a 506-taxon phylotranscriptomic study, in tandem with site heterogeneous models (including with PhyloBayes-mpi implementation) and investigations of the claim that genes supporting arachnid monophyly were less biased or "better" at inferring deep relationships. Ballesteros et al. [50] showed that rich sampling of Chelicerata only further undermined support for arachnid and acarine monophyly. Filtering for less saturated genes and use of site heterogeneous models (PMSF and CAT-GTR+Γ models) not only refuted arachnid monophyly, but also revealed that the monophyly of Acari was a long branch artifact, driven by the exclusion of slowly-evolving parasitiform taxa like Opilioacariformes in the analyses of Lozano-Fernández et al. [42] and Howard et al. [18] (see also reanalyses by Ballesteros et al. [44] and Ontano et al. [10] on the effect of including Opilioacariformes to their matrices). Ballesteros et al. [50] also dismantled the unsubstantiated notion that genes supporting the nested placement of Xiphosura exhibited artifacts; to the contrary, they were able to show that genes supporting arachnid monophyly, which were consistently in the minority across datasets, tended to be short and bear few parsimony-informative sites. Short genes with few informative sites have been closely linked to systematic artifacts and poor phylogenetic signal. Apropos, upon examining the distribution of signal across sites, they showed that sites supporting arachnid monophyly exhibited high levels of Shannon entropy, reflecting noise rather than signal. Tellingly, the number of sites supporting debunked, artificial groupings (e.g., Dromopoda, an erstwhile grouping of scorpions, pseudoscorpions, solifuges, and harvestmen, which has been refuted by rare genomic changes) exceeded the number of sites supporting Arachnida.

Simply stated, unbiased analyses of molecular data do not support arachnid monophyly. The ability to contrive matrices that can do so through cherry-picking of genes and taxa, as well as finding post hoc justifications for said matrices, is the modern equivalent of hunting for synapomorphies for preconceived groups. We submit that cherry-picking datasets and trees should be treated with the same level of skepticism as the obsolete practice of single-character systematics. In the best light, preferring only those trees that confirm morphology-based hypotheses of phylogeny reflects a form of confirmation bias that stems from a naïve misunderstanding of how signal and noise are distributed in molecular data at the genomic scale. In the worst light, this practice represents a gateway to pseudoscience.

5. Morphology in the Era of Phylogenomics

"He rain-made you. A guy says if you pay him, he can make it rain. You pay him. If and when it rains, he takes the credit. If and when it doesn't, he finds reasons for you to pay him more."

—Maurice Levy, *The Wire*, 2004

"The morphological data show ... very few nodes ... have significant resampling support and the strict consensus is poorly resolved. This is unquestionably a limitation of the taxon to character ratio used in the present study ... The character sample could potentially be bolstered studies [*sic*] on other character systems ... "

—Ganske et al., 2021 [60]

Before we address morphological support (or the lack thereof) for Arachnida, we consider here the broader role and value of morphological datasets in modern phylogenetics, with a critical eye toward the future of morphology in a phylogenomic era.

Over the course of the past 30 years, morphological data have declined in relevance as data sources for phylogenetics, and a paucity of research groups continue to examine both morphological and molecular datasets toward the goal of empirical systematics (e.g., [15,60–66]). Part of the reason for this, as diplomatically articulated by Giribet [67], is tied to declining costs of sequencing, as well as the difficulty of articulating clear homology statements for problematic morphological character systems. As a fuller (and blunter) answer, the decline of morphology's prominence in modern phylogenetics has just as much to do with the multifaceted superiority of molecular data as predictors of phylogeny. While these concepts below are fairly well known (if tacitly acknowledged) by the broader community of evolutionary biologists, we posit them here as a prelude to the debate over arachnid monophyly.

Firstly, molecular data present a universal character system. Molecular matrices contain an alphabet that is universally applicable to all cellular life, and thus any researcher can take a molecular matrix and work with it, for any group of organisms, with no barriers or steep learning curves for understanding the underlying data. Genomic sequencing can be performed with standardized approaches that are broadly transferable to all taxonomic groups. By contrast, understanding, interpreting, and analyzing morphological character systems requires familiarity with the morphological characters and states that underlie the matrix. This means that the data contained in a given morphological matrix can only be understood by taxonomic experts of that group (in some cases, a mere handful of individuals), barring a steep learning curve for novices to that taxon. At its core, a morphological character matrix is the product of one researcher (or team) coding subjective interpretations, often with little understanding of character dependencies or underlying costs of state transformations. For this reason, two different morphologists can (and routinely do) come up with markedly different interpretations, matrices, and character states for the same taxa. This makes the detection and correction of coding errors in historical morphological matrices more difficult (discussed below) and hinders the integration of different sets of morphological matrices with non-overlapping characters or character states.

Second, molecular sequences present the desirable quality of scalability. Molecular matrices greatly exceed morphological datasets in size and variance of evolutionary rates, particularly with respect to different parts of a genome and nucleotide versus peptide sequences. This means that molecular data can just as easily inform population genetics and microevolutionary processes, as the deep phylogenetic relationships and the origins of major taxonomic groups. By contrast, the informativeness of morphological data occupies a middle ground between the two extremes. Morphological characters do not evolve rapidly enough to inform population-level processes, which makes them irrelevant for such disciplines as population genetics and epidemiology (there are several good reasons why you do not see epidemiologists tracking the spread of COVID-19 using analyses of viral morphological trait matrices). At the same time, practicing phylogeneticists have broadly suspended efforts to create morphological datasets spanning pre-Paleozoic divergences, due to the paucity of available character systems and clear homology statements. As examples, the Metazoan Tree of Life team abandoned, after enormous investment of time and expense, efforts to establish the homology of characters that could apply to all extant animal phyla, because defining homologies at this scale had proven intractable (G. Giribet, personal communication). By the same token, constructing morphological datasets spanning even older lineages (Opisthokonta; Eukarya; the entire tree of life) is simply not feasible and our understanding of the basal-most splits in the Tree of Life today is predominantly informed by molecular sequence data.

Molecular sequence data can be collected inexpensively, reliably, reproducibly, and expeditiously. The collection of morphological data has remained precisely as painstaking, laborious, and slow as it was in the 1980s. While innovations like micro-CT scanning have improved the recovery and resolution of internal morphological characters, the coding of morphological matrices is simply not scalable to the level of genomic sequencing. The length of the operation aside, morphological data (from micro-CT or traditional microscopy) need to be interpreted and coded, which requires time, labor, and lineage-specific expertise. The coding itself demands questionable practices like subjective discretization of character states and arbitrary treatments of phenomena like phenotypic plasticity and polymorphism.

Thirdly, morphological data matrices have consistently struggled with the problem of character independence. Morphological characters are effectively black boxes, whose independence can be approximated only using congruence-based approaches (i.e., comparisons of state changes between characters on a tree), which is inherently circular. Molecular sequence data, in tandem with complete genomes, provide a straightforward and reliable way to assess the independence of loci, as a function of their distributions across the genome; in principle, character independence of loci can be quantitatively characterized.

A fourth consideration pertains to models of evolution. Due to the proliferation of molecular sequence data in the last half-century, substitution models available for both nucleotide and peptide alignments have become increasingly sophisticated, being informed by statistical and biochemical validation. This has enabled such applications as molecular divergence dating, assessment of compositionally heterogeneous evolution, assessment of heterotachous evolution, and measurement of incomplete lineage sorting. Even in cases where molecular phylogenies have initially faltered (as was the case of Myriochelata in arthropod phylogeny or the non-monophyly of Protobranchia in Sanger datasets of bivalves [68,69]), additional molecular sequence data, rare genomic changes, and/or improved substitution models accounting for asymmetrical branch lengths have facilitated the identification of the artifacts driving those results [70–72]. As stated above, morphological characters remain to this day a black box, rife with both theoretical and empirical pitfalls [73,74]. We lack reasonable models for their evolution, which has precipitated marked controversy over how morphological data matrices should be analyzed [75–80]. As a result, even in the context of molecular dating, morphological data have come to play a supporting role to molecular data in modern phylogenetics.

All these considerations bear upon the utility and accuracy of relationships predicted by morphological data matrices. Scores of systematic works have rejected or refuted traditional morphology-based relationships in the light of molecular data and subsequent reappraisals of morphological homology statements (e.g., Articulata; Aschelminthes; Coelenterata; Coelomata; Polychaeta). This is particularly the case for deeper nodes in the Tree of Life, groups that lack a large number of available character systems for scoring, and the higher-level relationships of groups prone to morphological convergence. As a result, most modern approaches to understanding the evolution of morphology in extant taxa will generate a molecular phylogeny first and map morphological character states onto the molecular tree thereafter. In the case of conflicts between morphological and molecular trees, the latter usually turn out to be more accurate, having prompted a reevaluation of morphological homology statements and character definitions. Prominent examples of these trends include the systematic history of Ecdysozoa, Pancrustacea, Gnathifera, and Ambulacraria [81–85].

Even in the case of groups with well-established morphological datasets and fairly clear evolutionary history, evidence for the superiority of molecular data is commonplace. In the case of Bivalvia, Bieler et al. [61] assessed the informativeness of different morphological character systems under a statistical framework, comparing the result to a total evidence tree of nine genes and morphology. They were able to show that more than 50% of characters coded exhibited phylogenetic signal indistinguishable from random structure, for all character systems except for external shell morphology and sperm ultrastructure. Upon extracting the 99 characters that exhibited any signal at all and computing a tree (Figure 35 of Bieler et al. [61]), they were able to recover the monophyly of only two of the six major bivalve lineages—a result that can be surpassed with a dataset of just four nuclear coding genes with a fraction of the effort, time, and expense required for collecting morphological data [70]. At a shallower phylogenetic node, Zou and Zhang [86] examined patterns of homoplasy in a mammal dataset of 3414 parsimony informative morphological characters and 5722 parsimony informative amino acid sites. They were able to show that morphological data were more prone to convergence than amino acid sites, in large part due to the small number of character states defined for discretized morphological characters. Similarly, in the case of biogeography, a recent meta-analysis of morphological and molecular phylogenies showed that molecular phylogenies exhibit better congruence to biogeographic distributions than their counterpart morphological trees [87], suggesting that homoplasy in morphological trees obscures inference of macroevolutionary processes.

The logical conclusion stemming from these trends should be that morphological data are non-universal and often unreliable arbiters of deep phylogenetic relationships, and therefore constitute an inferior data class for numerous taxa. Yet few published works arrive at this explicit conclusion. The archetypal conclusions postulated by phylogenetic workers that compare data classes to the detriment of morphology are (1) the importance of fossils and an integrated understanding of evolutionary history, which substantiates the continued relevance of morphology (e.g., [67,87,88]); and/or (2) a call for collecting more morphological data, despite the molecular data usually having resolved most of the relationships in question more efficiently and more reproducibly (e.g., [60]). Given the comparatively greater epistemological and practical challenges to collecting, interpreting, and analyzing morphological data in the context of phylogenetic inference, it is not surprising that morphological datasets have taken on an ancillary role by comparison to molecular data, in deciphering evolutionary relationships of extant taxa.

The considerations above bear directly upon the evaluation of morphological evidence in support of arachnid monophyly.

6. Morphological "Support" for Arachnida Is Grounded in Errors, Shared Absences, Homoplasy, and Circular Reasoning

> "Promoting or defending a specific phylogenetic hypothesis via lists of compatible synapomorphies is a common but problematic approach … A node supported by a long list of synapomorphies may seem convincing taken in isolation but may become less acceptable when its full phylogenetic implications are explored."
>
> —Jeffrey W. Shultz (2007) [2]

The base of Euchelicerata constitutes a recalcitrant node where molecular data do not yield clear answers, whereas morphological data are perceived to support arachnid monophyly unambiguously. But how strong is the actual evidence for this relationship? To examine this question, we revisited a series of morphological matrices recently produced by paleontologists and reanalyzed them using identical approaches as reported in their respective publications. We mapped onto the resulting trees all characters that constitute unambiguous synapomorphies for Arachnida. We additionally tabulated putative synapomorphies for Arachnida from historically significant works from the literature (Table 1).

We excluded from this analysis one dataset that was never published by the author [89] and another that could not recover arachnid monophyly without enforcing a topological constraint for this preconceived relationship [9].

For two datasets that we analyzed, one lacked a published character list accompanying the morphological matrix [90], and the second—a well-known and oft-cited work touting the reconciliation of morphology and molecules through the consideration of fossils—published an incorrect version of the morphological matrix with no correspondence to its character list [91]. Repeated requests to the author of this series of matrices [89–91] over the past two years for the correct versions of datasets elicited a bizarre online exchange that led us to conclude that the original matrices were lost or incorrectly analyzed. Documentation of this exchange is available upon request. The versions that were eventually and sporadically supplied to us did not match their published counterparts with respect to the number of taxa and characters analyzed. The author of the aforementioned matrices openly communicated to us their awareness of these issues over the past several years but has taken no steps to correct any of them. We are baffled as to how some researchers and editors in the field of arthropod paleontological systematics are able to operate with such lax standards and permissive requirements for science; for the rest of us, such oversights would be tantamount to trolling for journal retractions.

In any case, the series of matrices provided to us by the aforementioned author [89–91], as well as one that was correctly published [92], analyzed with or without a character list, recovered no characters constituting unambiguous synapomorphies for Arachnida.

Next, we examined a family of matrices emphasizing the sampling of aquatic euchelicerates (the merostomates) [5,8,93]. The character and taxon set for these studies exhibited broad overlap and all three recovered the same two characters as unreversed synapomorphies of Arachnida: (1) the identity of the first podomere of the fifth prosomal appendage that protrudes beyond the carapace, and (2) the condition of exopod armature, other than lamellae.

The first of these characters was coded as the second podomere (the trochanter) for all arachnids, a higher value for merostomate taxa (which exhibit comparatively greater lateral outgrowth of the carapace margin), and inapplicable for Pycnogonida (as sea spiders lack a carapace). The coding of this character is incorrect. In Palpigradi, Acariformes, and Solifugae, the carapace does not include the segments of the third or the fourth walking leg. In addition, in Solifugae and Palpigradi, the protruding coxae of these two leg-bearing segments are clearly visible in dorsal view; these arachnid orders should rightfully have been coded as inapplicable for this character, like the sea spiders (Figure 6a,b). Moreover, the designation of this character is arbitrary; comparable characters do not exist for any other prosomal appendages except for the fifth prosomal appendage in these matrices, suggesting that this character was erected for the express purpose of artificially distinguishing arachnids as a natural group.

The second character (exopod armature other than lamellae) is also incorrectly coded. It was scored as absent for all arachnids, present (setae or spines) for merostomates, and inapplicable for sea spiders. This coding is not logical; if coded as inapplicable in sea spiders because they lack exopods, then this character should also be coded as inapplicable in arachnid orders that unambiguously lack exopods as well (i.e., the non-arachnopulmonate arachnid orders). Alternatively, it should be coded as absent in sea spiders, following the same logic as the coding for arachnids. A broader issue with such character codings, recapitulating Ballesteros et al. [50], is that they group taxa based on shared absences that may be the result of convergence (especially in terrestrial habitats; see also [74]). Assuming for the sake of simplicity that this character was coded correctly (with only arachnids sharing a state of absence), it implies that the simplification or loss of exopods reflects a homologous condition. As revealed by the evolutionary history of hexapods, myriapods, and other terrestrial pancrustacean groups (e.g., terrestrial amphipods and isopods), the reduction or total loss of exites, rendering an arthropod limb uniramous, has evolved repeatedly, likely as a result of convergent adaptations to terrestrial habitats. Given the evolutionary history of the mandibulate limb in terrestrial taxa, the lack of exopods or exopod armature in terrestrial chelicerates does not offer compelling evidence uniting Arachnida.

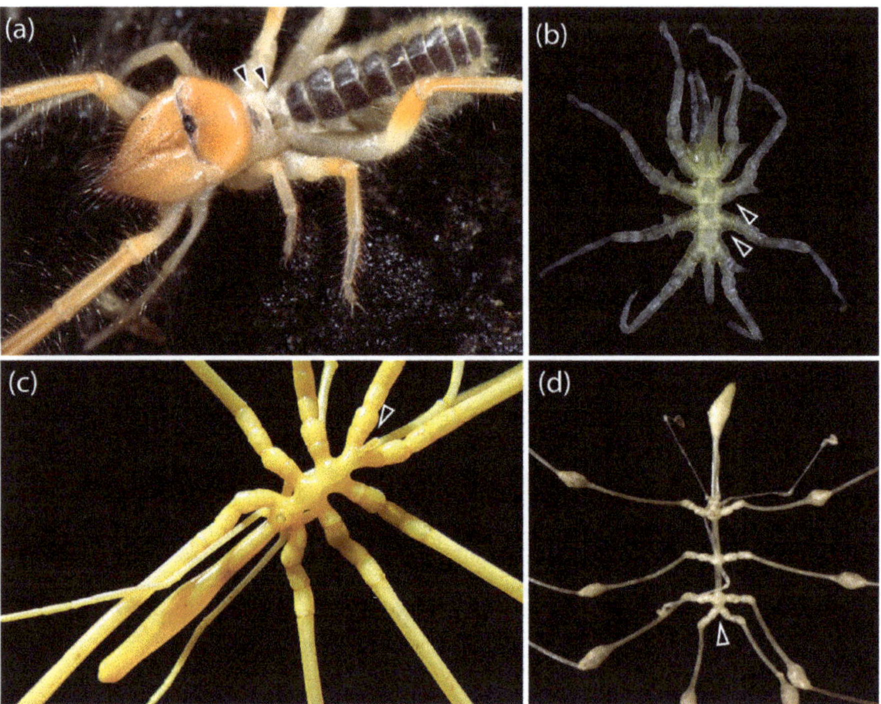

Figure 6. Incorrect codings of morphological character states underlie the recovery of Arachnida in morphological matrices. (**a**) Live habitus of Solifugae (*Eremobates* sp.). Arrowheads show the locations of the third and fourth walking leg segments, which are not part of the carapace (the protruding coxae are clearly visible from a dorsal view). The same condition is found in Palpigradi. (**b**) Micro-CT of an austrodecid sea spider (*Austrodecus glaciale*). Arrowheads indicate the visible coxae of walking leg segments. (**c**) Unsegmented and appendage-free anal tubercle (the remnant of the opisthosoma; arrowhead) of a colossendeid sea spider (*Colossendeis megalonyx*). (**d**) Dorsal view of the colossendeid *Rhopalorhynchus magdalena*. In such species, the opisthosomal rudiment has been lost altogether (arrowhead). Note as well the absence of exopods in sea spiders. Photographs: (**a**) G. Giribet; (**b**) G. Brenneis; (**c,d**) C.P. Arango.

The last series of morphological datasets we analyzed was drawn from a family of matrices originally generated by Shultz [2,39], who rooted Arachnida using Xiphosura, presuming arachnid monophyly. We analyzed derivations of this matrix, which subsequently underwent adaptations and expansions to include sea spiders and other phylogenetically significant lineages [94–97]. Of the morphological matrices we examined in the recent literature, this family of matrices represents the most intensive sampling of extant and fossil arachnids. In this family of matrices, we recovered the same two synapomorphies for Arachnida previously reported by Garwood and Dunlop [94]: (1) presence/absence of the appendages of the first opisthosomal segment (absent in arachnids; present in all other chelicerates), and (2) origin of the walking leg apotele depressor (tibial in arachnids; tarsal in horseshoe crabs and sea spiders).

Table 1. There is no compelling morphological support for the monophyly of Arachnida. Tabulation and critical evaluation of putative synapomorphies historically uniting Arachnida.

Source; Year	Character	Putative State Shared by Arachnida	Evaluation
Weygoldt and Paulus [1]; 1979	extraintestinal digestion	present	Absent in Opiliones, various Acariformes, and various Parasitiformes
	Malpighian tubules	present	Absent in Palpigradi, Pseudoscorpiones, Opiliones, and some groups of Acariformes
	eyes	disintegrated into five or fewer ocelli	Semi-aggregate eyes present in fossils of Scorpiones, Ricinulei, and Trigonotarbida
	slit sensilla	present	Absent in Palpigradi
Shultz [98]; 2001	reduced pleural fold (doublure) in the prosomal carapace	present	Also shared by Pycnogonida
	slit sensilla	present	Absent in Palpigradi
	anterodorsal rotation of anterior prosoma resulting in anteroventrally directed mouth	present	Same as the ancestral condition in the common ancestor of Pycnogonida
	appendages on somite VII (O1) in adults	absent	Shared absence; comparable to absence of tritocerebral appendages in hexapods and myriapods
	cardiac lobe or glabella on carapace	absent	Shared absence
	medial genital opening (paired in Xiphosura)	single	Initially paired in embryonic arachnids; if only scoring adults, character does not apply in Pycnogonida, so paired gonopores would be synapomorphic for merostomates
	appendages on somite XIII	absent	Shared absence if correctly scored; in fact, derivatives of O2 coxae form the genital operculum
	postcerebral crop and proventriculus	lost/reduced	Shared absence
	anterior oblique axial muscles	absent	Shared absence
	attachments of opisthosomal posterior oblique axial muscles	pleural	Cannot be polarized because sea spiders lack this character; tergal attachments in Xiphosura could represent a synapomorphy of the merostomates
	endosternal suspensors of somites I and II	absent/detached	Shared absence
Shultz [2]; 2007	carapacal pleural doublure	absent	Shared absence
	cardiac lobe	absent	Shared absence
	pedal gnathobases	absent	Shared absence
	moveable endites	absent	Shared absence
	aerial respiration	present	Prone to generating suites of morphological homoplasies (e.g., Atelocerata; Pulmonata); tautological
	anteriorly or anteroventrally directed mouth	present	Same as the ancestral condition in the common ancestor of Pycnogonida

Table 1. Cont.

Source; Year	Character	Putative State Shared by Arachnida	Evaluation
Legg et al. [91]; 2013	none	none	Note: Incorrectly published online; original version lost
Legg [89]; 2014	none	none	Note: Matrix never published; analyzed version provided to us mismatches publication
Briggs et al. [92]; 2016	none	none	Note: Character list not published
Siveter et al. [92]; 2017	none	none	
Lamsdell et al. [93]; 2015	first podomere of pa5 that fully projects beyond carapace	second podomere	Incorrectly scored for Solifugae, Palpigradi, Acariformes
	exopod armature other than lamellae	absent	Incorrectly scored for Pycnogonida
Lamsdell [5]; 2016	first podomere of pa5 that fully projects beyond carapace	second podomere	Incorrectly scored for Solifugae, Palpigradi, Acariformes
	exopod armature other than lamellae	absent	Incorrectly scored for Pycnogonida
Bicknell et al. [8]; 2019	first podomere of pa5 that fully projects beyond carapace	second podomere	Incorrectly scored for Solifugae, Palpigradi, Acariformes
	exopod armature other than lamellae	absent	Incorrectly scored for Pycnogonida
Garwood and Dunlop [94]; 2014	appendages on opishosomal segment 1	absent	Incorrectly scored for Pycnogonida
	origin of apotele depressor	tibia	Ambiguous to interpret for Xiphosura and Pycnogonida due to variation in podomere counts
Huang et al. [97]; 2018	appendages on opishosomal segment 1	absent	Incorrectly scored for Pycnogonida
	origin of apotele depressor	tibia	Ambiguous to interpret for Xiphosura and Pycnogonida due to variation in podomere counts

Again, both characters are ambiguously coded. For the former, sea spiders were scored as having appendages on the first opisthosomal segment—a problematic designation, given that sea spiders do not have an opisthosoma, but rather, an abdominal rudiment (at most) that is unsegmented and does not bear appendages (Figure 6c,d). The only fossil sea spider known to bear a segmented opisthosoma (*Flagellopantopus blocki*) also shows no appendages on this posterior tagma. Why then were sea spiders coded as bearing opisthosomal appendages?

Garwood and Dunlop [94] argued that the alignment of sea spider and arachnid body segments implies the positional homology of the first opisthosomal segment of arachnids and the fourth walking leg segment of sea spiders, due to the presence of an additional oviger-bearing segment in the sea spider cephalon. The problem with this coding is that it reflects an internal inconsistency across their matrices. Within chelicerates, the number of prosomal segments is generally fixed, but it exhibits additions in sea spiders (at least three origins of 10- and 12-legged genera [99,100]) and reductions or losses in groups like eriophyoid mites (four-legged upon hatching, due to early attenuation of the L3 and L4 segments during embryogenesis) [101]. In none of the sea spider genera with supernumerary appendages is the opisthosoma lost or otherwise affected, suggesting that the addition of prosomal segments does not require altering the boundary or identity of prosoma and opisthosoma (similarly, reduction of the posterior prosomal segments in eriophyoid mites does not impact the opisthosomal boundary). Opisthosomal segmentation is far more variable in extant chelicerates (e.g., between two and 13 segments across extant arachnids; unsegmented in sea spiders). The ensuing inconsistency is that other characters in the same matrix were coded as reflecting the absence of an opisthosoma in sea spiders altogether (e.g., characters pertaining to the condition of opisthosomal dorsal and ventral segmentation were coded as inapplicable), which conflicts with the definition of the fourth walking leg segment as "opisthosomal". A less tortuous coding would list the condition of the first opisthosomal appendage as absent in sea spiders, which renders this character as an invalid synapomorphy of arachnids.

The character of the apotele depressor stemmed from a comprehensive work by Shultz [98] on the musculoskeletal anatomy of *Limulus polyphemus*, wherein Shultz defined a series of putative synapomorphies of the arachnids (drawing upon his previous seminal work in 1990 [39]). Homologizing the podomeres and muscle attachment sites in extant chelicerates is complicated by the absence of the metatarsus in horseshoe crabs and the presence of additional podomeres in sea spiders—the same segmental landmarks do not exist in arachnid versus other chelicerate walking legs. Moreover, as previously argued by Ballesteros et al. [50], the use of this entire character system as a source of arachnid synapomorphies may be confounded by convergent evolution in terrestrial habitats, comparable to convergences in the musculoskeletal system of hexapods and myriapods. Arthropod appendages are highly adaptable structures that have undergone markedly different selective pressures in terrestrial versus aquatic environments, resulting in parallel evolution of numerous podomeres and rami [102].

An anemic defense of arachnid monophyly was recently mounted by Howard et al. [18] on morphological grounds. In their "holistic" treatment of chelicerate relationships, Howard et al. did not present a single morphological analysis, restricting their analytical contributions to a paralogy-riddled molecular analysis (discussed above; see also Figures S2 and S3 of Ontano et al. [10] and associated discussion), in addition to a discursive overview of a handful of morphological character systems in the context of arachnid evolution, such as the anatomy of the lateral eyes, the respiratory organs, and the feeding mouthparts. As previously rebutted by Ballesteros et al. [50], every one of the morphological character systems discussed by Howard et al. [18] has been shown to exhibit misleading levels of homoplasy in the phylogeny of Mandibulata, with many of these character systems contributing to the flawed traditional interpretation that hexapods and myriapods were sister groups (Atelocerata), representing a single colonization of land. The discourse of Howard et al. [18] consistently fails to grasp the implication of the repeated failure of morphological and paleontological datasets to recover the only higher-level chelicerate relationships that have been robustly supported by multiple data classes (i.e., Arachnopulmonata sensu Ontano et al., Panscorpiones; [10]): If morphological datasets are unable to recover these splits (all of which were accepted as valid by Howard et al. [18], and solely on the basis of molecular support), then other nodes recovered by morphological analyses of Chelicerata could just as easily be inaccurate as well.

This fundamental contradiction in reasoning aside, Howard et al. [18] cursorily pointed to the historical availability of "substantial lists of morphological autapomorphies ... proposed to support Arachnida" to promote the perception of unambiguous support for this grouping. Ironically, the 11 synapomorphies they refer to were previously defined in 2001 by Shultz [98], who later disparaged the practice of listing synapomorphies and character recycling [2]. As shown in Table 1, six of these 11 synapomorphies are shared absences, which do not constitute compelling apomorphies uniting any group (consider, for example, the patterns of shared absences observed in hexapods and myriapods for biramous appendages, gills, and tritocerebral appendages). Two others are predicating on ignoring character states in Pycnogonida that are shared with the arachnids (i.e., they reflect the practice of conceptualizing a tree of Arachnida rooted with Xiphosura a priori and dismissing the morphology of sea spiders as aberrant [2]). One character cannot be polarized due to the morphology of sea spiders and may in fact constitute a synapomorphy of merostomates. Another two characters still are in fact not arachnid synapomorphies at all but are restricted to subsets of terrestrial orders or to specific developmental stages. Shultz himself heavily revised the list of potential arachnid synapomorphies by 2007 and reduced it to six [2], recognizing that characters like slit sensilla and fluid feeding must have evolved within a subset of the arachnids (but see Table 1 for critical evaluation and rebuttals of all six characters). Even characters that clearly reflect functional adaptations to terrestrial habitats, like Malpighian tubules, are restricted to subsets of arachnid orders. Certain complex characters, like Malpighian tubules and tracheal tubules, are known to have evolved repeatedly across Panarthropoda, undercutting the value of excretory and respiratory organs as homoplasy-free character systems.

In reality, the "substantial lists" of arachnid synapomorphies alluded to by Howard et al. [18] do not exist, nor does the ability to conjure such characters (even if they existed) forfend the possibility of phylogenetic inaccuracy—consider, for example, the longer putative lists of synapomorphies once invoked to support the erstwhile groupings Articulata (Arthropoda + Annelida) and Atelocerata (Hexapoda + Myriapoda). There is little uniting Arachnida except for the condition of being mostly terrestrial and shared absences that likely stem from being terrestrial. Efforts to seek out support for this relationship with the goal of "rescuing morphology" risk incurring further circularity and confirmation bias.

To arachnologists—as well as to paleontologists with actual expertise in arachnids (e.g., [4,94])—none of this is new or surprising, given the discoveries and advances in arachnid biology in the past twenty years. Practicing arachnologists have understood for decades that morphological data are not strongly dispositive with respect to several competing hypotheses of various interordinal relationships (e.g., the placement of Opiliones, Palpigradi, and Pseudoscorpiones [15,44,103–105]), with various character systems demonstrably prone to convergent evolution at both deep and shallow taxonomic scales (e.g., respiratory organs) [10,106]. As a recent example, Ballesteros et al. [50] recently generated a comprehensive morphological matrix sampling 514 fossil and extant chelicerates, which they analyzed using both parsimony and Bayesian inference approaches. Neither analysis supported arachnid monophyly, with the base of Euchelicerata constituting a soft polytomy. When the morphological matrix was paired with their 506-taxon phylogenomic dataset, Merostomata was resurrected as a clade derived within Arachnida, with support. This total evidence analysis is the only case where morphological data, analyzed alone or in combination with molecules, have ever been able to recover relationships supported by rare genomic changes (Arachnopulmonata and Panscorpiones [10]). A more proximal placement of Merostomata to Arachnopulmonata may reconcile both the evolution of trabeculae in a derived group of chelicerates [6], as well as century-old observations of morphological correspondences between merostomates and scorpions—a stepwise colonization of land may indeed apply, but only for one derived group of arachnid orders..

To rule out the possibility that biased coding of the morphological matrix had driven this result, Ballesteros et al. [50] additionally paired their molecular dataset with two recent morphological matrices generated by paleontologists, with widely differing taxon sets. The inclusion of molecular data dissolved much of the higher-level structure of chelicerate relationships, but in all cases consistently refuted arachnid monophyly. In the case of a comprehensively sampled morphological matrix of Panarthropoda, the inclusion of molecular data again recovered Merostomata as nested within the arachnids as well. These analyses suggest that morphological support for traditional chelicerate relationships is not as robust as typically perceived, in the specific context of total evidence.

The perception that Arachnida is well-supported by morphology likely reflects a century-old recycling of an antiquated idea, predicated on the historical belief that terrestrialization was rare or irreversible in evolutionary history. This belief is shown to be groundless when examining the modern (i.e., molecular) phylogeny of Mandibulata.

Putting aside the myriapod and hexapod terrestrialization events, the history of decapod terrestrialization has been shown to result from at least ten colonizations of land, and a recent molecular work has even revealed the diphyly of the terrestrial isopods [107,108]. Thus, even on relatively shallow timescales, terrestrialization is common across Arthropoda, and a clear driver of morphological convergence, which obscures phylogenetic signal in morphological datasets. Similarly, a complex history of terrestrialization with repeated colonization of land has been revealed in other phyla, such as Nematoda and Mollusca (at least 12 colonizations of land in "Pulmonata"), through the lens of phylogenomics [109,110]. Arachnida is little more than the newest member of a growing list of erstwhile specious terrestrial groupings that reflects cases of multiple terrestrialization events.

7. Arthropod Systematic Paleontology and the Value of Validation

"Ever bought a fake picture, Toby? . . . The more you pay for it, the less inclined you are to doubt its authenticity."

—George Smiley, *Tinker Tailor Soldier Spy*

"There has been growing research interest on how people respond to corrections of misinformation . . . This body of research has converged on the conclusion that corrections are rarely fully effective: that is, despite being corrected, and despite acknowledging the correction, people by and large continue to rely at least partially on information they know to be false. This phenomenon is known as the continued-influence effect . . . In some circumstances, when the correction challenges people's worldviews, belief in false information may ironically even increase . . . "

—Stephan Lewandowsky, Ullrich K.H. Ecker, John Cook (2017) [111]

In his eulogy to the waning prominence of morphological data in 21st century phylogenetics, Giribet [67] articulated two broad reasons for maintaining research programs in morphology: an understanding of morphological evolution for its own sake, and molecular dating (with emphasis on total evidence dating, or tip-dating). Similar arguments were expounded by Lee and Palci [88], who extolled the availability of morphological characters as a source of validation for molecular clades (increasingly, a role superseded by analyses of genomic architecture and rare genomic changes), as well as their essential function in molecular dating. Indeed, in cases of taxa with (1) a richly detailed fossil record and (2) strong morphological signal in hard parts that are prone to fossilization, signals from morphological and molecular data partitions have been shown to complement one another and successfully yield an integrated dated resolution of extinct and extant lineages (e.g., [61,64,112]). Using taxon addition and deletion experiments for an array of morphological matrices, Mongiardino Koch and Parry [113] recently showed that fossil taxa can have profound impacts on tree topology, usually more so than extant taxa. Notably, however, Panarthropoda was one of two test cases where neontological data had greater impact than fossil data [113]. This may be because the authors of that Panarthropoda matrix coded molecular data, such as Hox gene expression, as morphological characters for extant taxa—a possibility we cannot test, given that the authors published an incorrect matrix version with a mismatching character list [91].

Overall, such works [67,88,113] make a compelling argument for the consideration of fossil taxa both for improved tree searches and for a more comprehensive understanding of character state evolution. In principle, the inclusion of fossil taxa in phylogenies toward a more comprehensive understanding of evolutionary history is not controversial and represents a desirable goal. The benefits of adding taxa to matrices is broadly recognized, particularly in the context of breaking up long-branch attraction artifacts (e.g., Acari; Pseudoscorpiones + Acariformes). If >95% of species have undergone extinction during the Phanerozoic, surely a comprehensive view of phylogenetic history must account for character states and histories of these groups. Would the integration of morphology and fossils not help to resolve the mystery of chelicerate relationships?

Putting aside the practical and algorithmic challenges of analyzing such combined datasets (specifically, the relative weights assigned to morphological versus molecular partitions), what proponents of the total evidence approach often overlook is that their favored test cases generally tend to be taxa that bear an exceptionally detailed, highly complete fossil record, with strong morphological signal in the body parts that fossilize in those groups. Examples of taxa that meet these criteria include the echinoderms, shelled mollusks, and various groups of vertebrates [61,62,112]. However, for many soft-bodied and/or small-bodied invertebrate lineages (e.g., Platyhelminthes; Placozoa; Cycliophora), the fossil record is prohibitively sparse for total evidence approaches, due to a lack of hard parts prone to fossilization. Most terrestrial lineages of animals (e.g., the arthropods) tend to have poorer fossil records than aquatic counterparts, save for cases of exceptional preservation (but for a counterexample, consider the fossil record of Pycnogonida [4,100]). As a result, empirical total evidence matrices for such groups tend to be flooded with missing entries for fossil taxa, which can result in the destabilization of the tree topology upon the inclusion of fossils. Among Chelicerata, the degree of missing data and uninformative characters for orders like Phalangiotarbida results in these groups acting as rogue taxa in Bayesian inference analyses [50,94,96].

Worse still, in some groups of invertebrates, external morphological characters that are observable in fossils may not be those that retain high phylogenetic signal. Such is the case for many groups of chelicerates. As examples, in Pycnogonida, it was recognized fairly early that the traditional morphological interpretation of a reductive trend in appendages across the sea spider tree was not substantiated by phylogenetic analysis. Subsequent comparisons of morphological and molecular data (either in isolation or in total evidence analyses) revealed marked incongruence in topology, partly attributable to the low number of codable characters in the sea spider bauplan, in addition to homoplasy [99,114]. In Solifugae, higher-level taxonomy is diagnosable only using a subset of feeding appendage characters, and in many cases, the diagnostic character states are only available in adult males. For Scorpiones, a higher-level phylogenomic analysis showed that external morphological characters were either uninformative or incongruent with higher level relationships inferred by thousands of genes [115]. Only a subset of internal morphological characters pertaining to embryonic developmental mode, the distribution of the digestive glands, and the morphology of the hemispermatophores is consistent with the molecular phylogeny and informative at higher taxonomic levels [115,116]. As with various arthropod taxa, numerous scorpion groupings based on traditional morphological analyses were shown to be non-monophyletic, requiring extensive systematic revision of higher-level relationships through the lens of phylogenomics [115,117–119]. Even at shallow taxonomic scales, parametric analyses of shape data have shown that structures traditionally held to harbor high phylogenetic information (e.g., carapace or pedipalpal podomere shape) exhibit extensive morphological convergence or broadly uninformative variation within derived scorpion groups [120]. Internal characters are virtually inaccessible to paleontologists, with the rare exceptions of some arachnid fossils (e.g., Opiliones) with exquisite preservation of intromittent organs [63,121].

These trends are uniquely problematic for chelicerate paleontologists, for whom fossilized traits available for coding occur in a small sample of specimens and constitute mostly external traits prone to morphological stasis or low phylogenetic signal. But limitations of morphological datasets with respect to phylogenetic signal are only heightened by the spareness of the arachnid fossil record. Putting the informativeness of morphological characters aside, the appearances of chelicerate orders cannot be interpreted as a sequence of divergences wherein arachnids are unambiguously at the younger end of the chelicerate stratigraphic range (implying a later origin than aquatic chelicerate orders), because the chelicerate fossil record is prohibitively incomplete.

As a point of comparison, the fossil record of Pancrustacea is consistent with the sequence of divergence wherein hexapods are nested within the "crustaceans"—the stratigraphic ranges of the marine pancrustaceans extend into the Cambrian, whereas Hexapoda do not appear until the Devonian. Within Hexapoda in turn, derived insect orders appear thereafter, in a window between the Jurassic and the Carboniferous. The temporal sequence of fossil appearances thus corroborates patterns predicted by molecular phylogenies of Pancrustacea (i.e., a series of nested relationships).

In Chelicerata, the oldest fossil horseshoe crabs (Xiphosura as well as synziphosurines) span the Ordovician in age (ca. 445–485 Myr old), whereas most fossil horseshoe crab taxa post-date the Ordovician-Silurian boundary [122]. The oldest unambiguous scorpion fossils (e.g., *Eramoscorpius brucensis*) are approximately 435 Myr old in age [123], occurring contemporaneously with Eurypterida (many of which postdate Silurian arachnid fossils). If scorpions are derived within Arachnopulmonata as the sister group of pseudoscorpions (again, an unambiguous split based on both rare genomic changes and phylogenomics [10,23]), then this phylogenomic placement, together with the age of *E. brucensis*, effectively guarantees that diversification of Arachnopulmonata must have predated 435 Myr. The age of *E. brucensis* cannot be taken as reflective of, much less synonymous with, the initial diversification of the arachnids—the group must be much older. Furthermore, the absences of other arachnopulmonate stem-group lineages (e.g., stem-Tetrapulmonata; stem-Pseudoscorpiones), as well as the apulmonate arachnid orders in Silurian strata only underscore the marked incompleteness of the chelicerate fossil record. Ordovician or older divergences of the arachnid assemblage would be more consistent with the derived placement of scorpions and molecular dating estimates of the basal arachnid diversification (regardless of the recovery of arachnid monophyly [18,116]) and would also accord with the estimated appearance of land plants by the Cambrian [54,124].

Thus, the temporal sequence of appearances of arachnid orders in the fossil record does not unambiguously substantiate inferences of arachnid monophyly. The distribution of chelicerate fossil ages is not consistent with the traditional interpretation of a gradual, stepwise colonization of land by a grade of merostomates, the scenario favored to this day by paleontologists [6]. This near-simultaneous appearance of major terrestrial and aquatic chelicerate groups in the fossil record just as easily accords with the short internodes at the base of the euchelicerate diversification in molecular phylogenies, in contradiction of the scenario of gradual and stepwise evolution of eurypterids and arachnids after their divergences from the horseshoe crab assemblage.

Given both low phylogenetic signal in morphological datasets, as well as the incompleteness of the chelicerate fossil record, it is not unexpected that chelicerate paleontologists have never successfully resolved a stable phylogeny of Chelicerata, despite the ability to sample extinct lineages and character states that are off-limits to molecular datasets. Recent paleontological phylogenies of chelicerates exhibit marked dissimilarity of topologies across different families of data matrices and analytical approaches, with broad disagreement on the basally branching placement of scorpions; the monophyly of Euchelicerata; and the monophyly of Tetrapulmonata. The strict consensus of chelicerate morphological phylogenies published in the past five years (after reanalyzing one study to remove an a priori constraint for arachnid monophyly) constitutes a total polytomy (Figure 1). The only relationships that are somewhat consistently recovered by morphological phylogenies (e.g., a basally branching placement of scorpions at the base of Arachnida; Pseudoscorpiones + Solifugae) happen to be exactly those that have been wholly refuted by evidence based on phylogenomics and genome architecture [10,21–23]. There is no evidence that the addition of fossils rescues chelicerate phylogeny from topological instability at the base of Euchelicerata, nor that morphological phylogenetic signal is consistent and unambiguous across datasets.

Given the topological uncertainty surrounding chelicerate relationships, molecular dating—an oft-cited justification for valuing morphological datasets—for the group may be a premature, error-prone exercise. Recent efforts to date the chelicerate tree of life have been based upon paralogy-riddled datasets [18,125], one of which went as far as estimating rates of cladogenesis on a phylogeny that undersampled the higher-level diversity of diverse groups of arachnids (Acariformes and Parasitiformes) and did not include all extant arachnid orders [125]. These analyses provide little to testing the hypotheses of chelicerate relationships and evolution; the ability to place precise confidence intervals on the ages of unstable nodes that may not exist is not synonymous with evolutionary insight. In a group that bears over 130,000 described species, measures of diversification dynamics using a dataset of less than 100 exemplars that omit much of the higher-level ordinal diversity are meaningless. If the vertebrate literature serves as any guide, robust hypothesis-testing using diversification rate analyses and comparative methods requires many hundreds of terminals with dense (ideally, complete) species-level sampling. The authors of these premature analyses of chelicerate diversification dynamics [125] have repeatedly mistaken the ability to perform an analysis and obtain a palatable result for the notion of achieving "consilience". What they call consilience is more akin to stacking suppositions on top of other suppositions.

Consilience, strictly defined, refers to independent lines of scientific evidence arriving at the same conclusion. As explained above, independent lines of evidence (e.g., phylogenomic analyses; rare genomic changes; gene expression patterns) have repeatedly rejected the validity of several relationships postulated by morphological data partitions. The robust support for groups like Arachnopulmonata and Panscorpiones from independent lines of phylogenomic, developmental genetic, and genomic analyses underscore one of the most valuable principles of the scientific enterprise: validation.

In the context of phylogenetics, independently evolving partitions in molecular datasets can be identified and compared for validation of phylogenetic signal; genes and sites can be examined and subsampled to assess sources of systematic bias; whole genomes can be sequenced to search for rare genomic changes like microRNAs and transposon insertions; internal and external anatomical character systems can be surveyed and scrutinized for evidence of evolutionary history. In principle, for the case of ambiguous morphological characters, approaches in functional genetics and comparative development (evo-devo) can be used to understand the developmental genetic basis for trait convergence and thereby evaluate the validity of homology statements. The collection, comparison, and evaluation of all these data sources that substantiate or refute hypotheses of evolutionary relationships is integral to validation and the achievement of consilience. In Chelicerata, it was through such integration of different datasets that a robust phylogeny of enigmatic groups like sea spiders was recently obtained and competing hypotheses for the placement of the long-branch taxon Pseudoscorpiones were arbitered with rare genomic changes [10,100].

The ability to perform such validation through diverse and independent sources of phylogenetic data, however, is the domain of living organisms. With the exception of recent fossil taxa that can be sequenced (e.g., Neanderthals; moas; mammoths [126–128]) and applications of metazoan biomarkers for the interpretation of fossils (e.g., [129,130]), the study of paleontological systematics usually features access only to a single data class: morphology. As discussed above, the collection, interpretation, and codification of morphological data is fraught with subjectivity and inherent limitations, even for extant lineages. Tree topologies for the internal relationships of ancient extinct groups (e.g., Eurypterida; Synziphosurina) effectively lack validation by other data classes altogether. While methods exist to address homoplasy in morphological datasets (e.g., implied weighting), these congruence-based approaches rely upon the same morphological dataset that is being evaluated, and their interpretation is often subjective or circular.

There is usually no way for the practicing paleontologist to validate the relationships of fossils using data sources other than morphology. While this is not a barrier for the study of diverse assemblages with strong phylogenetic signal and large sample sizes (perhaps best exemplified by the and shelled mollusk and Mesozoic vertebrate fossil record), it represents a fundamental epistemological limitation of paleontological systematics for groups like chelicerates. When the fossil record is sparse and the parts prone to fossilization lack strong phylogenetic signal, how does the systematic paleontologist distinguish signal from investigator bias? How do they test the accuracy of the sole data class available to them?

And what happens if the systematic paleontologist has (as in the case of Chelicerata) a priori reason to doubt the informativeness of that sole available data class?

For neontologists, the decision would be simple: Abandon morphology, or alternatively, retain only the morphological character systems that clearly exhibit strong phylogenetic signal (by whatever criterion signal is evaluated [60,86,115]). Similar decisions are routinely made by molecular phylogeneticists for data partitions that are uninformative, noisy, data-poor, or low-quality—neontologists have access to various data classes and types, and thus typically place little investment in specific data partitions. Chelicerate paleontologists do not have this option—for some, admitting that arachnids may not be monophyletic is tantamount to conceding that their only available data class is unreliable. The downstream implication is that placements of other Paleozoic fossil groups could be little more than untestable conjecture.

As far as we can tell, the only communities that benefit from the perpetuation of arachnid monophyly are those with a vested interest in the validity of morphological phylogenies—principally, paleontologists and "molecular paleobiologists". Considering how much time, training, and investment has been poured into a specific scenario of chelicerate terrestrialization and morphological evolution over the past decades by adherents of these approaches, much is at stake for this traditionalist community when the nearly monolithic interpretation of arachnid terrestrialization is questioned. More generally, entrenched resistance to alternative evolutionary scenarios and hypotheses of character state transformations is a historically recurring phenomenon among some groups of morphologists and paleontologists, as exemplified by the history of Ecdysozoa and Pancrustacea.

It is for this reason that "molecular paleobiological" works like those of Lozano-Fernández et al. [42] and Howard et al. [18], which grasp for the defense of traditional relationships, do not represent balanced investigations grounded in consilience. They represent an advocacy, one that aims to defend a particular discipline, overestimates informativeness and completeness of a preferred data class, and overvalues a preferred set of preconceived hypotheses, in contradiction of the actual data, the sum of analyses, and the literature. Such an advocacy is irrevocably tainted with investigator bias, inasmuch as it is unwilling a priori to consider the conclusion that a preferred data class (e.g., morphology; the fossil record) could be flawed, uninformative, or incomplete, despite the weight of empirical evidence.

As is often the case when molecular data overturn long-held morphological hypotheses of relationships, adherents of Arachnida (principally, some morphologists and paleontologists) will likely continue to mount a spirited, but ultimately fruitless, effort to defend their preconceived vision of chelicerate phylogeny. As underscored by our previous works, accruing data from genome-scale analyses with comprehensive taxonomic sampling, site heterogeneous models, filtering for systematic biases, and total evidence approaches (that is, concomitant application of every approach embraced by "molecular paleobiologists") are overturning arachnid monophyly, the notion that arachnid paraphyly is attributable to an artifact, and an array of relationships previously thought to be robustly supported by morphological data (e.g., scorpions sister group to the remaining arachnids; pseudoscorpions + solifuges). As shown above, Arachnida has never been robustly supported even by morphological data.

Given that there are more characters supporting spurious, debunked relationships than those supporting arachnid monophyly, both in molecular datasets (e.g., Dromopoda) and in morphological ones (e.g., Atelocerata), we simply have to ask: What value is there in continuing to mainstream the delusion that arachnid monophyly is a certainty?

8. Last One Out, Get the Lights: The Future of Chelicerate Phylogeny

"Articulata . . . is based on much better morphological evidence [than] Ecdysozoa."

—Wägele and Misof (2001) [131]

" . . . morphological evidence suggests that myriapods are the sister group to Hexapoda."

—Wägele and Kück (2014) [132]

"I agree with you that the morphological support for Atelocerata [the debunked grouping of Hexapoda + Myriapoda] is stronger than that for Arachnida."

—Gregory D. Edgecombe (9 October 2019, personal communication to PPS)

In the interval since we first encountered the parable of the toucans, we have collectively engaged in a broad array of scientific approaches, such as taxonomic description and revision (including the description of fossils), bioinformatics, morphological and molecular phylogenetics, population genetics, phylogenomics, molecular dating, comparative genomics, developmental genetics, and evolutionary developmental biology (note, however, that we do not identify as "molecular paleobiologists", and we take serious exception to being labeled as such). While we recognize that every discipline bears its share of in-built assumptions and logical bridges, of all the approaches and methods we have pursued, the coding of morphological character matrices has been the most subjective, assumption-riddled, and validation-free activity we have engaged in, particularly as it impacts the placement of fossils. The characters and states we have encountered in chelicerate morphological datasets (even for extant taxa) are steeped in questionable assumptions of homology and artificial discretization of phenotypic traits, respectively. Given that the developmental genetic basis for the majority of arachnid morphological traits remains unknown, there is little validation available for the majority of traits that putatively inform chelicerate phylogeny. While we do not doubt that robust morphological matrices and sufficiently detailed fossil histories, with internal validation, are available for better test cases of the total evidence approach (e.g., echinoderms; shelled mollusks; vertebrates), we submit that chelicerates are simply and clearly not among this group.

Morphology has essentially outlived its usefulness as a phylogenetic data class for the higher-level relationships of numerous taxa that do not exhibit a large number of morphological features to score, lack detailed fossil records, exhibit a propensity for morphological convergence, or lack clear phylogenetic signal in anatomical character systems (e.g., Annelida; Nematoda; Sipuncula; Platyhelminthes; Cycliophora; Gnathostomulida; Chaetognatha). Despite the exhortations of a few adherents of the total evidence approach to collect more morphological data for combined analyses—even when molecular data have superiorly resolved the phylogenetic questions at hand in the same study (e.g., [60])—the notion that the next generation of evolutionary biologists will turn away in droves from fields like genomics and molecular evolution to dedicate time for coding morphological matrices is not realistic. It is detrimental to the next generation of arthropod morphologists to advertise the continuation of questionable phylogenetic practices, as well as to pass on the burden of defending unsupported, century-old hypotheses, as a viable course of scholarly pursuit for the future—particularly given that new finds and advanced imaging techniques have actively transformed our understanding of arthropod paleontology, outdated homology statements, and the evolution of character states in the past decade alone [63,96,133–135]. Paleontology has the potential to contribute greatly to the understanding of chelicerate evolution, but that potential is a function of how objectively fossil data and competing evolutionary scenarios are interpreted. It is far better to approach new discoveries in paleontology and morphology with an open mind than to cudgel their interpretations to accord with obsolete ideas and ossified biases.

Bluntly stated, morphology has nothing left to contribute to the defense of Arachnida, and unbiased analyses of molecular data outright refute it. Morphologists have had centuries to uncover characters for chelicerate relationships. We long ago hit the point of diminishing returns with regard to new character systems to score. Recent exceptions that have leveraged high-end imaging approaches to survey internal characters are certainly uncovering new character systems [136–139], but these only reinforce support for new hypotheses of relationships derived from phylogenomics, thereby further undermining morphological topologies that continue to be produced by chelicerate paleontologists (e.g., [5,7,9,89,91]). At some point, adherents of Arachnida need to weigh objectively whether their goal is to test a systematic hypothesis using unbiased assessments of available evidence and balanced treatment of new ideas, or simply to affirm a traditional worldview—an affirmation that seems to require cherry-picking of taxa, matrices, characters, and substitution models, in tandem with shallow analytical interrogations of error-riddled datasets.

As was the case for proponents of Articulata and Atelocerata, adherents of Arachnida will gradually dwindle in number as their case inexorably erodes. The field will continue to move away from the limitations of subjective character interpretations and untested assumptions passed down through time, leaving Arachnida behind in the dustbin of obsolete morphology-based groupings. After a time, the post hoc realignment of morphological interpretations to accord with the new outcomes of molecular phylogenies will bury an older generation of paleontology- and morphology-driven hypotheses of chelicerate relationships, paralleling the history of Ecdysozoa and Pancrustacea [81,91]. In the case of Arachnopulmonata and Panscorpiones, this process is already well underway [10,137]. Inevitably, when this post hoc realignment is completed (usually, about a decade after molecular datasets have cracked the case), some paleontologists and "molecular paleobiologists" will proclaim its success as a victory and a vindication of morphology, without a trace of irony (e.g., [91]).

Propitious developments in chelicerate genomics foretell a bright future for the study of chelicerate phylogeny and evolution. The discovery of rare genomic changes in the chelicerate tree of life has offered a new and powerful class of data for testing competing tree topologies and validating inferential approaches to phylogenomics—enabling arachnologists to break century-old impasses in the placement of groups like scorpions and pseudoscorpions [10,20]. The advent of "molecular morphology" in the form of analyses of genomic architecture, together with independent evaluations of various data classes (e.g., gene expression patterns; microRNA incidence and enrichment; gene family duplications; [10,20–27]), offers a robust vehicle for the achievement of consilience in chelicerate evolutionary relationships. The parallel trend of accruing functional genetic tools for various chelicerate model systems offers new routes to understanding morphological traits and their evolution, as well as discovering functional links between genomes and phenotypes [20,29,34,140–145]. It is in such vehicles that future scholars of chelicerate evolution should direct their efforts, energy, and attention.

Author Contributions: Conceptualization, P.P.S., J.A.B. and C.E.S.-L.; methodology and analysis, P.P.S. and J.A.B.; writing—original draft preparation, P.P.S.; writing—review and editing, J.A.B. and C.E.S.-L.; visualization, P.P.S.; funding acquisition, P.P.S. All authors have read and agreed to the published version of the manuscript.

Funding: This research was funded by the National Science Foundation, grant numbers IOS-1552610 and IOS-2016141. The APC was funded by IOS-2016141.

Data Availability Statement: Datasets corresponding to Figures 3–5 are available on FigShare: https://figshare.com/articles/dataset/Supplementary_material_for_Sharma_et_al_2021_-_Diversity/16913386.

Acknowledgments: We are indebted to Matjaž Kuntner for the invitation to contribute to this special issue. Gonzalo Giribet, Georg Brenneis, and Claudia P. Arango provided photographs in Figure 6. Comments from Matjaž Kuntner and two anonymous reviewers improved a previous draft of the work.

References

1. Weygoldt, P.; Paulus, H.F. Untersuchungen zur Morphologie, Taxonomie und Phylogenie der Chelicerata II. Cladogramme und die Entfaltung der Chelicerata. *J. Zool. Syst. Evol. Res.* **1979**, *17*, 177–200. [CrossRef]
2. Shultz, J.W. A phylogenetic analysis of the arachnid orders based on morphological characters. *Zool. J. Linn. Soc.* **2007**, *150*, 221–265. [CrossRef]
3. Giribet, G. Current views on chelicerate phylogeny—A tribute to Peter Weygoldt. *Zool. Anz.* **2018**, *273*, 7–13. [CrossRef]

4. Dunlop, J.A. Geological history and phylogeny of Chelicerata. *Arthropod Struct. Dev.* **2010**, *39*, 124–142. [CrossRef]

5. Lamsdell, J.C. Horseshoe crab phylogeny and independent colonizations of fresh water: Ecological invasion as a driver for morphological innovation. *Palaeontology* **2016**, *59*, 181–194. [CrossRef]

6. Lamsdell, J.C.; McCoy, V.E.; Perron-Feller, O.A.; Hopkins, M.J. Air breathing in an exceptionally preserved 340- million-year-old sea scorpion. *Curr. Biol.* **2020**, *30*, 4316–4321.e2. [CrossRef]

7. Wolfe, J.M. Metamorphosis is ancestral for crown euarthropods, and evolved in the Cambrian or earlier. *Integr. Comp. Biol.* **2017**, *57*, 499–509. [CrossRef] [PubMed]

8. Bicknell, R.D.C.; Lustri, L.; Brougham, T. Revision of *"Bellinurus" carteri* (Chelicerata: Xiphosura) from the Late Devonian of Pennsylvania, USA. *C. R.—Palevol* **2019**, *18*, 967–976. [CrossRef]

9. Aria, C.; Caron, J.-B. A Middle Cambrian arthropod with chelicerae and proto-book gills. *Nature* **2019**, *573*, 586–589. [CrossRef]

10. Ontano, A.Z.; Gainett, G.; Aharon, S.; Ballesteros, J.A.; Benavides, L.R.; Corbett, K.F.; Gavish-Regev, E.; Harvey, M.S.; Monsma, S.; Santibáñez-López, C.E.; et al. Taxonomic sampling and rare genomic changes overcome long-branch attraction in the phylogenetic placement of pseudoscorpions. *Mol. Biol. Evol.* **2021**, *38*, 2446–2467. [CrossRef] [PubMed]

11. Wheeler, W.; Hayashi, C. The phylogeny of the extant chelicerate orders. *Cladistics* **1998**, *14*, 173–192. [CrossRef]

12. Giribet, G.; Edgecombe, G.; Wheeler, W. Arthropod phylogeny based on eight molecular loci and morphology. *Nature* **2001**, *413*, 157–161. [CrossRef] [PubMed]

13. Masta, S.; Longhorn, S.; Boore, J. Arachnid relationships based on mitochondrial genomes: Asymmetric nucleotide and amino acid bias affects phylogenetic analyses. *Mol. Phylogenet. Evol.* **2009**, *50*, 117–128. [CrossRef]

14. Masta, S.E.; McCall, A.; Longhorn, S.J. Rare genomic changes and mitochondrial sequences provide independent support for congruent relationships among the sea Spiders (Arthropoda, Pycnogonida). *Mol. Phylogenet. Evol.* **2010**, *57*, 59–70. [CrossRef] [PubMed]

15. Pepato, A.R.; Rocha, C.E.F.d.; Dunlop, J.A. Phylogenetic position of the acariform mites: Sensitivity to homology assessment under total evidence. *BMC Evol. Biol.* **2010**, *10*, 235. [CrossRef] [PubMed]

16. Regier, J.C.; Shultz, J.W.; Zwick, A.; Hussey, A.; Ball, B.; Wetzer, R.; Martin, J.W.; Cunningham, C.W. Arthropod relationships revealed by phylogenomic analysis of nuclear protein-coding sequences. *Nature* **2010**, *463*, 1079–1083. [CrossRef]

17. Sharma, P.P.; Kaluziak, S.T.; Perez-Porro, A.R.; Gonzalez, V.L.; Hormiga, G.; Wheeler, W.C.; Giribet, G. Phylogenomic interrogation of Arachnida reveals systemic conflicts in phylogenetic signal. *Mol. Biol. Evol.* **2014**, *31*, 2963–2984. [CrossRef]

18. Howard, R.J.; Puttick, M.N.; Edgecombe, G.D.; Lozano-Fernandez, J. Arachnid Monophyly: Morphological, palaeontological and molecular support for a single terrestrialization within Chelicerata. *Arthropod Struct. Dev.* **2020**, *59*, 100997. [CrossRef]

19. Scholtz, G.; Kamenz, C. The Book Lungs of Scorpiones and Tetrapulmonata (Chelicerata, Arachnida): Evidence for homology and a single terrestrialisation event of a common arachnid ancestor. *Zoology* **2006**, *109*, 2–13. [CrossRef] [PubMed]

20. Sharma, P.P.; Schwager, E.E.; Extavour, C.G.; Wheeler, W.C. Hox gene duplications correlate with posterior heteronomy in scorpions. *Proc. R. Soc. B Biol. Sci.* **2014**, *281*, 20140661. [CrossRef] [PubMed]

21. Leite, D.J.; Ninova, M.; Hilbrant, M.; Arif, S.; Griffiths-Jones, S.; Ronshaugen, M.; Mcgregor, A.P. Pervasive MicroRNA duplication in chelicerates: Insights from the embryonic microRNA repertoire of the spider *Parasteatoda tepidariorum*. *Genome Biol. Evol.* **2016**, *8*, 2133–2144. [CrossRef] [PubMed]

22. Leite, D.J.; Baudouin-Gonzalez, L.; Iwasaki-Yokozawa, S.; Lozano-Fernandez, J.; Turetzek, N.; Akiyama-Oda, Y.; Prpic, N.-M.; Pisani, D.; Oda, H.; Sharma, P.P.; et al. Homeobox gene duplication and divergence in arachnids. *Mol. Biol. Evol.* **2018**, *35*, 2240–2253. [CrossRef] [PubMed]

23. Schwager, E.E.; Sharma, P.P.; Clarke, T.; Leite, D.J.; Wierschin, T.; Pechmann, M.; Akiyama-Oda, Y.; Esposito, L.; Bechsgaard, J.; Bilde, T.; et al. The house spider genome reveals an ancient whole-genome duplication during arachnid evolution. *BMC Biol.* **2017**, *15*, 62. [CrossRef] [PubMed]

24. Schwager, E.E.; Schoppmeier, M.; Pechmann, M.; Damen, W.G. Duplicated hox genes in the spider *Cupiennius salei*. *Front. Zool.* **2007**, *4*, 10. [CrossRef]

25. Pechmann, M.; Prpic, N.-M. Appendage patterning in the South American bird spider *Acanthoscurria geniculata* (Araneae: Mygalomorphae). *Dev. Genes Evol.* **2009**, *219*, 189–198. [CrossRef] [PubMed]

26. Nolan, E.D.; Santibáñez-López, C.E.; Sharma, P.P. Developmental gene expression as a phylogenetic data class: Support for the monophyly of Arachnopulmonata. *Dev. Genes Evol.* **2020**, *230*, 137–153. [CrossRef] [PubMed]

27. Turetzek, N.; Pechmann, M.; Schomburg, C.; Schneider, J.; Prpic, N.-M. Neofunctionalization of a duplicate dachshundgene underlies the evolution of a novel leg segment in arachnids. *Mol. Biol. Evol.* **2015**, *33*, 109–121. [CrossRef]

28. Setton, E.V.W.; Hendrixson, B.E.; Sharma, P.P. Embryogenesis in a Colorado population of *Aphonopelma hentzi* (Girard, 1852) (Araneae: Mygalomorphae: Theraphosidae): Establishing a promising system for the study of mygalomorph development. *J. Arachnol.* **2019**, *47*, 209–216. [CrossRef]

29. Gainett, G.; Sharma, P.P. Genomic resources and toolkits for developmental study of whip spiders (Amblypygi) provide insights into arachnid genome evolution and antenniform leg patterning. *EvoDevo* **2020**, *11*, 18. [CrossRef]

30. Gainett, G.; Ballesteros, J.A.; Kanzler, C.R.; Zehms, J.T.; Zern, J.M.; Aharon, S.; Gavish-Regev, E.; Sharma, P.P. Systemic paralogy and function of retinal determination network homologs in arachnids. *BMC Genom.* **2020**, *21*, 811. [CrossRef]

31. Harper, A.; Gonzalez, L.B.; Schönauer, A.; Janssen, R.; Seiter, M.; Holzem, M.; Arif, S.; McGregor, A.P.; Sumner-Rooney, L. Widespread retention of ohnologs in key developmental gene families following whole genome duplication in arachnopulmonates. *G3 Genes Genomes Genet.* **2021**. [CrossRef]

32. Grbić, M.; Leeuwen, T.V.; Clark, R.M.; Rombauts, S.; Rouzé, P.; Grbić, V.; Osborne, E.J.; Dermauw, W.; Ngoc, P.C.T.; Ortego, F.; et al. The genome of *Tetranychus urticae* reveals herbivorous pest adaptations. *Nature* **2011**, *479*, 487–492. [CrossRef] [PubMed]

33. Gulia-Nuss, M.; Nuss, A.B.; Meyer, J.M.; Sonenshine, D.E.; Roe, R.M.; Waterhouse, R.M.; Sattelle, D.B.; Fuente, J.e.d.l.; Ribeiro, J.M.; Megy, K.; et al. Genomic insights into the *Ixodes scapularis* tick vector of Lyme disease. *Nat. Commun.* **2016**, *7*, 10507. [CrossRef] [PubMed]

34. Gainett, G.; González, V.L.; Ballesteros, J.A.; Setton, E.V.W.; Baker, C.M.; Barolo Gargiulo, L.; Santibáñez-López, C.E.; Coddington, J.A.; Sharma, P.P. The genome of a daddy-long-legs (Opiliones) illuminates the evolution of arachnid appendages. *Proc. R. Soc. B Biol. Sci.* **2021**, *288*, 20211168. [CrossRef]

35. Kenny, N.J.; Chan, K.W.; Nong, W.; Qu, Z.; Maeso, I.; Yip, H.Y.; Chan, T.F.; Kwan, H.S.; Holland, P.W.H.; Chu, K.H.; et al. Ancestral whole-genome duplication in the marine chelicerate horseshoe crabs. *Heredity* **2016**, *116*, 190–199. [CrossRef]

36. Shingate, P.; Ravi, V.; Prasad, A.; Tay, B.-H.; Garg, K.M.; Chattopadhyay, B.; Yap, L.-M.; Rheindt, F.E.; Venkatesh, B. Chromosome-level assembly of the horseshoe crab genome provides insights into its genome evolution. *Nat. Commun.* **2020**, *11*, 2322. [CrossRef]

37. Shingate, P.; Ravi, V.; Prasad, A.; Tay, B.-H.; Venkatesh, B. Chromosome-level genome assembly of the coastal horseshoe crab (*Tachypleus gigas*). *Mol. Ecol. Resour.* **2020**, *20*, 1748–1760. [CrossRef]

38. Nong, W.; Qu, Z.; Li, Y.; Barton-Owen, T.; Wong, A.Y.P.; Yip, H.Y.; Lee, H.T.; Narayana, S.; Baril, T.; Swale, T.; et al. Horseshoe crab genomes reveal the evolution of genes and microRNAs after three rounds of whole genome duplication. *Commun. Biol.* **2021**, *4*, 83. [CrossRef] [PubMed]

39. Shultz, J.W. Evolutionary morphology and phylogeny of Arachnida. *Cladistics* **1990**, *6*, 1–38. [CrossRef]

40. Dunlop, J. The origins of tetrapulmonate book lungs and their significance for chelicerate phylogeny. In *Proceedings of the 17th European Colloquium of Arachnology*; British Arachnological Society: Edinburgh, UK, 1998; pp. 9–16.

41. Ballesteros, J.A.; Sharma, P.P. A critical appraisal of the placement of Xiphosura (Chelicerata) with account of known sources of phylogenetic error. *Syst. Biol.* **2019**, *68*, 896–917. [CrossRef]

42. Lozano-Fernandez, J.; Tanner, A.R.; Giacomelli, M.; Carton, R.; Vinther, J.; Edgecombe, G.D.; Pisani, D. Increasing species sampling in chelicerate genomic-scale datasets provides support for monophyly of Acari and Arachnida. *Nat. Commun.* **2019**, *10*, 2295. [CrossRef]

43. Ballesteros, J.A.; Hormiga, G. A New orthology assessment method for phylogenomic data: Unrooted Phylogenetic Orthology. *Mol. Biol. Evol.* **2016**, *33*, 2117–2134. [CrossRef] [PubMed]

44. Ballesteros, J.A.; Santibáñez-López, C.E.; Kováč, Ľ.; Gavish-Regev, E.; Sharma, P.P. Ordered phylogenomic subsampling enables diagnosis of systematic errors in the placement of the enigmatic arachnid order Palpigradi. *Proc. R. Soc. B Biol. Sci.* **2019**, *286*, 20192426. [CrossRef]

45. Altenhoff, A.M.; Schneider, A.; Gonnet, G.H.; Dessimoz, C. OMA 2011: Orthology inference among 1000 complete genomes. *Nucleic Acids Res.* **2010**, *39*, D289–D294. [CrossRef] [PubMed]

46. Lartillot, N.; Philippe, H. A Bayesian mixture model for across-site heterogeneities in the amino-acid replacement process. *Mol. Biol. Evol.* **2004**, *21*, 1095–1109. [CrossRef]

47. Klompen, H.; Lekveishvili, M.; IV, W.C.B. Phylogeny of parasitiform mites (Acari) based on rRNA. *Mol. Phylogenet. Evol.* **2007**, *43*, 936–951. [CrossRef] [PubMed]

48. Lozano-Fernandez, J.; Giacomelli, M.; Fleming, J.F.; Chen, A.; Vinther, J.; Thomsen, P.F.; Glenner, H.; Palero, F.; Legg, D.A.; Iliffe, T.M.; et al. Pancrustacean evolution illuminated by taxon-rich genomic-scale data sets with an expanded remipede sampling. *Genome Biol. Evol.* **2019**, *11*, 2055–2070. [CrossRef]

49. Whelan, N.V.; Halanych, K.M. Who let the CAT out of the bag? Accurately dealing with substitutional heterogeneity in phylogenomic analyses. *Syst. Biol.* **2017**, *66*, 232–255. [CrossRef]

50. Ballesteros, J.A.; Santibáñez-López, C.E.; Baker, C.M.; Benavides, L.R.; Cunha, T.J.; Gainett, G.; Ontano, A.Z.; Setton, E.V.W.; Arango, C.P.; Gavish-Regev, E.; et al. Comprehensive species sampling and sophisticated algorithmic approaches refute the monophyly of Arachnida. *bioRxiv* **2021**. [CrossRef]

51. Shen, X.-X.; Hittinger, C.T.; Rokas, A. Contentious relationships in phylogenomic studies can be driven by a handful of genes. *Nat. Ecol. Evol.* **2017**, *1*, 1–10. [CrossRef]

52. Maxmen, A. The sea spider's contribution to T.H. Morgan's (1866–1945) development. *J. Exp. Zool. Part B Mol. Dev. Evol.* **2008**, *310B*, 203–215. [CrossRef] [PubMed]

53. Mongiardino Koch, N. Phylogenomic subsampling and the search for phylogenetically reliable loci. *Mol. Biol. Evol.* **2021**, *38*, 4025–4038. [CrossRef] [PubMed]

54. Hedges, S.B.; Tao, Q.; Walker, M.; Kumar, S. Accurate timetrees require accurate calibrations. *Proc. Natl. Acad. Sci. USA* **2018**, *115*, E9510–E9511. [CrossRef] [PubMed]

55. Laumer, C.E.; Fernández, R.; Lemer, S.; Combosch, D.; Kocot, K.M.; Riesgo, A.; Andrade, S.C.S.; Sterrer, W.; Sørensen, M.V.; Giribet, G. Revisiting metazoan phylogeny with genomic sampling of all phyla. *Proc. R. Soc. B Biol. Sci.* **2019**, *286*, 20190831. [CrossRef] [PubMed]

56. Szucsich, N.U.; Bartel, D.; Blanke, A.; Böhm, A.; Donath, A.; Fukui, M.; Grove, S.; Liu, S.; Macek, O.; Machida, R.; et al. Four myriapod relatives—But who are sisters? No end to debates on relationships among the four major myriapod subgroups. *BMC Evol. Biol.* **2020**, *20*, 144. [CrossRef] [PubMed]

57. Li, Y.; Shen, X.-X.; Evans, B.; Dunn, C.W.; Rokas, A. Rooting the animal Tree of Life. *Mol. Biol. Evol.* **2021**, *38*, 4322–4333. [CrossRef] [PubMed]

58. Gustafson, G.T.; Miller, K.B.; Michat, M.C.; Alarie, Y.; Baca, S.M.; Balke, M.; Short, A.E.Z. The enduring value of reciprocal illumination in the era of insect phylogenomics: A response to Cai et al. (2020). *Syst. Entomol.* **2021**, *46*, 473–486. [CrossRef]

59. Telford, M.J.; Moroz, L.L.; Halanych, K.M. A sisterly dispute. *Nature* **2016**, *529*, 286–287. [CrossRef]

60. Ganske, A.-S.; Vahtera, V.; Dányi, L.; Edgecombe, G.D.; Akkari, N. Phylogeny of Lithobiidae Newport, 1844, with emphasis on the megadiverse genus *Lithobius* Leach, 1814 (Myriapoda, Chilopoda). *Cladistics* **2021**, *37*, 162–184. [CrossRef] [PubMed]

61. Bieler, R.; Mikkelsen, P.M.; Collins, T.M.; Glover, E.A.; González, V.L.; Graf, D.L.; Harper, E.M.; Healy, J.; Kawauchi, G.Y.; Sharma, P.P. Investigating the bivalve Tree of Life–an exemplar-based approach combining molecular and novel morphological characters. *Invertebr. Syst.* **2014**, *28*, 32–115. [CrossRef]

62. Mongiardino Koch, N.; Thompson, J.R. A total-evidence dated phylogeny of Echinoidea combining phylogenomic and paleontological data. *Syst. Biol.* **2021**, *70*, 421–439. [CrossRef]

63. Garwood, R.J.; Sharma, P.P.; Dunlop, J.A.; Giribet, G. A Paleozoic stem group to mite harvestmen revealed through integration of phylogenetics and development. *Curr. Biol. CB* **2014**, *24*, 1017–1023. [CrossRef] [PubMed]

64. Sharma, P.P.; Giribet, G. A revised dated phylogeny of the arachnid order Opiliones. *Front. Genet.* **2014**, *5*, 255. [CrossRef] [PubMed]

65. Chen, A.; White, N.D.; Benson, R.B.J.; Braun, M.J.; Field, D.J. Total-evidence framework reveals complex morphological evolution in nightbirds (Strisores). *Diversity* **2019**, *11*, 143. [CrossRef]

66. Paterson, R.S.; Rybczynski, N.; Kohno, N.; Maddin, H.C. A total evidence phylogenetic analysis of pinniped phylogeny and the possibility of parallel evolution within a monophyletic framework. *Front. Ecol. Evol.* **2020**, *7*, 457. [CrossRef]

67. Giribet, G. Morphology should not be forgotten in the era of genomics–A phylogenetic perspective. *Zool. Anz.—J. Comp. Zool.* **2015**, *256*, 96–103. [CrossRef]

68. Pisani, D.; Poling, L.L.; Lyons-Weiler, M.; Hedges, S.B. The colonization of land by animals: Molecular phylogeny and divergence times among arthropods. *BMC Biol.* **2004**, *2*, 1. [CrossRef]

69. Giribet, G.; Wheeler, W. On bivalve phylogeny: A high-level analysis of the Bivalvia (Mollusca) based on combined morphology and DNA sequence data. *Invertebr. Biol.* **2002**, *121*, 271–324. [CrossRef]

70. Sharma, P.P.; González, V.L.; Kawauchi, G.Y.; Andrade, S.C.S.; Guzmán, A.; Collins, T.M.; Glover, E.A.; Harper, E.M.; Healy, J.M.; Mikkelsen, P.M.; et al. Phylogenetic analysis of four nuclear protein-encoding genes largely corroborates the traditional classification of Bivalvia (Mollusca). *Mol. Phylogenet. Evol.* **2012**, *65*, 64–74. [CrossRef]

71. Rota-Stabelli, O.; Campbell, L.; Brinkmann, H.; Edgecombe, G.D.; Longhorn, S.J.; Peterson, K.J.; Pisani, D.; Philippe, H.; Telford, M.J. A congruent solution to arthropod phylogeny: Phylogenomics, microRNAs and morphology support monophyletic Mandibulata. *Proc. R. Soc. B Biol. Sci.* **2011**, *278*, 298–306. [CrossRef] [PubMed]

72. Gonzalez, V.L.; Andrade, S.C.S.; BIELER, R.; Collins, T.M.; Dunn, C.W.; Mikkelsen, P.M.; Taylor, J.D.; Giribet, G. A Phylogenetic backbone for Bivalvia: An RNA-Seq approach. *Proc. R. Soc. B Biol. Sci.* **2015**, *282*, 20142332. [CrossRef] [PubMed]

73. Jenner, R.A. Bilaterian Phylogeny and uncritical recycling of morphological data sets. *Syst. Biol.* **2001**, *50*, 730–742. [CrossRef]

74. Jenner, R. Boolean logic and character state identity: Pitfalls of character coding in metazoan cladistics. *Contrib. Zool.* **2002**, *71*, 67–91. [CrossRef]

75. Wright, A.M.; Hillis, D.M. Bayesian analysis using a simple likelihood model outperforms parsimony for estimation of phylogeny from discrete morphological data. *PLoS ONE* **2014**, *9*, e109210. [CrossRef]

76. O'Reilly, J.E.; Puttick, M.N.; Parry, L.; Tanner, A.R.; Tarver, J.E.; Fleming, J.; Pisani, D.; Donoghue, P.C.J. Bayesian methods outperform parsimony but at the expense of precision in the estimation of phylogeny from discrete morphological data. *Biol. Lett.* **2016**, *12*, 20160081. [CrossRef]

77. O'Reilly, J.E.; Puttick, M.N.; Pisani, D.; Donoghue, P.C.J. Probabilistic methods surpass parsimony when assessing clade support in phylogenetic analyses of discrete morphological data. *Palaeontology* **2018**, *61*, 105–118. [CrossRef]

78. Goloboff, P.A.; Pittman, M.; Pol, D.; Xu, X. Morphological data sets fit a common mechanism much more poorly than dna sequences and call into question the Mkv model. *Syst. Biol.* **2019**, *68*, 494–504. [CrossRef] [PubMed]

79. Goloboff, P.A.; Galvis, A.T.; Arias, J.S. Parsimony and model-based phylogenetic methods for morphological data: Comments on O'Reilly et al. *Palaeontology* **2018**, *61*, 625–630. [CrossRef]

80. Goloboff, P.A.; Arias, J.S. Likelihood approximations of implied weights parsimony can be selected over the Mk model by the Akaike Information Criterion. *Cladistics* **2019**, *35*, 695–716. [CrossRef] [PubMed]

81. Giribet, G.; Edgecombe, G.D. Current understanding of Ecdysozoa and its internal phylogenetic relationships. *Integr. Comp. Biol.* **2017**, *57*, 455–466. [CrossRef]

82. Regier, J.C.; Shultz, J.W.; Kambic, R.E. Pancrustacean phylogeny: Hexapods are terrestrial crustaceans and maxillopods are not monophyletic. *Proc. R. Soc. B Biol. Sci.* **2005**, *272*, 395–401. [CrossRef]

83. Marlétaz, F.; Peijnenburg, K.T.C.A.; Goto, T.; Satoh, N.; Rokhsar, D.S. A new spiralian phylogeny places the enigmatic arrow worms among gnathiferans. *Curr. Biol.* **2019**, *29*, 312–318.e3. [CrossRef] [PubMed]

84. Fröbius, A.C.; Funch, P. Rotiferan Hox genes give new insights into the evolution of metazoan bodyplans. *Nat. Commun.* **2017**, *8*, 9. [CrossRef]
85. Kapli, P.; Natsidis, P.; Leite, D.J.; Fursman, M.; Jeffrie, N.; Rahman, I.A.; Philippe, H.; Copley, R.R.; Telford, M.J. Lack of support for Deuterostomia prompts reinterpretation of the first Bilateria. *Sci. Adv.* **2021**, *7*, eabe2741. [CrossRef] [PubMed]
86. Zou, Z.; Zhang, J. Morphological and molecular convergences in mammalian phylogenetics. *Nat. Commun.* **2016**, *7*, 12758. [CrossRef] [PubMed]
87. Oyston, J.; Wilkinson, M.; Ruta, M.; Wills, M. Molecular phylogenies map to biogeography better than morphological ones. *ResearchSquare* **2021**. preprint. [CrossRef]
88. Lee, M.S.Y.; Palci, A. Morphological phylogenetics in the genomic age. *Curr. Biol.* **2015**, *25*, R922–R929. [CrossRef]
89. Legg, D.A. *Sanctacaris uncata*: The oldest chelicerate (Arthropoda). *Naturwissenschaften* **2014**, *101*, 1065–1073. [CrossRef]
90. Briggs, D.E.G.; Siveter, D.J.; Siveter, D.J.; Sutton, M.D.; Legg, D. Tiny individuals attached to a new Silurian arthropod suggest a unique mode of brood care. *Proc. Natl. Acad. Sci. USA* **2016**, *113*, 4410–4415. [CrossRef]
91. Legg, D.A.; Sutton, M.D.; Edgecombe, G.D. Arthropod fossil data increase congruence of morphological and molecular phylogenies. *Nat. Commun.* **2013**, *4*, 2485. [CrossRef]
92. Siveter, D.J.; Briggs, D.E.G.; Siveter, D.J.; Sutton, M.D.; Legg, D. A new crustacean from the Herefordshire (Silurian) Lagerstätte, UK, and its significance in malacostracan evolution. *Proc. R. Soc. B Biol. Sci.* **2017**, *284*, 20170279. [CrossRef]
93. Lamsdell, J.C.; Briggs, D.E.G.; Liu, H.P.; Witzke, B.J.; McKay, R.M. A new Ordovician arthropod from the Winneshiek Lagerstätte of Iowa (USA) reveals the ground plan of eurypterids and chasmataspidids. *Naturwissenschaften* **2015**, *102*, 63. [CrossRef]
94. Garwood, R.J.; Dunlop, J. Three-dimensional reconstruction and the phylogeny of extinct chelicerate orders. *PeerJ* **2014**, *2*, e641. [CrossRef] [PubMed]
95. Garwood, R.J.; Dunlop, J.A.; Selden, P.A.; Spencer, A.R.T.; Atwood, R.C.; Vo, N.T.; Drakopoulos, M. Almost a spider: A 305-million-year-old fossil arachnid and spider origins. *Proc. R. Soc. B Biol. Sci.* **2016**, *283*, 20160125. [CrossRef]
96. Wang, B.; Dunlop, J.A.; Selden, P.A.; Garwood, R.J.; Shear, W.A.; Müller, P.; Lei, X. Cretaceous Arachnid *Chimerarachne yingi* gen. et sp. nov. illuminates spider origins. *Nat. Ecol. Evol.* **2018**, *2*, 614–622. [CrossRef] [PubMed]
97. Huang, D.; Hormiga, G.; Cai, C.; Su, Y.; Yin, Z.; Xia, F.; Giribet, G. Origin of spiders and their spinning organs illuminated by mid-Cretaceous amber fossils. *Nat. Ecol. Evol.* **2018**, *2*, 623–627. [CrossRef]
98. Shultz, J.W. Gross muscular anatomy of *Limulus polyphemus* (Xiphosura, Chelicerata) and its bearing on evolution in the Arachnida. *J. Arachnol.* **2001**, *29*, 283–303. [CrossRef]
99. Arango, C.P. Morphological phylogenetics of the sea spiders (Arthropoda: Pycnogonida). *Org. Divers. Evol.* **2002**, *2*, 107–125. [CrossRef]
100. Ballesteros, J.A.; Setton, E.V.W.; Santibáñez-López, C.E.; Arango, C.P.; Brenneis, G.; Brix, S.; Corbett, K.F.; Cano-Sánchez, E.; Dandouch, M.; Dilly, G.F.; et al. Phylogenomic resolution of sea spider diversification through integration of multiple data classes. *Mol. Biol. Evol.* **2021**, *38*, 686–701. [CrossRef]
101. Chetverikov, P.E.; Desnitskiy, A.G. A study of embryonic development in eriophyoid mites (Acariformes, Eriophyoidea) with the use of the fluorochrome DAPI and confocal microscopy. *Exp. Appl. Acarol.* **2015**, *68*, 97–111. [CrossRef]
102. Boxshall, G.A. The evolution of arthropod limbs. *Biol. Rev.* **2004**, *79*, 253–300. [CrossRef] [PubMed]
103. Giribet, G.; Edgecombe, G.D.; Wheeler, W.C.; Babbitt, C. Phylogeny and systematic position of Opiliones: A combined analysis of chelicerate relationships using morphological and molecular data. *Cladistics* **2002**, *18*, 5–70. [CrossRef] [PubMed]
104. Dunlop, J.A. The epistomo-labral plate and lateral lips in solifuges, pseudoscorpions and mites. *Ekologia* **2000**, *19*, 67–78.
105. Dunlop, J.A.; Krüger, J.; Alberti, G. The sejugal furrow in camel spiders and acariform mites. *Arachnol. Mitt.* **2012**, *43*, 29–36. [CrossRef]
106. Ramírez, M.J.; Magalhaes, I.L.F.; Derkarabetian, S.; Ledford, J.; Griswold, C.E.; Wood, H.M.; Hedin, M. Sequence capture phylogenomics of true spiders reveals convergent evolution of respiratory systems. *Syst. Biol.* **2021**, *70*, 14–20. [CrossRef]
107. Watson-Zink, V.M. Making the grade: Physiological adaptations to terrestrial environments in decapod crabs. *Arthropod Struct. Dev.* **2021**, *64*, 101089. [CrossRef]
108. Dimitriou, A.C.; Taiti, S.; Sfenthourakis, S. Genetic evidence against monophyly of Oniscidea implies a need to revise scenarios for the origin of terrestrial isopods. *Sci. Rep.* **2019**, *9*, 18508. [CrossRef]
109. Jörger, K.M.; Stöger, I.; Kano, Y.; Fukuda, H.; Knebelsberger, T.; Schrödl, M. On the origin of Acochlidia and other enigmatic euthyneuran gastropods, with implications for the systematics of Heterobranchia. *BMC Evol. Biol.* **2010**, *10*, 323. [CrossRef] [PubMed]
110. Smythe, A.B.; Holovachov, O.; Kocot, K.M. Improved phylogenomic sampling of free-living nematodes enhances resolution of higher-level nematode phylogeny. *BMC Evol. Biol.* **2019**, *19*, 1–15. [CrossRef]
111. Lewandowsky, S.; Ecker, U.K.H.; Cook, J. Beyond misinformation: Understanding and coping with the "post-truth" era. *J. Appl. Res. Mem. Cogn.* **2017**, *6*, 353–369. [CrossRef]
112. Pyron, R.A. Divergence time estimation using fossils as terminal taxa and the origins of Lissamphibia. *Syst. Biol.* **2011**, *60*, 466–481. [CrossRef] [PubMed]
113. Koch, N.M.; Parry, L.A. Death is on our side: Paleontological data drastically modify phylogenetic hypotheses. *Syst. Biol.* **2020**, *69*, 1052–1067. [CrossRef] [PubMed]

114. Arango, C.P.; Wheeler, W.C. Phylogeny of the sea spiders (Arthropoda, Pycnogonida) based on direct optimization of six loci and morphology. *Cladistics* **2007**, *23*, 255–293. [CrossRef]
115. Sharma, P.P.; Fernández, R.; Esposito, L.A.; González-Santillán, E.; Monod, L. Phylogenomic resolution of scorpions reveals multilevel discordance with morphological phylogenetic signal. *Proc. R. Soc. B Biol. Sci.* **2015**, *282*, 20142953. [CrossRef] [PubMed]
116. Monod, L.; Cauwet, L.; González-Santillán, E.; Huber, S. The male sexual apparatus in the order Scorpiones (Arachnida): A comparative study of functional morphology as a tool to define hypotheses of homology. *Front. Zool.* **2017**, *14*, 1–48. [CrossRef] [PubMed]
117. Santibáñez-López, C.E.; Graham, M.R.; Sharma, P.P.; Ortiz, E.; Possani, L.D. Hadrurid scorpion toxins: Evolutionary conservation and selective pressures. *Toxins* **2019**, *11*, 637. [CrossRef]
118. Santibáñez-López, C.E.; González-Santillán, E.; Monod, L.; Sharma, P.P. Phylogenomics facilitates stable scorpion systematics: Reassessing the relationships of Vaejovidae and a new higher-level classification of scorpiones (Arachnida). *Mol. Phylogenet. Evol.* **2019**, *135*, 22–30. [CrossRef]
119. Santibáñez-López, C.E.; Ojanguren-Affilastro, A.A.; Sharma, P.P. Another one bites the dust: Taxonomic sampling of a key genus in phylogenomic datasets reveals more non-monophyletic groups in traditional scorpion classification. *Invertebr. Syst.* **2020**, *34*, 133–143. [CrossRef]
120. Santibáñez-López, C.E.; Kriebel, R.; Sharma, P.P. *eadem figura manet:* Measuring morphological convergence in diplocentrid scorpions (Arachnida: Scorpiones: Diplocentridae) under a multilocus phylogenetic framework. *Invertebr. Syst.* **2017**, *31*, 233–248. [CrossRef]
121. Dunlop, J.; Anderson, L.; Kerp, H.; Hass, H. A harvestman (Arachnida: Opiliones) from the Early Devonian Rhynie Cherts, Aberdeenshire, Scotland. *Trans. R. Soc. Edinb. Earth Sci.* **2004**, *94*, 341–354. [CrossRef]
122. Bicknell, R.D.C.; Pates, S. Pictorial atlas of fossil and extant horseshoe crabs, with focus on Xiphosurida. *Front. Earth Sci.* **2020**, *8*, 98. [CrossRef]
123. Waddington, J.; Rudkin, D.M.; Dunlop, J.A. A new mid-Silurian aquatic scorpion—One step closer to land? *Biol. Lett.* **2015**, *11*, 20140815. [CrossRef]
124. Morris, J.L.; Puttick, M.N.; Clark, J.W.; Edwards, D.; Kenrick, P.; Pressel, S.; Wellman, C.H.; Yang, Z.; Schneider, H.; Donoghue, P.C.J. The timescale of early land plant evolution. *Proc. Natl. Acad. Sci. USA* **2018**, *115*, E2274–E2283. [CrossRef] [PubMed]
125. Lozano-Fernandez, J.; Tanner, A.R.; Puttick, M.N.; Vinther, J.; Edgecombe, G.D.; Pisani, D. A Cambrian–Ordovician terrestrialization of arachnids. *Front. Genet.* **2020**, *11*, 182. [CrossRef]
126. Prüfer, K.; Racimo, F.; Patterson, N.; Jay, F.; Sankararaman, S.; Sawyer, S.; Heinze, A.; Renaud, G.; Sudmant, P.H.; de Filippo, C.; et al. The complete genome sequence of a Neanderthal from the Altai Mountains. *Nature* **2014**, *505*, 43–49. [CrossRef]
127. Phillips, M.J.; Gibb, G.C.; Crimp, E.A.; Penny, D. Tinamous and moa flock oogether: Mitochondrial genome sequence analysis reveals independent losses of flight among ratites. *Syst. Biol.* **2010**, *59*, 90–107. [CrossRef] [PubMed]
128. Van der Valk, T.; Pečnerová, P.; Díez-del-Molino, D.; Bergström, A.; Oppenheimer, J.; Hartmann, S.; Xenikoudakis, G.; Thomas, J.A.; Dehasque, M.; Sağlıcan, E.; et al. Million-year-old DNA sheds light on the genomic history of mammoths. *Nature* **2021**, *591*, 265–269. [CrossRef]
129. Bobrovskiy, I.; Hope, J.M.; Ivantsov, A.; Nettersheim, B.J.; Hallmann, C.; Brocks, J.J. Ancient steroids establish the ediacaran fossil *Dickinsonia* as one of the earliest animals. *Science* **2018**, *361*, 1246–1249. [CrossRef]
130. Bailleul, A.M.; Zheng, W.; Horner, J.R.; Hall, B.K.; Holliday, C.M.; Schweitzer, M.H. Evidence of proteins, chromosomes and chemical markers of DNA in exceptionally preserved dinosaur cartilage. *Natl. Sci. Rev.* **2020**, *7*, 815–822. [CrossRef] [PubMed]
131. Wägele, J.W.; Misof, B. On quality of evidence in phylogeny reconstruction: A reply to Zrzavý's defence of the 'Ecdysozoa' hypothesis. *J. Zool. Syst. Evol. Res.* **2001**, *39*, 165–176. [CrossRef]
132. Wägele, J.; Kück, P. Arthropod phylogeny and the origin of Tracheata (= Atelocerata) from Remipedia–like Ancestors. In *Deep Metazoan Phylogeny: The Backbone of the Tree of Life*, 1st ed.; Wägele, J.W., Bartholamaeus, T., Eds.; De Gruyter: Berlin, Germany, 2014; pp. 285–341.
133. Ma, X.; Hou, X.; Edgecombe, G.D.; Strausfeld, N.J. Complex brain and optic lobes in an early Cambrian arthropod. *Nature* **2013**, *490*, 258–261. [CrossRef]
134. Tanaka, G.; Hou, X.; Ma, X.; Edgecombe, G.D.; Strausfeld, N.J. Chelicerate neural ground pattern in a Cambrian great appendage arthropod. *Nature* **2013**, *502*, 364–367. [CrossRef]
135. Briggs, D.E.G.; Siveter, D.J.; Siveter, D.J.; Sutton, M.D.; Garwood, R.J.; Legg, D. Silurian horseshoe crab illuminates the evolution of arthropod limbs. *Proc. Natl. Acad. Sci. USA* **2012**, *109*, 15702–15705. [CrossRef] [PubMed]
136. Klußmann-Fricke, B.J.; Pomrehn, S.W.; Wirkner, C.S. A wonderful network unraveled-detailed description of capillaries in the prosomal ganglion of scorpions. *Front. Zool.* **2014**, *11*, 28. [CrossRef] [PubMed]
137. Klußmann-Fricke, B.J.; Wirkner, C.S. Comparative morphology of the hemolymph vascular system in Uropygi and Amblypygi (Arachnida): Complex correspondences support Arachnopulmonata. *J. Morphol.* **2016**, *277*, 1084–1103. [CrossRef] [PubMed]
138. Lehmann, T.; Melzer, R.R. The visual system of Thelyphonida (whip scorpions): Support for Arachnopulmonata. *Arthropod Struct. Dev.* **2019**, *51*, 23–31. [CrossRef]
139. Lehmann, T.; Melzer, R.R. Also looking like *Limulus*?—Retinula axons and visual neuropils of Amblypygi (whip spiders). *Front. Zool.* **2018**, *15*, 52. [CrossRef]

140. Sharma, P.P.; Schwager, E.E.; Giribet, G.; Jockusch, E.L.; Extavour, C.G. *Distal-less* and *dachshund* pattern both plesiomorphic and apomorphic structures in chelicerates: RNA interference in the harvestman *Phalangium opilio* (Opiliones). *Evol. Dev.* **2013**, *15*, 228–242. [CrossRef]
141. Akiyama-Oda, Y.; Oda, H. Hedgehog signaling controls segmentation dynamics and diversity via *msx1* in a spider embryo. *Sci. Adv.* **2020**, *6*, eaba7261-17. [CrossRef]
142. Schönauer, A.; Paese, C.L.B.; Hilbrant, M.; Leite, D.J.; Schwager, E.E.; Feitosa, N.M.; Eibner, C.; Damen, W.G.M.; Mcgregor, A.P. The Wnt and Delta-Notch signalling pathways interact to direct pair-rule gene expression via *caudal* during segment addition in the spider *Parasteatoda tepidariorum*. *Development* **2016**, *143*, 2455–2463. [CrossRef]
143. Pechmann, M. Embryonic development and secondary axis induction in the Brazilian white knee tarantula *Acanthoscurria geniculata*, C.L. Koch, 1841 (Araneae; Mygalomorphae; Theraphosidae). *Dev. Genes Evol.* **2020**, *230*, 75–94. [CrossRef] [PubMed]
144. Brenneis, G.; Scholtz, G. Serotonin-immunoreactivity in the ventral nerve cord of Pycnogonida—Support for individually identifiable neurons as ancestral feature of the arthropod nervous system. *BMC Evol. Biol.* **2015**, *15*, 136. [CrossRef] [PubMed]
145. Brenneis, G.; Scholtz, G.; Beltz, B.S. Comparison of ventral organ development across Pycnogonida (Arthropoda, Chelicerata) provides evidence for a plesiomorphic mode of late neurogenesis in sea spiders and myriapods. *BMC Evol. Biol.* **2018**, *18*, 47. [CrossRef] [PubMed]

Article

Improving Taxonomic Practices and Enhancing Its Extensibility—An Example from Araneology

Jason E. Bond [1,*], Rebecca L. Godwin [1], Jordan D. Colby [1], Lacie G. Newton [1], Xavier J. Zahnle [1], Ingi Agnarsson [2], Chris A. Hamilton [3] and Matjaž Kuntner [4,5]

[1] Department of Entomology and Nematology, University of California Davis, Davis, CA 95616, USA; rgodwin@ucdavis.edu (R.L.G.); jdcolby@ucdavis.edu (J.D.C.); lgnewton@ucdavis.edu (L.G.N.); xjzahnle@ucdavis.edu (X.J.Z.)

[2] Faculty of Life and Environmental Sciences, University of Iceland, Sturlugata 7, 102 Reykjavik, Iceland; iagnarsson@gmail.com

[3] Department of Entomology, Plant Pathology & Nematology, University of Idaho, Moscow, ID 83844, USA; hamiltonlab@uidaho.edu

[4] Evolutionary Zoology Laboratory, Department of Organisms and Ecosystems Research, National Institute of Biology, SI-1000 Ljubljana, Slovenia; matjaz.kuntner@nib.si

[5] Jovan Hadži Institute of Biology, Research Centre of the Slovenian Academy of Sciences and Arts, SI-1000 Ljubljana, Slovenia

* Correspondence: jbond@ucdavis.edu

Abstract: Planetary extinction of biodiversity underscores the need for taxonomy. Here, we scrutinize spider taxonomy over the last decade (2008–2018), compiling 2083 published accounts of newly described species. We evaluated what type of data were used to delineate species, whether data were made freely available, whether an explicit species hypothesis was stated, what types of media were used, the sample sizes, and the degree to which species constructs were integrative. The findings we report reveal that taxonomy remains largely descriptive, not integrative, and provides no explicit conceptual framework. Less than 4% of accounts explicitly stated a species concept and over one-third of all new species described were based on 1–2 specimens or only one sex. Only ~5% of studies made data freely available, and only ~14% of all newly described species employed more than one line of evidence, with molecular data used in ~6% of the studies. These same trends have been discovered in other animal groups, and therefore we find it logical that taxonomists face an uphill challenge when justifying the scientific rigor of their field and securing the needed resources. To move taxonomy forward, we make recommendations that, if implemented, will enhance its rigor, repeatability, and scientific standards.

Keywords: Araneae; taxonomy; taxonomic crisis; species concepts; data management; monographic research

Citation: Bond, J.E.; Godwin, R.L.; Colby, J.D.; Newton, L.G.; Zahnle, X.J.; Agnarsson, I.; Hamilton, C.A.; Kuntner, M. Improving Taxonomic Practices and Enhancing Its Extensibility—An Example from Araneology. *Diversity* **2022**, *14*, 5. https://doi.org/10.3390/d14010005

Academic Editor: Luc Legal

Received: 11 November 2021
Accepted: 21 December 2021
Published: 23 December 2021

Publisher's Note: MDPI stays neutral with regard to jurisdictional claims in published maps and institutional affiliations.

1. Introduction

The biological field of taxonomy and systematics, the science of describing and classifying species, is often maligned as merely descriptive [1–5]. Despite this characterization, taxonomic products play a pivotal role by providing the underlying framework for every biological study [6]—rigorous and repeatable ecological, biochemical, comparative, evolutionary, and physiological studies would be impossible without accurate species delimitation. Nevertheless, taxonomy is typically regarded as a science in crisis [5,7,8]. Fewer students are being trained in organismal expertise, taxonomic works are under cited, funding for taxonomic research is limited, and the paucity of professional taxonomic positions at academic institutions portends unimportance among fellow researchers [7–11]. Arguments that the number of taxonomists has increased in recent years [12] are debatable [8] and likely reflect changes in scientific publication practices rather than increasing taxonomic expertise [13]. Recent reliance on citation metrics and journal impact factors for making hiring decisions, promotions, and other rewards [14] reinforces that taxonomic work is undervalued [9]. Citations and the correlated impact of taxonomic journals [15] mirrors the bias against proper full citation of scientific names [4,7,16–18]. The perceived diminished impact of taxonomy becomes self-perpetuating. The latest example of this dangerous trend was the proposed suppression (and subsequent reversal, due to backlash) in early 2020 of impact factors from journals with high instances of self-citations—oftentimes taxonomic journals, a direct consequence of the tradition not to cite taxonomic authorities—as expressed in an announcement from Clarivate (https://jcr.help.clarivate.com/Content/title-suppressions.htm (accessed on 28 June 2020)).

Contemporary taxonomists find themselves at the forefront of the battle lines drawn by human-driven climate change and mass extinction due to habitat loss and other factors [19,20]. As this large and rapid extinction event unfolds [21], taxonomists are thrust into the unsustainable position of documenting this monumental historical loss of biodiversity [22] and, in some cases, grimly identifying and naming new species already extinct or destined thusly. One of the authors of this paper (JEB) has first-hand experience describing new species (Californian trapdoor spiders) after their extinction [23]. Taxonomy alone stands between a species being lost to both extinction and obscurity. As such, one could argue that never before has the discipline been so important; it is impossible to "save," conserve, and/or inventory undiscovered species. Yet, lag times between when a new species is collected and when it is described remain on average as high as 35 years or more with only 15% of all species described within five years of collection [13]. The implications of the Anthropocene epoch will never fully be understood without an exhaustive inventory of Earth's biodiversity—a task where taxonomists will play fundamental roles. Taxonomic work is an undeniably critical component to saving our planet.

The arguments above notwithstanding, it would be fair to say that the taxonomic literature, from nearly any organismal field, represents works that span a very broad range, from those that are very descriptive (e.g., taxonomic descriptions of a single species that lack broader context) to large-scale phylogenetically-informed monographs that explicitly contain many elements of hypothesis testing and experimentation. Both types of work have utility and value. Moreover, considerable variation across many taxonomic groups exists as a function of species diversity and taxonomic "maturity". For example, taxonomically-biased groups such as birds and mammals [24], in which the majority of species have likely been described, can afford now to be broadly integrative. In contrast, megadiverse arthropod groups with many undescribed species remaining [25,26] likely lag behind collectively in terms of highly sophisticated approaches to species delimitation, in part because taxonomists must find the balance between the urgent need to rapidly describe species and the production of more integrative works evaluating species using other approaches (e.g., molecular taxonomy). That said, we probably know very little about the progress in most taxonomic fields of study, and what we do know is based largely on anecdotal observations of individual works with which we are most familiar.

To our knowledge, there are few, if any, examples of multi-year surveys within any one taxonomic field evaluating the data and rigor of the species hypotheses being formulated (but see Liu et al. 2019). As such, we aimed to evaluate the data being collected and the species hypotheses being proposed across the large taxonomic field of spider systematics (Order: Araneae). Spiders are a hyper-diverse group comprising approximately 50,000 species parceled among >4000 genera and 129 families (World Spider Catalog 2021; WSC [27]). By some estimates, there are likely well over 120,000–200,000 species [28], an estimate supported by non-asymptotic new species discovery (WSC). Progress in spider taxonomy is documented via the WSC where each year hundreds of species descriptions are catalogued with the accompanying literature. The WSC is an information-rich resource, a rare database when compared against most other major lineages on the Tree of Life, that facilitates questions such as the ones we pose herein. Using this unique dataset, we surveyed the spider taxonomic literature from 2008 to 2018 (11 years). As part of our survey, we tabulated 22 parameters that included the type of data used in species delimitation, the species concept employed (if stated), and the number of specimens available for each species described. The results we outline below present a sobering overview of spider taxonomy (and perhaps of taxonomy in general): few papers describe an explicit species hypothesis, very few studies are integrative with the vast majority relying on one morphological data point, and a large proportion of species are based on two or fewer specimens.

We, like nearly all biologists, agree that taxonomy is important and more is needed, but it does not seem like anyone is really willing to ask hard questions about why it has a less than desirable reputation. If it is important, relevant, hypothesis-driven work, we see no reason why it should garner less respect now, or in the past, than any other field of biological science. As a model system that likely reflects the state of the science, we aim to take a critical look at the field of spider taxonomy by asking some difficult questions about the nature of the work we are doing with the hope of provoking change. We feel it is best that this criticism be honest and direct, and we acknowledge openly that it certainly applies to some of the works of this paper's authors.

2. Methods

We downloaded nearly all the taxonomic works documented in the WSC during the time period of 2008–2018. Each investigator documented authorship and the number of new species described per publication; our review focused exclusively on newly described taxa during the study time period. Only a few non-English works were omitted from this study for which we were unable to find a translation that allowed confident data scoring. Table 1 below lists the parameters reviewed and how they were scored. Binary scorings were based on interrogative NO/YES (0/1) responses, whereas others were quantitative (documenting absolute numbers of observations or counts). Generally, we assessed the following: (1) the type of data used to establish species constructs; (2) how species were illustrated; (3) whether raw specimen data were downloadable; (4) how many specimens were examined for each species; and (5) what sexes were available for each species. The number of specimens available was tabulated as (1), (2), or (>2); >2 is somewhat arbitrary and underestimates the paucity of data associated with some species, but objectively captures the variation in the dataset without documenting the absolute number of specimens for all species. A study was classified as 'integrative' if morphological data (i.e., genitalic or other) was used in combination with at least one other data source. For species concept, we assessed each paper to determine whether the author stated explicitly what species concept they used to delineate taxa. The data were tabulated in a MS Excel spreadsheet; summary statistics and bar graphs were produced using the base R statistics packages and carried out in R-Studio [29].

Table 1. Parameters evaluated and how they were assessed/scored.

General Descriptor	Parameters	Scoring Method
General information	(1) Authorship; (2) year; (3) country/region	Recorded directly
Species	(4) Number of new species	Exact count
Type of data used to delimit species	(5) Morphology—genitalic; (6) morphology—other; (7) molecular; (8) ecological; (9) behavioral	Binary scoring NO/YES (0/1)
Integrative	(10) Did species delimitation employ more than one data type?	Binary scoring NO/YES (0/1)
Species concept	(11) Was a species concept explicitly stated? (12) If species concept stated then which concept used?	(11) Binary scoring NO/YES (0/1) and (12) recorded directly
Illustration types used	(13) Drawings; (14) digital images; (15) scanning electron microscopy	Binary scoring NO/YES (0/1)
Data	(16) Are the raw data (measurements, material examined, etc.) available as a downloadable resource?	Binary scoring NO/YES (0/1)

Table 1. *Cont.*

General Descriptor	Parameters	Scoring Method
Sex if specimens examined for new species description	(17) Males only; (18) females only; (19) both sexes available	Exact counts
Number of specimens examined for new species description	(20) 1 specimen; (21) 2 specimens; (22) 3 or more specimens	Exact counts

Although most of the parameters evaluated can be objectively scored, the role that a particular data type played in species delimitation might be viewed as a matter of opinion *sans* asking the author directly. Generally, the team erred on the side of inclusivity—for example, if a researcher documented having collected ecological or behavioral data, we assumed those data played a role in delimiting species. For molecular studies where multiple populations were sequenced, if not stated it was considered implicit that the researcher employed a phylogenetic lineage species concept (indeed in nearly every case monophyly or exclusivity was stated as the criterion), otherwise a species concept was only noted if explicitly stated.

3. Results

Spanning an 11 year time period (2008–2018), we evaluated 2083 spider taxonomic works that described 8433 new species. The data are summarized in Tables 2–4 and Figure 1; the raw data can be downloaded as a .csv (comma separated) file at https://doi.org/10.6084/m9.figshare.17263835.v1. Nearly half the papers surveyed (>990) described a single new species with a median and average of two and four new species per publication, respectively. Note that *some* of these papers describing new species were part of a larger work or revision that also redescribed existing nominal taxa. Because we were primarily interested in the data used to describe new species, per se, we did not evaluate the context. In retrospect, this would have been a worthwhile parameter to assess and as such may extend the data set to include this at some point in the future.

Table 2. The 2008–2018 summary data showing numbers of species described, numbers of papers examined, types of data considered, numbers of studies scored as integrative, and the number of papers that communicated an explicit species concept.

Year	# Species	# of Papers Surveyed	Genitalia	Other	Ecology	Behavior	Molecular	Integrative	Concept?
2008	574	165	158	67	21	10	6	25	1
2009	705	195	193	110	36	16	7	41	0
2010	762	179	177	92	24	10	3	26	2
2011	884	177	171	110	32	16	8	41	5
2012	970	188	186	124	46	13	12	56	7
2013	916	187	181	116	33	11	6	38	5
2014	763	189	181	129	38	15	10	47	6
2015	691	205	196	121	14	9	11	27	10
2016	678	194	193	103	1	7	14	19	2
2017	691	199	187	127	4	3	24	28	27
2018	799	207	193	116	7	3	29	35	28
Totals	8433	2083	2016	1215	256	113	130	298	93

Table 3. The 2008–2018 summary data showing the type of imaging used to illustrate species and number of papers with electronic data available in a downloaded format.

Year	Drawings	Digital Images	SEM	Data Availability
2008	159	74	46	4
2009	182	102	41	0
2010	160	114	37	5
2011	143	112	33	3
2012	152	148	44	17
2013	139	153	52	6
2014	143	170	40	11
2015	164	180	50	9
2016	145	181	43	17
2017	144	183	60	18
2018	138	195	61	28
Totals	1669	1612	508	118

Table 4. The 2008–2018 summary data showing numbers of species for which males and females were described and the number of specimens associated with each new species description. Percentages (of Totals) are the percentage of each column relative to the total number of species evaluated during the project time period.

Year	Male Specimens Only	Female Specimens Only	Both Sexes	% Species Based on One Sex	Based on 1 Specimen	Based on 2 Specimens	Based on >2 Specimens	% Based on 2 or Fewer Specimens
2008	139	107	328	42%	140	90	344	40%
2009	125	140	440	37%	152	90	463	34%
2010	140	114	508	33%	140	94	528	31%
2011	188	134	562	36%	172	132	580	34%
2012	197	149	624	35%	197	142	631	35%
2013	175	126	615	32%	157	144	615	33%
2014	180	103	480	37%	164	91	508	33%
2015	139	115	437	36%	148	100	443	36%
2016	113	93	472	30%	123	71	484	29%
2017	128	127	436	36%	156	97	438	37%
2018	186	87	526	34%	129	129	541	32%
Totals	1710 20%	1295 15%	5428 64%	35.6%	1678 (20%)	1180 (14%)	5575 66%	34%

3.1. How Integrative Is Spider Taxonomy?

Most surveyed studies rely on morphology alone (Figure 1A), with few including additional types of evidence. Only 14% of taxonomic works were classified as integrative using the criterion requiring two or more data types, whereas all studies classified as integrative included morphology as one of the data sources. Our evaluation of the degree to which taxonomy is integrative is liberal because most instances of "integrative taxonomy" we documented did not invoke an explicit statement of a predefined integrative approach, e.g., [29–31], by the author(s). Consequently, our results report a best-case scenario for integrative taxonomy. As shown in Figure 1A, genitalic morphology is the dominant data type followed by 'morphology—other', the latter representing morphological differences such as coloration, patterning, setal differences, and measurements. Ecological and behavioral data were included to a modest degree from 2008 to 2014, though diminish in prevalence thereafter.

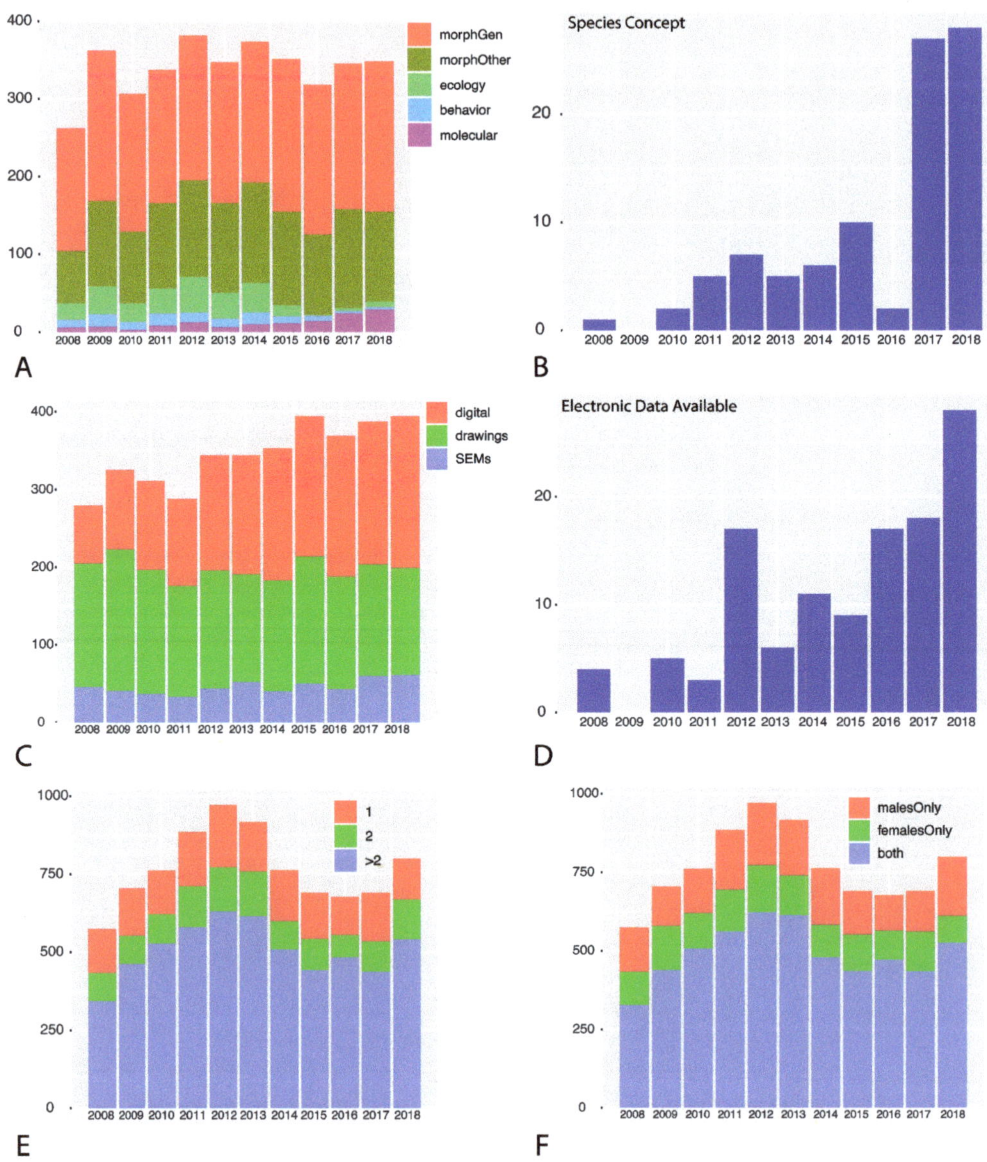

Figure 1. Summary of data per year 2008–2018. (**A**) Number of species based on each data type; (**B**) number of publications explicitly stating a species concept; (**C**) illustration type used in each publication; (**D**) prevalence of downloadable electronic data; (**E**) number of specimens examined per new species; (**F**) prevalence of both male and female specimens per new species described.

At this time, molecular data have not been heavily involved in spider taxonomy (Figure 1A), though of the molecular studies we surveyed, almost all were integrative;

we defined 98% as integrative with only two ostensibly not including other lines of evidence. Studies were considered to have a molecular component if they included any molecular data and thus spanned a wide range of data types from COI barcoding to multi-locus genomic studies. Molecular data were used in ~43% of all studies we considered to be integrative. Despite the increased access to cheaper molecular tools over the last decade, e.g., [32], molecules remain surprisingly scarce in spider taxonomic studies (see below for potential explanations). We document 130 occurrences of molecular usage over the 11 years, accounting for ~6% of the total publications surveyed, with nearly 2/3s of the molecular-based studies published within the last three surveyed years (2016–2018).

3.2. Are Species Constructs Defined Conceptually?

The vast majority of spider taxonomic publications fail to articulate any defined species concept (Figure 1B); only 93 of 2083 (4%) publications defined an operational species criterion or concept. From our survey, approximately five concepts are employed—morphological, phylogenetic, cohesion, general lineage, and the unified species concept.

3.3. What Data Are Being Used? What Is Available, and How Much?

Modern day spider species descriptions are primarily illustrated via drawings, scanning electron microscopy (SEM), and digital images (Figure 1C). From 2008 to 2012, drawings accounted for the greatest proportion of illustrations, whereas in subsequent years digital images using light microscopy and other techniques (e.g., image stacking) have become more prevalent [33]. The proportion of SEM illustration has remained relatively constant and likely reflects the need for fine-scale detail not possible using light microscopy. We did not collect data regarding the deposition of images in online databases; however, anecdotally it seems that very few authors archive images in any online repositories such as Morphbank (http://www.morphbank.net/). We observe a transition to high resolution light photographic images, a positive trend, given the recent push to digitize collections. Although drawings certainly have value as interpretive images, high quality photographs of actual specimens (e.g., holotypes) are potentially data rich and in some cases may abrogate the need for borrowing material from collections [34].

Raw measurement data, included in every spider species description, is seldom available as searchable supplementary data (Figure 1D). We found only 118 (~5%) documented occurrences where raw measurement data were available in a downloadable format (e.g., as a spreadsheet or database). Our survey shows a minor sustained upward trend in increasing data availability after 2015. For 2018, 28 of 207 papers, approximately 14%, provided data available for download. As discussed by Bik [6], availability of traditional taxonomic data would "fundamentally improve database resources for all scientific disciplines".

Although taxonomic works are potentially data rich, a non-trivial number of spider species are based on little material and thus sample sizes are small. During the time period reviewed, 34% were based on two or fewer specimens (Figure 1E) and 35% of all species were based on one sex (Figure 1F). Twenty percent (20%) of all spider species were based on a single specimen (i.e., the sample size for a new species is $n = 1$). In numerous instances, a publication consisted entirely of the description of only one new species, based on only one specimen (185 in total, 8.5%).

4. Discussion

Current trends in spider taxonomy are predominantly of non-revisionary nature and non-integrative approaches, findings that seem consistent with trends reported in other groups (e.g., insects—[10]). This suggests that taxonomy has been slow to improve upon the traditional morphology-only approach despite rapid development and easy availability of a vast number of other data sources—a finding likely shared within taxonomic groups across the Tree of Life. Such a slow pace of transition to more inclusive taxonomy seems to characterize not only spiders, but also other hyper-diverse taxa, with calls to devise somewhat controversial methods [35] to drastically shortcut species descriptions for groups containing many undescribed species (e.g., barcode and rapid digital imaging of ichneumonoid wasps, [36].

4.1. How Integrative Is Taxonomy?

The spider taxonomic literature, like that of many other hyper-diverse taxa and 'non-charismatic microbiota', appears to be largely dominated by traditional morphological taxonomy. If relying on a single source of data, it can be argued that morphology is more important than other data sources, for example, by providing essential information on taxa and the necessary link to taxonomy going back to its beginnings [37]. Nevertheless, the field overwhelmingly lacks an explicit integrative perspective and the sole reliance on morphology restricts the field. From the typical number of new species described per paper (1–2 species), one can probably infer the field is not generally comparative (note that this was not an evaluative aim of our study), and one might also infer anecdotally that the literature is dominated by alpha taxonomy, with the majority of papers being non-monographic [7]. This is a trend that has been countered by the creation of taxonomic initiatives, for example the USA National Science Foundation PEET program (Partnership for Enhancing Expertise in Taxonomy [11]. However, more such efforts, especially international and global initiatives, are necessary.

The scarcity of molecular data is likely the consequence of a number of factors. First, many species are known from only one to a few specimens and thus taxonomists may hesitate damaging type specimens for DNA extraction. Likewise, while it is possible to extract DNA from older museum specimens using techniques that allow for non-destructive sampling [38], the same reluctance to damage or destroy precious material likely prevails. Further, morphological taxonomy can be produced on a salary alone, whereas molecular data often require additional funding.

All things considered, why have molecular and/or integrative approaches not become more common in spider taxonomy, particularly the former? All other fields in the life sciences have been methodologically transformed over the past decade by the major technological innovation of high-throughput sequencing technologies. Yet, this does not seem to be true for the taxonomy of spiders and other megadiverse organisms see [39] for a summary of how taxonomy could be advanced using genomics, likely in part due to the challenges in striking a balance between the needs for rapid discovery on one hand and taxonomic rigorousness on the other. That notwithstanding, rapid technological advances have essentially rejuvenated molecular phylogenetic systematics and have an amazing potential to invigorate taxonomy as well [40]. One likely general explanation for the slow adoption of molecular methods is that the majority of traditional taxonomists, in particular those from countries with little financial aid for scientific research available, are resource limited—a point that directly links with the problematic impact-factor decision by Clarivate (mentioned above) and has significant potential to harm our colleagues in these countries. However, the cost of obtaining molecular data is ever decreasing. The argument could further be made that morphological data are sufficient, yet numerous molecular studies at multiple hierarchical levels consistently show instances where morphology underestimates diversity and/or fails to accurately delimit species [41–44]; discussed in more detail below.

4.2. Are Species Constructs Defined Conceptually?

As we discussed above, taxonomy provides a potentially rich hypothetic-deductive framework for testing species boundaries [2,5,7]. Yet, spider taxonomists generally fail to convey any information on how they conceptualized the species they are describing, including some authors of this paper. Additionally, while we are certainly a part of this past, we argue, however, that the well-worn argument that taxonomic species constructs are rigorous species boundary hypotheses is vastly strengthened by placing the work in the context of an operational species construct. Otherwise, we believe taxonomists have played a role in diminishing their own work's rigor. Arguably, taxonomic works that lack an explicit hypothetical framework could be branded as purely descriptive, and essentially unfalsifiable. We do believe that good 'diagnoses' sections found in many papers potentially provide some testable hypotheses, as do papers that clearly place species in a phylogenetic context and those using molecular data (e.g., ability to demonstrate monophyly, and lack of gene flow, important elements of most species concepts, at least intrinsically). However, the explicit use of species concepts as the underlying hypothetical framework of taxonomy, would vastly improve and clearly place taxonomic works on par with any other hypothesis-based sciences.

The absence of a stated conceptual species framework may provide some insight into why the value of taxonomy is diminished as a largely "descriptive science"; that is, why taxonomy is viewed as non-experimental observation and thus not as intellectually informed as other scientific disciplines perceived to be more hypothesis-driven. Although descriptive science should have merit in its capacity to discover and illuminate novel phenomena, a number of authors posit that taxonomy is just as much a hypothesis-driven science as any other [2]. Nicely articulated by Haszpruner [45], taxonomic works formulate and implicitly convey tests of three or more hypotheses: (1) species delimitation; (2) species classification; and (3) homology statements (often multiple hierarchical hypotheses) regarding organismal form and function. In short, a taxonomist with a collection of specimens postulates tests of what constitutes the limits of a species, where in the hierarchy of life that species is placed, and what synapomorphies support their hypotheses. The latter constitutes a complex nested set of homology statements that represent hypotheses based on, for example, anatomy, function, and ontogeny. Moreover, broad-scale studies (e.g., a family-level taxonomic revision, monograph with good diagnoses, and morphological studies in a phylogenetic framework) could be characterized as carefully designed experiments that include informed taxon and character selection used to construct phylogenetic inferences specifically aimed to test species, genera, and family-level limits (e.g., tests of monophyly). Taxonomy is potentially both descriptive and experimental, by employing levels of experimentation, sophistication, and knowledge that are equal among its counterpart disciplines in the life sciences. "That a taxonomic study is hypothesis-driven and analytical from its very beginning is not obvious to the uninitiated" [5]—for taxonomy to increase potential to receive its due credit, this needs to become explicit and obvious.

4.3. Species and Data: What Is Being Used? What Is Available, and How Much?

In general, we find that during the surveyed period, studies only employ a small portion of the available relevant data used to formulate and test the taxonomic hypotheses. In a large proportion of studies, the data presented barely meet an acceptable minimum; indeed we can think of no other field of biology (or other scientific discipline, in general) where such a paucity of data would be the acceptable basis for scientific inference.

Regardless, the historical taxonomic literature is (potentially) incredibly data rich, containing valuable information related to geographic distributions, temporal occurrence patterns, and morphology that captures a vast wealth of quantitative and qualitative observations. Because these data are generally inaccessible, one could argue that they simply do not exist—"If it isn't online, it might as well not exist" [1,6] p. 2. Although some taxonomic journals such as Zookeys (Magnolia Press) use XML markup to facilitate downstream data extraction, there seems little reason for not making taxonomic data available in electronic form. For example, with the recent advances to spider phylogeny [46–49], accessible morphological data could be used to enhance comparative evolutionary studies that are de rigueur in other more exhaustively studied groups such as mammals, birds, and amphibians and reptiles.

Additionally, rare, singleton species (species known from only one specimen) are not uncommon in biodiversity samples, representing over 30% of all species found in tropical arthropod inventories [50]. In a 10 year survey (2000–2010) of American Museum of Natural History publications, Lim et al. [51] found that 17.7% of new species were known from only one specimen. Although very rare species are expected in collections and biodiversity surveys, some authors have proposed that new species should never be described on the basis of a single specimen. Species hypotheses using so little data are prone to anatomical mistakes and lack thorough evaluation of infra- and interspecific variation [29]. Lim et al. [51] suggest that these problems are not germane to only morphological species delimitation because molecular approaches such as GMYC and PAA only accommodate singleton species to some degree; many species are indeed rare and current molecular analytical methods are severely limited in treating these situations. In general, we believe such species descriptions in isolation, and absent a context should cease to exist, but clearly species based on single specimens may be very appropriate for inclusion in broader revisions. That is, single species and/or single specimen descriptions may be justifiable in cases given a very clear and well-defined context. These could include biological (e.g., unique species biology, part of a thorough biodiversity inventory), revisionary (e.g., adding a new species to a recent revision), or phylogenetic (e.g., critical phylogenetic taxon) contexts where their description serves a clear purpose despite the limitations detailed here.

We would also like to explore here the notion that many singleton species are unfalsifiable and represent a particularly poor inductive argument. As discussed by Wheeler [2], "species hypotheses are not efficiently tested in isolation" and "to critically test the distribution of attributes defining one species it is necessary to examine variation within that and all nearly related species". The singleton specimen species problem often conveys no such test or distribution of attributes described by Wheeler; it is not a hypothesis that can be falsified based on a broader context character diagnosis—the essence of the species hypothesis. Alternatively, in the absence of any description of variation, the species is potentially typological and thus conceivably falsified upon discovery of any additional material showing variation. Second, the inductive assumptions of species based on one specimen are severe; it essentially assumes the uniformity of all individuals based on a single observation. It is not that we believe that the diagnosis of all species based on a single specimen are false, they may very well be "good species", but they are certainly weak hypotheses, trivially falsifiable by a single additional datum. This merits acknowledging the flaws inherent in such logic and why the quality of such science might be viewed as wanting.

4.4. Recommendations

Below, we make four recommendations we think need to be quickly adopted by the taxonomic community. The first three are pragmatic changes to taxonomic publications that entail how species hypotheses are stated, data are made available, and provide minimum data standard guidelines. The fourth recommendation relates to an aspirational goal of ultimately achieving a more integrative taxonomy in the coming years. The International Code of Zoological Nomenclature (ICZN) establishes the rules for naming species but is silent with respect to what constitutes good scientific practice. Codes governing the naming of other taxa such as plants, fungi, and viruses similarly focus on the naming rules rather than scientific practice. If the field of taxonomy is going to gain the respect it deserves in the scientific community, then we need to move past simple compliance with the ICZN and other naming rules and adopt practices that are more in line with other scientific disciplines. Our recommendations are as follows:

(1) All taxonomic works should clearly state the species concept being applied. Failing to state what epistemological and/or conceptual-based criteria are being used to distinguish one specimen/species from the next should become an important factor for reviewers and editors in assessing manuscript quality. We emphasize that some of our own past works would have greatly benefited from doing so and we fail to see any valid reason for future taxonomic work not doing so. The notion that some underlying yet unstated species concept/hypothesis is implicit in every taxonomic work [sensu 5] should be no more acceptable to practicing taxonomists than would an experimental ecologist failing to describe experimental design, define hypotheses tested, and elaborate a statistical alpha level. Taxonomic journals, subject editors, and reviewers must start demanding that authors include such a statement. Such a change would benefit taxonomists in a number of ways including requiring authors to more carefully consider the criteria they are applying to differentiate taxa, by further enhancing the objectivity of species determination with explicit criteria, and by emphasizing to others that species descriptions are conceptually formulated and hypothesis driven.

(2) With increasing databank availability and ease of use, all data should be made electronically accessible. The demand for electronic data should include all quantitative, qualitative, specimen (e.g., geographic and collection event), and original (unaltered) image data. Nearly every journal that publishes papers containing genetic data have required authors for decades to deposit data in databases such as NCBI or EMBL. Although morphological databases with similar governmental support do not currently exist (they should!), archives such as Dryad (http://dryad.org), Symbiota (https://symbiota.org/), figshare (https://figshare.com/) are available and easy to use, and GBIF is online and free for archiving locality data (https://www.gbif.org/). Because some databases such as Dryad are fee based, for-profit journals capitalizing on the hard work of taxonomist authors, editors, and reviewers must work to ensure that free or minimal cost options are available to investigators that may not have adequate funding. The mandate to electronically archive all taxonomic associated data should begin immediately. Digitally archived morphological data should conform to field/organismal anatomy ontologies whenever possible, e.g., [52], the spider anatomy ontology (SPD) [53]. It follows that systematics and taxonomy need to also focus attention and computational resources to begin harvesting critical legacy morphological data but, in the meantime, cease accumulating data that are not electronically accessible. Efforts to employ computational approaches such as natural language and machine learning, coupled with well-developed ontologies, should be further exploited to facilitate recovery of 200+ years of data embedded in publications. As noted by Godfray [1], "the quantity of taxonomic information available on the web is pitiful"—a statement that remains as true today as it did over 18 years ago—that needs to change.

(3) Descriptions of new species based on singleton specimen data that are described outside a very clear context (see above) should stop. This practice strikes a balance between the positive aspects of describing rare new taxa with Dayrat's [29] recommendation to preclude these altogether. Such data-deficient hypotheses should be as critically scrutinized in taxonomy as they would be in any other field of science. This should not necessarily stymie efforts describing rare taxa, but simply require that the taxonomic description thoroughly document interspecific variation and demonstrate sufficient evidence that a new, rare species is warranted. To be fair, some (but not all) singleton species descriptions typically examine types of related species, other material, or digital images of congeners, but we would argue that the bar needs to be higher. Unfortunately, this may leave some new candidate species waiting description (although it should motivate larger-scale studies, particularly international collaboration). As such, these specimens could be documented as candidate species in published taxonomic notes or using online data narratives. Nevertheless, the practice of describing a new taxon on the basis of very little data needs critical reevaluation. In general, it is not good scientific practice because it is a weakly formulated hypothesis (at best), logically flawed, and caters to the notion that taxonomy lacks robust data and rigorous publication standards.

(4) Finally, taxonomy must aspire to become more integrative and do so quickly [31]. Other fields in the life sciences have been transformed over the past decade via quantum leaps in genomics, proteomics, and biological imaging (to be fair most other fields do not have the task of describing hundreds of thousands of new entities, many of which are rare and difficult to discover, in a very short timeline). Taxonomy does not appear to have overwhelmingly capitalized on these advances. That said, it is important to recognize that not all taxonomists have access to technology and sufficient funding, thus the aspiration of a fully integrative taxonomy that includes genomics, for example, is primarily targeted at those labs able to do that sort of work (but see below)—with an emphatic call for global collaboration and an end to the colonialist mindset of "parachuting" into a region or country, collecting specimens, and describing that diversity without the help (and co-authorship) of local colleagues. We follow Dayrat's [29] definition of integrative taxonomy "as the science that aims to delimit the units of life's diversity from multiple and complementary perspectives (phylogeography, comparative morphology, population genetics, ecology, development, behaviour, etc.)"; therefore, integrative taxonomy is not limited to technology per se but fundamentally incorporates multiple data types. Whenever possible, future generations of taxonomists need to be trained in morphological taxonomy as well as modern techniques that capitalize on next generation technologies, thereby extending their potential to gain "complementary perspectives"—training only in the former severely limits marketability for jobs. Adopting an integrative approach is not just good practice because more modern techniques are in vogue, it is good practice because it is good science. Evolutionary biologists have long acknowledged the limitations of morphology in species delimitation. Phenomena such as convergent and parallel evolution, phenotypic plasticity, and morphological stasis (species crypsis) can obscure species boundaries requiring multiple lines of evidence to accurately resolve. If taxonomists are interested in "getting it right", we need to consider other data related to the species origin and evolutionary trajectory [31]. The majority of spider taxonomic studies employing molecular data were ironically integrative for exactly these reasons; that is, molecular systematists have long acknowledged that one data point, one gene, one data type, is not sufficient to confidently assign populations to species. Species diagnosed using a single character system should be just as questionable as a species or phylogenetic hypothesis based on one gene, yet they are not. Taxonomy needs to genuinely transform itself as a collaborative, integrative information science [1]. Finally, we will also add that a recent review of the taxonomic literature characterized as "integrative" [54] seems to indicate that these works may "open doors" to top ranked journals and enhance citation performance.

The first three of the recommendations can easily be implemented through enhanced editorial practices and vigilant peer review that enforces a set of community standards. Recommendation #4 is a little more complicated in its implementation and should not be confused as a decree that a new integrative taxonomy be prohibitively costly and only technology driven. Although we do see an aspirational integrative taxonomy as taking full advantage of many of the methodological innovations available in a modern life sciences tool kit, this does not preclude integration of traditional morphology with other types of natural history data that can be collected at much lower or no cost (oftentimes just as valuable, e.g., differences in mating rituals and temporally different mating activity patterns). Moreover, such an aspiration should encourage partnerships and collaborations among scientists with access to those tools. There is a lot of quality taxonomy being done in parts of the world that must continue; however, if the field of taxonomy is going to advance it cannot remain stagnant for the purpose of holding to a standard that every practicing taxonomist can easily achieve. We certainly do not see other fields of biology only adhering to older practices as a means of accommodating all researchers.

5. Conclusions

The data we have presented here document a rather sobering depiction of the state of spider taxonomy, consistent with the sparse available evidence on other hyper-diverse taxonomic groups, e.g., [10]. We find that much of the work over the last decade is generally not integrative, not accessible electronically, and based on very few specimens and consequently little data. Spider taxonomists seldom state a species hypothesis or concept. Although there are certainly exceptional works that are broad in scope (monographic), integrative, and data rich/accessible, unfortunately such studies are the minority. As such, it would seem that many of the criticisms leveled at taxonomists including the stereotype that the work we do is largely descriptive, are justified. These wounds are to some extent self-inflicted. Despite the importance and relevance of the work we do, these data either bely the sophistication and intellectual underpinnings of taxonomy or, accurately depict the work being done. We believe that for many professional taxonomists, it is the former rather than the latter. Nevertheless, the majority of taxonomists do a very poor job emphasizing the tremendous intellectual contributions of the work they are doing within the context of those publications. As discussed recently by Wheeler [3], it is time that taxonomy got an "image makeover" and recognized for the "incomparable benefits to other sciences and society". The recommendations we make above are aimed toward exactly such a makeover because it is simply not enough to just extol the virtues and intellectual content implicit in every taxonomic work—the intellectual content must be made explicit in every work, otherwise taxonomy will continue to dwindle in perception as a bona fide discipline along with its funding and academic positions. As a world community, we must heavily invest now in training modern integrative taxonomic specialists who take full advantage of all the tools modern biology has to offer. Currently, the resources needed to effectively discover and describe new species are seriously lacking, despite the fact that understanding the true diversity on our planet and finding answers to how that diversity has evolved is perhaps one of the most important endeavors we can answer as biologists. The resources needed for taxonomists to effectively do their jobs require immense person power and funding for field work, identification, data collection (e.g., morphological and DNA), specimen storage, and training the next generation. There is clear evidence that efforts such as the USA National Science Foundation's PEET program, can dramatically advance the field. We strongly advocate that the NSF urgently consider reinstating the PEET program, and that similar efforts be initiated and/or sustained through other national, and especially international programs. It is not surprising that quick-fix approaches such as DNA barcoding [55] that propose to "democratize" taxonomy [56], garner funding and attention because their proponents have seemingly done a better job advocating for their science. DNA barcoding outwardly seems modern, rigorous and objective, and hypothesis driven. Consequently, governments continue to invest millions of dollars into this methodology—with little regard for the fact that for this approach to be truly effective, the species (and their boundaries) it seeks to identify need to be identified and diagnosed by trained taxonomists.

In closing, our planet is facing an extinction crisis. Taxonomy is integral to solving that crisis. It is time for taxonomists to stop complaining about being disrespected, underfunded, and not cited. Instead, we must acknowledge the problems with how we have been working on and presenting our science, as well as the role that we have played in fomenting negative perceptions. We must change some of our practices and demand the respect and resources that our noble field of taxonomy requires and deserves.

Author Contributions: Conceptualization, J.E.B.; methodology, J.E.B.; validation, all authors; formal analysis, J.E.B.; investigation, all authors; data curation, J.E.B.; writing—original draft preparation, J.E.B., I.A., M.K.; writing—review and editing, all authors; visualization, J.E.B.; supervision, J.E.B.; project administration, J.E.B.; funding acquisition, J.E.B., M.K. All authors have read and agreed to the published version of the manuscript.

Funding: This work was supported by USA National Science Foundation grants AF 1401176 and DEB 1937604 to JEB and the Evert and Marion Schlinger Foundation. MK was supported by the Slovenian Research Agency, grants P1-0236, J1-9163.

Data Availability Statement: The raw data can be downloaded at https://doi.org/10.6084/m9.figshare.17263835.v1.

Acknowledgments: We thank Joel Ledford, Lynn Kimsey, Vera Opatova, Laura Montes de Oca, and two reviewers for critical comments on an earlier version of this manuscript (having commented does not imply agreement with all of the content!).

Conflicts of Interest: The authors declare no conflict of interest.

References

1. Godfray, H.C.J. Challenges for Taxonomy. *Nature* **2002**, *417*, 17–19. [CrossRef] [PubMed]
2. Wheeler, Q.D. Taxonomic Triage and the Poverty of Phylogeny. *Philos. Trans. R. Soc. Lond. B Biol. Sci.* **2004**, *359*, 571–583. [CrossRef] [PubMed]
3. Wheeler, Q. A Taxonomic Renaissance in Three Acts. *Megataxa* **2020**, *1*, 4–8. [CrossRef]

4. Wägele, H.; Klussmann-Kolb, A.; Kuhlmann, M.; Haszprunar, G.; Lindberg, D.; Koch, A.; Wägele, J.W. The Taxonomist—An Endangered Race. A Practical Proposal for Its Survival. *Front. Zool.* **2011**, *8*, 25. [CrossRef]
5. Sluys, R. The Unappreciated, Fundamentally Analytical Nature of Taxonomy and the Implications for the Inventory of Biodiversity. *Biodivers. Conserv.* **2013**, *22*, 1095–1105. [CrossRef]
6. Bik, H.M. Let's Rise up to Unite Taxonomy and Technology. *PLoS Biol.* **2017**, *15*, e2002231. [CrossRef]
7. Agnarsson, I.; Kuntner, M. Taxonomy in a Changing World: Seeking Solutions for a Science in Crisis. *Syst. Biol.* **2007**, *56*, 531–539. [CrossRef]
8. Bacher, S. Still Not Enough Taxonomists: Reply to Joppa et al. *Trends Ecol. Evol.* **2012**, *27*, 65–66. [CrossRef]
9. Drew, L.W. Are We Losing the Science of Taxonomy? *BioScience* **2011**, *61*, 942–946. [CrossRef]
10. Liu, Y.; Dietrich, C.; Braxton, S.; Wang, Y. Publishing Trends and Productivity in Insect Taxonomy from 1946 through 2012 Based on an Analysis of the Zoological Record for Four Species-Rich Families. *Eur. J. Taxon.* **2019**, *504*, 1–24. [CrossRef]
11. Rodman, J.E.; Cody, J.H. The Taxonomic Impediment Overcome: NSF's Partnerships for Enhancing Expertise in Taxonomy (PEET) as a Model. *Syst. Biol.* **2003**, *52*, 428–435. [CrossRef] [PubMed]
12. Joppa, L.N.; Roberts, D.L.; Pimm, S.L. The Population Ecology and Social Behaviour of Taxonomists. *Trends Ecol. Evol.* **2011**, *26*, 551–553. [CrossRef] [PubMed]
13. Bebber, D.P.; Wood, J.R.I.; Barker, C.; Scotland, R.W. Author Inflation Masks Global Capacity for Species Discovery in Flowering Plants. *New Phytol.* **2014**, *201*, 700–706. [CrossRef] [PubMed]
14. Gruber, T. Academic Sell-out: How an Obsession with Metrics and Rankings Is Damaging Academia. *J. Mark. High. Educ.* **2014**, *24*, 165–177. [CrossRef]
15. Bilton, D.T. What's in a Name? What Have Taxonomy and Systematics Ever Done for Us? *J. Biol. Educ.* **2014**, *48*, 116–118. [CrossRef]
16. Vink, C.; Paquin, P.; Cruickshank, R. Taxonomy and Irreproducible Biological Science. *BioScience* **2012**, *62*, 451–452. [CrossRef]
17. Packer, L.; Monckton, S.K.; Onuferko, T.M.; Ferrari, R.R. Validating Taxonomic Identifications in Entomological Research. *Insect Conserv. Divers.* **2018**, *11*, 1–12. [CrossRef]
18. Monckton, S.K.; Johal, S.; Packer, L. Inadequate Treatment of Taxonomic Information Prevents Replicability of Most Zoological Research. *Can. J. Zool.* **2020**, 633–642. [CrossRef]
19. Ceballos, G.; Ehrlich, P.R.; Dirzo, R. Biological Annihilation via the Ongoing Sixth Mass Extinction Signaled by Vertebrate Population Losses and Declines. *Proc. Natl. Acad. Sci. USA* **2017**, *114*, E6089–E6096. [CrossRef]
20. Young, H.S.; McCauley, D.J.; Galetti, M.; Dirzo, R. Patterns, Causes, and Consequences of Anthropocene Defaunation. *Annu. Rev. Ecol. Evol. Syst.* **2016**, *47*, 333–358. [CrossRef]
21. Régnier, C.; Achaz, G.; Lambert, A.; Cowie, R.H.; Bouchet, P.; Fontaine, B. Mass Extinction in Poorly Known Taxa. *Proc. Natl. Acad. Sci. USA* **2015**, *112*, 7761–7766. [CrossRef]
22. Dubois, A. Zoological Nomenclature in the Century of Extinctions: Priority vs. 'Usage'. *Org. Divers. Evol.* **2010**, *10*, 259–274. [CrossRef]
23. Bond, J. Phylogenetic Treatment and Taxonomic Revision of the Trapdoor Spider Genus *Aptostichus* Simon (Araneae, Mygalomorphae, Euctenizidae). *ZooKeys* **2012**, *252*, 1–209. [CrossRef]
24. Troudet, J.; Vignes-Lebbe, R.; Grandcolas, P.; Legendre, F. The Increasing Disconnection of Primary Biodiversity Data from Specimens: How Does It Happen and How to Handle It? *Syst. Biol.* **2018**, *67*, 1110–1119. [CrossRef] [PubMed]
25. Mora, C.; Tittensor, D.P.; Adl, S.; Simpson, A.G.B.; Worm, B. How Many Species Are There on Earth and in the Ocean? *PLoS Biol.* **2011**, *9*, e1001127. [CrossRef]
26. Stork, N.E. How Many Species of Insects and Other Terrestrial Arthropods Are There on Earth? *Annu. Rev. Entomol.* **2018**, *63*, 31–45. [CrossRef] [PubMed]
27. World Spider Catalog. Version 20.5. Natural History Museum Bern. Available online: http://wsc.nmbe.ch (accessed on 3 November 2021).
28. Agnarsson, I.; Coddington, J.A.; Kuntner, M. Systematics: Progress in the Study of Spiders and Evolution. In *Spider Research in the 21st Century: Trends and Perspectives*; Siri Scientific Press: Manchester, UK, 2013; pp. 58–1111. ISBN 978-0-9574530-1-2.
29. Dayrat, B. Towards Integrative Taxonomy: Integrative Taxonomy. *Biol. J. Linn. Soc.* **2005**, *85*, 407–415. [CrossRef]
30. DeSalle, R.; Egan, M.G.; Siddall, M. The Unholy Trinity: Taxonomy, Species Delimitation and DNA Barcoding. *Philos. Trans. R. Soc. B Biol. Sci.* **2005**, *360*, 1905–1916. [CrossRef]
31. Padial, J.M.; Miralles, A. TRehvieew Integrative Future of Taxonomy. *Front. Zool.* **2010**, *7*, 1–4. [CrossRef]
32. Pomerantz, A.; Peñafiel, N.; Arteaga, A.; Bustamante, L.; Pichardo, F.; Coloma, L.A.; Barrio-Amorós, C.L.; Salazar-Valenzuela, D.; Prost, S. Real-Time DNA Barcoding in a Rainforest Using Nanopore Sequencing: Opportunities for Rapid Biodiversity Assessments and Local Capacity Building. *GigaScience* **2018**, *7*, giy033. [CrossRef]
33. Mertens, J.; Van Roie, M.; Merckx, J.; Dekoninck, W. The Use of Low Cost Compact Cameras with Focus Stacking Functionality in Entomological Digitization Projects. *ZooKeys* **2017**, *712*, 141–154. [CrossRef] [PubMed]
34. Brecko, J.; Mathys, A.; Dekoninck, W.; Leponce, M.; VandenSpiegel, D.; Semal, P. Focus Stacking: Comparing Commercial Top-End Set-Ups with a Semi-Automatic Low Budget Approach. A Possible Solution for Mass Digitization of Type Specimens. *ZooKeys* **2014**, *464*, 1–23. [CrossRef] [PubMed]

35. Zamani, A.; Vahtera, V.; Sääksjärvi, I.E.; Scherz, M.D. The Omission of Critical Data in the Pursuit of 'Revolutionary' Methods to Accelerate the Description of Species. *Syst. Entomol.* **2020**, *46*, 1–4. [CrossRef]
36. Meierotto, S.; Sharkey, M.J.; Janzen, D.H.; Hallwachs, W.; Hebert, P.D.N.; Chapman, E.G.; Smith, M.A. A Revolutionary Protocol to Describe Understudied Hyperdiverse Taxa and Overcome the Taxonomic Impediment. *Dtsch. Entomol. Z.* **2019**, *66*, 119–145. [CrossRef]
37. Clerck, C. *Svenska Spindlar, Uti Sina Hufvud-Slågter Indelte Samt under Några Och Sextio Särskildte Arter Beskrefne Och Med Illuminerade Figurer Uplyste*; Stockholmiae: Stockholm, Sweden, 1757; pp. 1–154.
38. Patzold, F.; Zilli, A.; Hundsdoerfer, A.K. Advantages of an Easy-to-Use DNA Extraction Method for Minimal-Destructive Analysis of Collection Specimens. *PLoS ONE* **2020**, *15*, e0235222. [CrossRef]
39. Coates, D.J.; Byrne, M.; Moritz, C. Genetic Diversity and Conservation Units: Dealing With the Species-Population Continuum in the Age of Genomics. *Front. Ecol. Evol.* **2018**, *6*, 165. [CrossRef]
40. Gómez Daglio, L.; Dawson, M.N. Integrative Taxonomy: Ghosts of Past, Present and Future. *J. Mar. Biol. Assoc. UK* **2019**, *99*, 1237–1246. [CrossRef]
41. Hamilton, C.A.; Formanowicz, D.R.; Bond, J.E. Species Delimitation and Phylogeography of Aphonopelma Hentzi (Araneae, Mygalomorphae, Theraphosidae): Cryptic Diversity in North American Tarantulas. *PLoS ONE* **2011**, *6*, e26207. [CrossRef]
42. Hedin, M. High-Stakes Species Delimitation in Eyeless Cave Spiders (*Cicurina*, Dictynidae, Araneae) from Central Texas. *Mol. Ecol.* **2015**, *24*, 346–361. [CrossRef]
43. Opatova, V.; Bond, J.E.; Arnedo, M.A. Ancient Origins of the Mediterranean Trap-Door Spiders of the Family Ctenizidae (Araneae, Mygalomorphae). *Mol. Phylogenet. Evol.* **2013**, *69*, 1135–1145. [CrossRef]
44. Thomas, S.M.; Hedin, M. Multigenic Phylogeographic Divergence in the Paleoendemic Southern Appalachian Opilionid Fumontana Deprehendor Shear (Opiliones, Laniatores, Triaenonychidae). *Mol. Phylogenet. Evol.* **2008**, *46*, 645–658. [CrossRef] [PubMed]
45. Haszprunar, G. Species Delimitations—Not 'Only Descriptive'. *Org. Divers. Evol.* **2011**, *11*, 249–252. [CrossRef]
46. Bond, J.E.; Garrison, N.L.; Hamilton, C.A.; Godwin, R.L.; Hedin, M.; Agnarsson, I. Phylogenomics Resolves a Spider Backbone Phylogeny and Rejects a Prevailing Paradigm for Orb Web Evolution. *Curr. Biol.* **2014**, *24*, 1765–1771. [CrossRef] [PubMed]
47. Fernández, R.; Hormiga, G.; Giribet, G. Phylogenomic Analysis of Spiders Reveals Nonmonophyly of Orb Weavers. *Curr. Biol.* **2014**, *24*, 1772–1777. [CrossRef]
48. Garrison, N.L.; Rodriguez, J.; Agnarsson, I.; Coddington, J.A.; Griswold, C.E.; Hamilton, C.A.; Hedin, M.; Kocot, K.M.; Ledford, J.M.; Bond, J.E. Spider Phylogenomics: Untangling the Spider Tree of Life. *PeerJ* **2016**, *4*, e1719. [CrossRef]
49. Opatova, V.; Hamilton, C.A.; Hedin, M.; De Oca, L.M.; Král, J.; Bond, J.E. Phylogenetic Systematics and Evolution of the Spider Infraorder Mygalomorphae Using Genomic Scale Data. *Syst. Biol.* **2020**, *69*, 671–707. [CrossRef]
50. Coddington, J.A.; Agnarsson, I.; Miller, J.A.; Kuntner, M.; Hormiga, G. Undersampling Bias: The Null Hypothesis for Singleton Species in Tropical Arthropod Surveys. *J. Anim. Ecol.* **2009**, *78*, 573–584. [CrossRef]
51. Lim, G.S.; Balke, M.; Meier, R. Determining Species Boundaries in a World Full of Rarity: Singletons, Species Delimitation Methods. *Syst. Biol.* **2012**, *61*, 165–169. [CrossRef]
52. Vogt, L.; Nickel, M.; Jenner, R.A.; Deans, A.R. The Need for Data Standards in Zoomorphology. *J. Morphol.* **2013**, *274*, 793–808. [CrossRef]
53. Ramírez, M.J.; Michalik, P. The Spider Anatomy Ontology (SPD)—A Versatile Tool to Link Anatomy with Cross-Disciplinary Data. *Diversity* **2019**, *11*, 202. [CrossRef]
54. Vinarski, M.V. Roots of the Taxonomic Impediment: Is the "Integrativeness" a Remedy? *Integr. Zool.* **2020**, *15*, 2–15. [CrossRef] [PubMed]
55. Hebert, P.D.N.; Cywinska, A.; Ball, S.L.; deWaard, J.R. Biological Identifications through DNA Barcodes. *Proc. R. Soc. Lond. B Biol. Sci.* **2003**, *270*, 313–321. [CrossRef] [PubMed]
56. Ellis, R.; Waterton, C.; Wynne, B. Taxonomy, Biodiversity and Their Publics in Twenty-First-Century DNA Barcoding. *Public Underst. Sci.* **2010**, *19*, 497–512. [CrossRef] [PubMed]

Article

In Silico Assessment of Probe-Capturing Strategies and Effectiveness in the Spider Sub-Lineage Araneoidea (Order: Araneae)

Yi-Yen Li [1,2], Jer-Min Tsai [3], Cheng-Yu Wu [1], Yi-Fan Chiu [1], Han-Yun Li [1], Natapot Warrit [4], Yu-Cen Wan [1], Yen-Po Lin [2,5], Ren-Chung Cheng [2,*] and Yong-Chao Su [1,*]

[1] Department of Biomedical Science and Environmental Biology, Kaohsiung Medical University, Kaohsiung 80708, Taiwan; yiyenl30@gmail.com (Y.-Y.L.); u109551003@kmu.edu.tw (C.-Y.W.); ivan06513i@gmail.com (Y.-F.C.); hanyun1012@gmail.com (H.-Y.L.); u109551007@gap.kmu.edu.tw (Y.-C.W.)

[2] Department of Life Sciences, National Chung Hsing University, Taichung 40227, Taiwan; yplin@tesri.gov.tw

[3] Department of Information and Communication, Kun Shan University, Tainan 710, Taiwan; tjm@fhl.net

[4] Center of Excellence in Entomology and Department of Biology, Faculty of Science, Chulalongkorn University, Bangkok 10330, Thailand; natapot.w@chula.ac.th

[5] Taiwan Endemic Species Research Institute, No.1, Minsheng East Rd., Jiji Township, Nantou 55244, Taiwan

[*] Correspondence: bolasargiope@email.nchu.edu.tw (R.-C.C.); ycsu527@kmu.edu.tw (Y.-C.S.); Tel.: +886-4-228-40416 (ext. 707) (R.-C.C.); +886-7-312-1101 (ext. 6983) (Y.-C.S.)

Abstract: Reduced-representation sequencing (RRS) has made it possible to identify hundreds to thousands of genetic markers for phylogenomic analysis for the testing of phylogenetic hypotheses in non-model taxa. The use of customized probes to capture genetic markers (i.e., ultraconserved element (UCE) approach) has further boosted the efficiency of collecting genetic markers. Three UCE probe sets pertaining to spiders (Araneae) have been published, including one for the suborder Mesothelae (an early diverged spider group), one for Araneae, and one for Arachnida. In the current study, we developed a probe set specifically for the superfamily Araneoidea in spiders. We then combined the three probe sets for Araneoidea, Araneae, and Arachnid into a fourth probe set. In testing the effectiveness of the 4 probe sets, we used the captured loci of the 15 spider genomes in silico (6 from Araneoidea). The combined probe set outperformed all other probe sets in terms of the number of captured loci. The Araneoidea probe set outperformed the Araneae and Arachnid probe sets in most of the included Araneoidea species. The reconstruction of phylogenomic trees using the loci captured from the four probe sets and the data matrices generated from 50% and 75% occupancies indicated that the node linked to the *Stegodyphus* + RTA (retrolateral tibial apophysis) clade has unstable nodal supports in the bootstrap values, gCFs, and sCFs. Our results strongly indicate that developing ad hoc probe sets for sub-lineages is important in the cases where the origins of a lineage are ancient (e.g., spiders ~380 MYA).

Keywords: target sequencing; reduced representation sequencing (RRS); spider phylogenomics; deep phylogeny

Citation: Li, Y.-Y.; Tsai, J.-M.; Wu, C.-Y.; Chiu, Y.-F.; Li, H.-Y.; Warrit, N.; Wan, Y.-C.; Lin, Y.-P.; Cheng, R.-C.; Su, Y.-C. In Silico Assessment of Probe-Capturing Strategies and Effectiveness in the Spider Sub-Lineage Araneoidea (Order: Araneae). *Diversity* **2022**, 14, 184. https://doi.org/10.3390/d14030184

Academic Editor: Luc Legal

Received: 18 December 2021
Accepted: 28 February 2022
Published: 3 March 2022

Publisher's Note: MDPI stays neutral with regard to jurisdictional claims in published maps and institutional affiliations.

1. Introduction

High-throughput sequencing is widely used for the generation of genomic data in phylogenomic research [1–4]. Reduced-representation sequencing (RRS) methods [5] have made it possible to collect hundreds to thousands of genetic markers at a fraction of the cost of whole-genome sequencing [6]. The ultraconserved elements approach (UCE approach), a form of target DNA sequencing, is becoming particularly prevalent [7–9]. The UCE approach using customized probes makes it possible for researchers to capture thousands of genetic markers from non-model taxa, thereby making it possible to test hypotheses about phylogeny from shallow (e.g., <5 MYA) to deep (e.g., >200 MYA) divergence times [10]. Despite the importance of the UCE approach in phylogenomics, the design of ad hoc probe sets remains a technical gap such that many researchers are forced to use probe sets designed for similar taxa or for different taxonomic levels. In the current study, we compared the effectiveness of an ad hoc probe set for spiders in the superfamily Araneoidea to the existing probe sets that are known to be applicable to higher taxonomic levels in arachnids [9,11].

UCEs are non-variable genomic fragments that occur across species in a given taxonomic group [12]. These genomic fragments, which are often in >200 bps conserved regions [13], have been detected in a variety of taxa [14]. The functions of these UCEs are unknown [15], and the types of UCEs vary among taxonomic groups [16]. Hedin et al. [17] showed that the spider UCEs mostly correspond to exons. In a pioneering work, Faircloth et al. [18] captured 854 UCE loci to reconstruct the phylogenomic tree of birds. Subsequent research assessed the utility of the UCE approach in applying phylogenomic hypotheses to taxa dated from 5 MYA to 200 MYA [7,19,20]. The UCE approach has also been extended to the reconstruction of species trees and coalescent methods [21,22]. Recent advances in the UCE approach have strengthened phylogenetic hypothesis testing and phylogenetic tree reconstruction, including for arthropods [9,10,23].

The UCE approach was first applied to arachnids by Starrett et al. [23], and to Araneae by Kulkarni et al. [11]. Note that the order Araneae includes 49,877 species [24] in three subclades, suborder Mesothelae, infraorder Mygalomorphae, and infraorder Araneomorphae, with evolutionary time extending back to more than 300 MYA [25,26]. The application of the probe sets designed for the higher levels (for order and class) could be problematic because the probes may not be fully targeted, thus reducing the number of captured loci when testing the hypotheses of the phylogeny within suborders and lower taxonomic levels. Xu et al. [27] tested a customized probe set in the suborder Mesothelae. Hedin et al. [23] reconstructed the UCE phylogenomic tree in the Mygalomorphae species. In the current study, we developed an ad hoc UCE probe set for the superfamily Araneoidea, which contains 17 families, 25% spider diversity, and a variety of web architectures [28].

We developed the Araneoidea probe set in accordance with the pipeline outlined by Faircloth [9]. We then compared the effectiveness of our Araneoidea probe set with those for arachnid and Araneae. Finally, we combined these three probe sets as the fourth probe set. Note that we did not assess the Mesothelae probe set because it is clearly applicable at that suborder level [27]. We evaluated the effectiveness of the four probe sets in two schemes. (1) We performed in silico testing on the number of captured UCE loci in 15 genomes of Araneoidea and other spider species. (2) We compared the phylogenomic trees reconstructed using the concatenation and gene–tree–species–tree approaches with various data matrices to compare the tree topologies and node supports.

2. Materials and Methods

2.1. Data Sources of UCE Loci

As data sources for our in silico testing, we employed two published probe sets for ultraconserved elements [9,11], including 14 published genomes (Table 1) and 1 de novo assembled genome (*Argyrodes miniaceus*).

2.2. Genome Assembly

We assembled the genomes de novo using the procedure below. We used TRIMMOMATIC [29] for raw read trimming and adaptor removal. KMERGENIE [30] was then used to estimate the optimal k-mer length for genomic assembly. Finally, ABYSS 2.0 [31] was used to assemble the genome for *Argyrodes miniaceus* using the following settings: k = 55, B = 30 G. ABYSS–FAC was used to evaluate the quality of the genome assembly.

Table 1. Genomes fetched from GenBank. Information for all 14 genomes used in this research, which were fetched from GenBank. This table displays species name, assembly accession, assembly level, assembly submission date, N50 of contigs, coverage rate, and references for each genome. Adding *Argyrodes miniaceus*, 15 genomes are included in this study.

Organism Name	Assembly Accession	Total Sequence Length	Assembly Level	Submission Date	Contig N50	Coverage	Reference
Acanthoscurria geniculata	GCA_000661875.1	7,178,402,394	Contig	2014-04-29	541	21.5×	[32]
Anelosimus studiosus	GCA_008297655.1	2,033,432,615	Scaffold	2019-09-05	1132	79.0×	[33]
Araneus ventricosus	GCA_013235015.1	3,656,621,265	Scaffold	2019-08-02	22,999	70×	[34]
Argiope bruennichi	GCA_015342795.1	1,670,285,661	Chromosome	2020-11-16	284,772	70×	[35]
Dolomedes plantarius	GCA_907164885.1	2,381,335,874	Chromosome	2021-05-16	292,830	19.4×	[36]
Dysdera silvatica	GCA_006491805.2	1,365,686,336	Scaffold	2021-07-07	21,954	96.9×	[37]
Latrodectus hesperus	GCA_000697925.2	1,233,806,489	Scaffold	2018-02-05	15,961	80.0×	[38]
Loxosceles reclusa	GCA_001188405.1	3,262,478,678	Contig	2015-04-27	1834	55×	[38]
Oedothorax gibbosus	GCA_019343175.1	821,427,276	Chromosome	2021-08-05	979,336	14.0×	[39]
Parasteatoda tepidariorum	GCA_000365465.3	1,228,972,128	Scaffold	2019-06-14	66,479	48.0×	[40]
Pardosa pseudoannulata	GCA_008065355.1	4,207,954,893	Scaffold	2019-08-22	23,226	423.95×	[41]
Stegodyphus dumicola	GCF_010614865.1	2,551,871,595	Scaffold	2020-02-14	254,130	49.0×	[42]
Stegodyphus mimosarum	GCA_000611955.2	2,738,704,917	Scaffold	2014-08-01	40,146	86.0×	[32]
Trichonephila clavipes	GCA_002102615.1	2,439,301,466	Scaffold	2017-04-20	7993	140.0×	[43]

2.3. Design of UCE Probe Set for Araneidae

The design of UCE probes was based on the PHYLUCE pipeline [9,44]. The genomes of *Argyrodes miniaceus, Latrodectus hesperus, Loxosceles reclusa, Trichonephila clavipes, Parasteatoda tepidariorum,* and *Stegodyphus mimosarum* were used in UCE probe design as follows: (1) ART v2016.06.05 [45] was used to simulate genomic fragments of *Argyrodes miniaceus, Latrodectus hesperus, Loxosceles reclusa, Trichonephila clavipes,* and *Parasteatoda tepidariorum* into 100-bps reads. (2) Simulated short reads were aligned to *Stegodyphus mimosarum* (i.e., the base genome [32]) using STAMPY (substitution rate = 0.05 and insert size = 400) [46]. Misaligned fragments were removed using SAMTOOLS [47], and the aligned fragments were combined in the browser-extensible data (BED) format using BEDTOOLS [48]. (3) Duplicated genomic fragments were removed using PHYLUCE script (phyluce_probe_strip_masked_loci_from_set) to detect and remove fragments that were mapped but designated too short (<80 bps) or within masked regions of the *Stegodyphus mimosarum* genome (more than 25%). (4) SQLITE v 3.34.0 [49] was used to construct a database of candidate UCE sites for *Argyrodes miniaceus, Latrodectus hesperus, Loxosceles reclusa, Trichonephila clavipes,* and *Parasteatoda tepidariorum* to determine the shared conserved regions. PHYLUCE was then used to remove duplicated candidate probes, and LASTZ [50] was used to align the candidate probes with a given genome to enable the extraction of UCE sites for a given species. (5) Finally, we relaxed the similarity to 50% and reconstructed the database in SQLITE to create the final UCE probes, whereupon we repeated the duplicate-probe removal process in PHYLUCE. The final probe length was 120 bps with tiling, with 60 bps overlapping (thus covering 180 bps) per target locus.

2.4. In Silico Simulation of Probe Sets Aimed at Capturing Affinity

Simulations were conducted using four probe sets for Arachnids (Arachnid probe set [9]), Araneae (Araneae probe set [11]), Araneoidea (Araneoidea probe set), and a combination of these three (combined probe set). In generating the combined probe set, we compiled the probe sets for Arachnid, Araneae, and Araneoidea and removed potential duplicated probes using LASTZ Python script (phyluce_probe_remove_duplicate_hits_from_probes_using_lastz [44]). Each probe set was tested on 15 genomes using standardized testing procedures. (1) The probes were aligned with the targeted genome using LASTZ [50]. (2) The probes were then aligned and mapped to the targeted genome using PHYLUCE [44] to extract the 500-bps regions on both sides of the UCE probe sites. Note that our objective was to simulate the fragment length when conducting in-solution capture. (3) The probes were aligned with each extracted sequence using LASTZ by running phyluce_assembly_match_contigs_to_probes to identify which loci sequences belonged. The duplicates were again sorted out. Following the capture and filtering of fragments from each probe set and each targeted genome, the number of captured loci and proportion of identified loci (defined as the capture rate) in all simulated contigs were calculated per genome per probe set. The captured loci per probe set were then used to reconstruct the phylogenomic tree per data matrix from each probe set.

2.5. Reconstruction of Phylogenomic Tree Using the Captured Data Matrix for Each Probe Set

We generated the data matrices for different occupancy (the smallest percentage of data per locus in a matrix). We counted the number of in silico captured loci in each genome under occupancies from 10% to 100%, with 10% as the increment. We then decided the occupancies to use in the final tree-reconstruction analyses. The script phyluce_align_get_only_loci_with_min_taxa was used to output the locus matrices of the probe set using occupancies 50% and 75%, respectively. This allowed the omission of up to 50% and up to 25% missing taxa per locus, which resulted in two matrices per probe set. In total, we used eight locus matrices in reconstructing the phylogenomic trees of available spider species.

The MAFFT [51] script, phyluce_align_seqcap_align, was used to align each UCE locus, whereupon the ends of the aligned fragments were trimmed using GBLOCKS (default arguments of PHYLUCE: −b1 = 0.5, −b2 = 0.85, −b3 = 8, and −b4 = 10) [52] via the script phyluce_align_get_gblocks_trimmed_alignments_from_untrimmed. We then concatenated the aligned loci using phyluce_align_concatenate_alignments to produce a matrix for each probe-set per occupancy and then output the matrices in PHYLIP format (partition scheme) and in NEXUS format. After detecting the models with the best fit in each locus using MODELFINDER [53], IQTREE-2.0.3 [54] was used to reconstruct the phylogenomic trees via concatenation involving 1000 bootstrap operations. We also used the gene–tree–species–tree approach in IQTREE-2.0.3 to infer the gene trees and calculate the gene concordance factors (gCFs) and site concordance factors (sCFs) for the nodes associated with species tree [55]. In accordance with the methods outlined by Wheeler et al. [56], *Acanthoscurria geniculata* (Theraphosidae) was used as an outgroup. Finally, FIGTREE [57] was used to visualize phylogenomic trees.

3. Results

3.1. De Novo Genome Assembly

Using Illumina Hi-seq short-read sequences, we assembled a genome, *Argyrodes miniaceus*. From 7,051,281 contigs in *Argyrodes miniaceus*, we obtained a total assembled length of 35.51×10^6 bps with N50 = 618 bps and a maximum assembled contig of 5834 bps (for genome assembly statistics, see Table S1). This de novo assembled draft genome was then intended to be used to detect UCE probes.

3.2. Probe Detection

Using *Stegodyphus mimosarum* as the base genome in accordance with the methods outlined by Faircloth et al. [9], we detected 12,679 probes related to 1374 UCE loci using *Argyrodes miniaceus, Latrodectus hesperus, Loxosceles reclusa, Trichonephila clavipes,* and *Parasteatoda tepidariorum.*

3.3. In Silico Testing of Capture Efficiency

We used four probe sets for the in silico capture of targeted loci from 15 genomes. We detected a total of 7357 loci using the newly designed Araneoidea probe set (Figure 1 and Tables S2–S5). From the Arachnid probe set, we detected a total of 4579 loci. From the Araneae probe set, we detected a total of 9103 loci. Accordingly, even though we mostly used the genomes in Araneoidea in this study, we collected fewer loci compared with the Araneae probe set, which mostly included the well-assembled genomes fetched from GenBank (Table 2). From the combined probe set, we detected 14,271 loci.

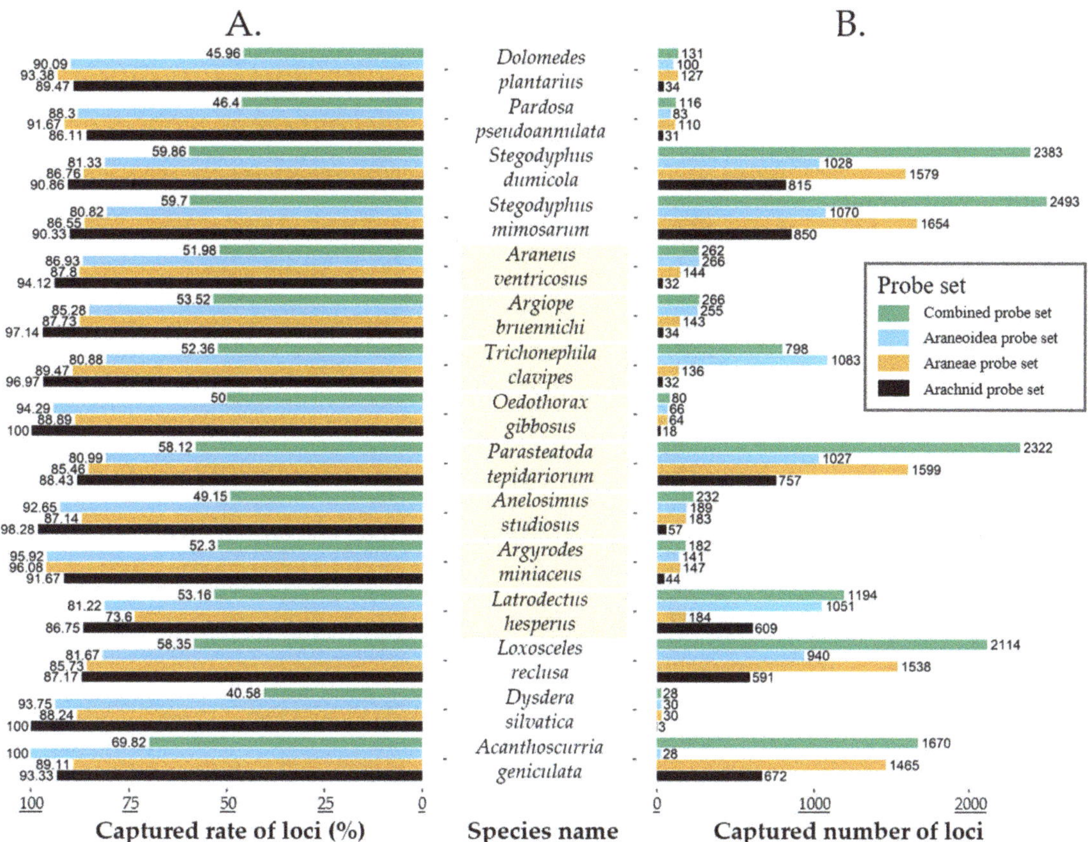

Figure 1. Results of in silico tests. Capture rate of loci (**A**) and captured loci number (**B**) in each probe set. The Araneoidea species are shaded in orange.

The performance of each probe set in capturing the loci in each genome varied as a function of the taxonomic group. In eight of the genomes, the Araneae probe set outperformed the Araneoidea probe set in the capture of loci (Figure 1). The Arachnid probe set outperformed the Araneoidea probe set in only one species, *Acanthoscurria geniculata* (Mygalomorphae). Nonetheless, the Araneoidea probe set outperformed two published probe sets, mostly in the Araneoidea species (e.g., *Trichonephila clavipes*, *Latrodectus hesperus*, *Argiope bruennichi*, *Anelosimus studiosus*, *Oedothorax gibbosus*, and *Araneus ventricosus*). The combined probe set outperformed all probe sets in most species except *Araneus ventricosus*, *Argiope bruennichi*, and *Dysdera silvatica*, which presented very few loci (3–30 loci) in each probe set (Figure 1).

The recovery rates of the probe sets were assessed using the combined probe set as targeted contigs to determine the number of captures and capture rates. The re-capture number and re-capture rates were lowest in the Arachnid probe set, followed by the Araneoidea probe set and the Araneae probe set (Figure 2).

Table 2. List of the genomes for probe-set design and the numbers of captured loci. All probe sets used in this research are listed below, ordered by published year. This table displays the targeted taxa of probe-set design, species names of the genomes used to identify UCE loci, species names of the genomes used to design probes, number of UCE loci, number of probes, and the published year of the probe set.

Target Taxon	Genomes Used to Identify UCEs	Genomes Used to Design Probes	Number of UCE Loci	Number of Probes	Publication Year	Reference
Arachnida	*Trithyreus pentapaltis, Atypoides riversi, Phrynus marginemaculatus, Cryptocellus goodnighti, Mitopus morio, Bothriurus keyserlingi, Pseudouroctonus apacheanus, Hadogenes troglodytes, Vaejovis deboerae, Ixodes scapularis, Limulus polyphemus*	*Ixodes scapularis, Limulus polyphemus, Acanthoscurria geniculata, Centruroides exilicauda, Latrodectus hesperus, Mesobuthus martensii, Parasteatoda tepidariorum, Stegodyphus mimosarum, Amblyomma americanum*	1120	14,799	2017	[9]
Araneae	*Parasteatoda tepidariorum, Acanthoscurria geniculata, Stegodyphus mimosarum*	*Parasteatoda tepidariorum, Acanthoscurria geniculata, Stegodyphus mimosarum*	2021	15,051	2020	[11]
Araneoidea	*Argyrodes miniaceus, Latrodectus hesperus, Loxosceles reclusa, Trichonephila clavipes, Parasteatoda tepidariorum, Stegodyphus mimosarum*	*Argyrodes miniaceus, Latrodectus hesperus, Loxosceles reclusa, Trichonephila clavipes, Parasteatoda tepidariorum, Stegodyphus mimosarum*	1374	12,679	2021	This article
-	-	-	3344	30,379	2021	This article

3.4. Capture Rates and Number of Loci in Various Occupancies

We present the number of captured loci in each genome under occupancies from 10% to 100%, in increments of 10%. The numbers of captured loci were higher in the combined probe set and Araneae probe set under occupancies of 10% to 30%. The numbers of captured loci did not vary considerably under occupancies of >50%. We observed similar trends in the retention ratio, with the highest retention in the Araneoidea probe set, and a merging of results at occupancies of >50% (Figure 3). Thus, in accordance with the UCE-phylogenomic results published earlier, we used occupancies of 50% and 75% in reconstructing the phylogenomic trees [11,25]. Note, however, that this strategy reduced the in silico capture number to less than 350 loci in each genome (see Figure 4; for other trees, see Figures S1–S7).

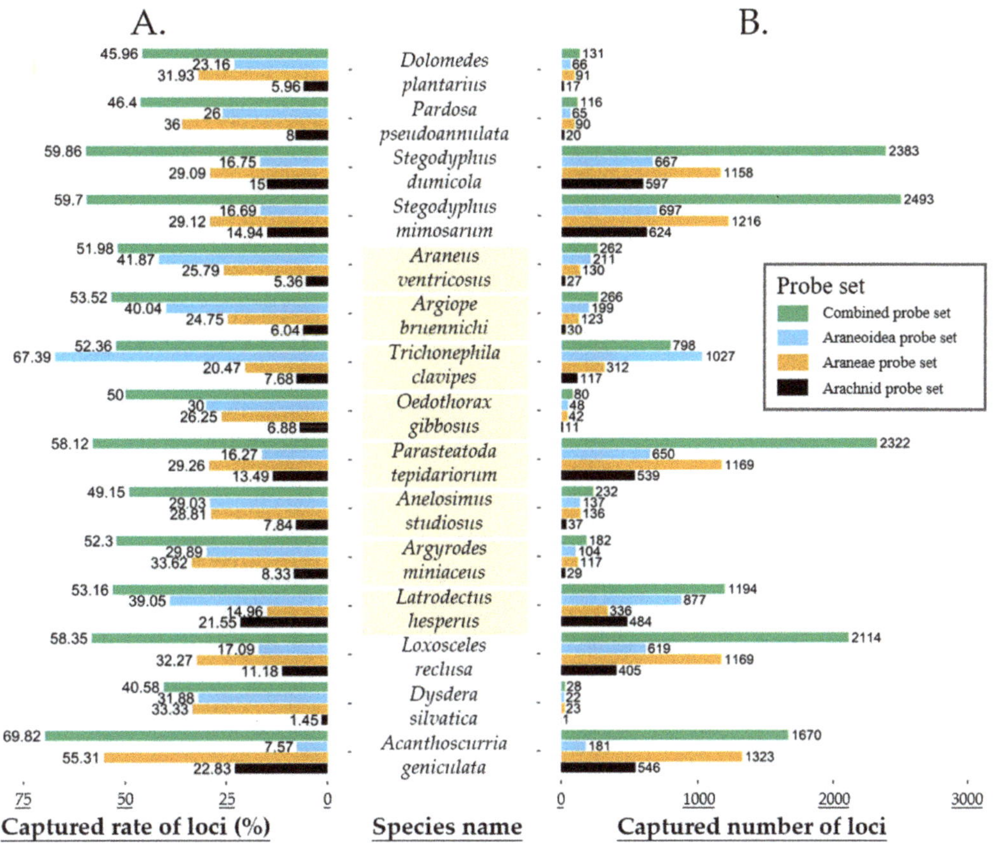

Figure 2. Comparison of all three probe sets with the combined probe set. Capture rate of loci (**A**) and captured loci number (**B**) when each probe set aligned with simulated contigs using the sequences of the combined probe set.

3.5. Tree Reconstruction Using Simulated Captured Loci

We reconstructed the phylogenomic trees using two occupancies (50% and 75%) for the loci captured from the four probe sets, thereby resulting in eight data matrices. The resulting topologies were similar to previous findings (e.g., Kulkarni et al. [58]), and the supports (i.e., bootstrap, gCF, and sCF) of each node were similar between the results of these two datasets (shown in Figure 4B,C).

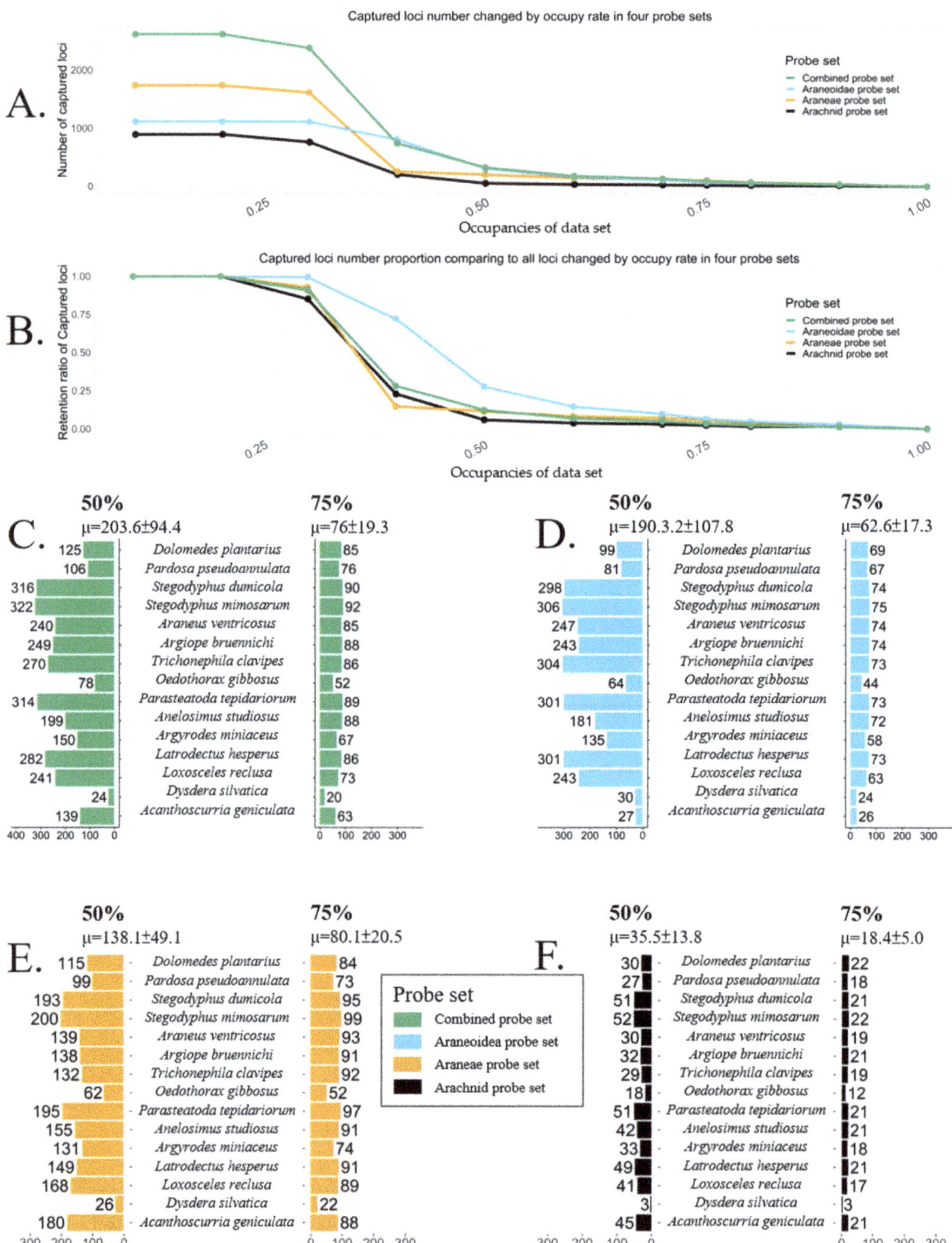

Figure 3. Loci numbers of data sets. (**A**,**B**) Number of loci changed by filtered increment of 10% (from 10% to 100%) occupancies (for data, see Table S6). (**C–F**) Number of loci used in tree reconstruction of each data set (Table S7).

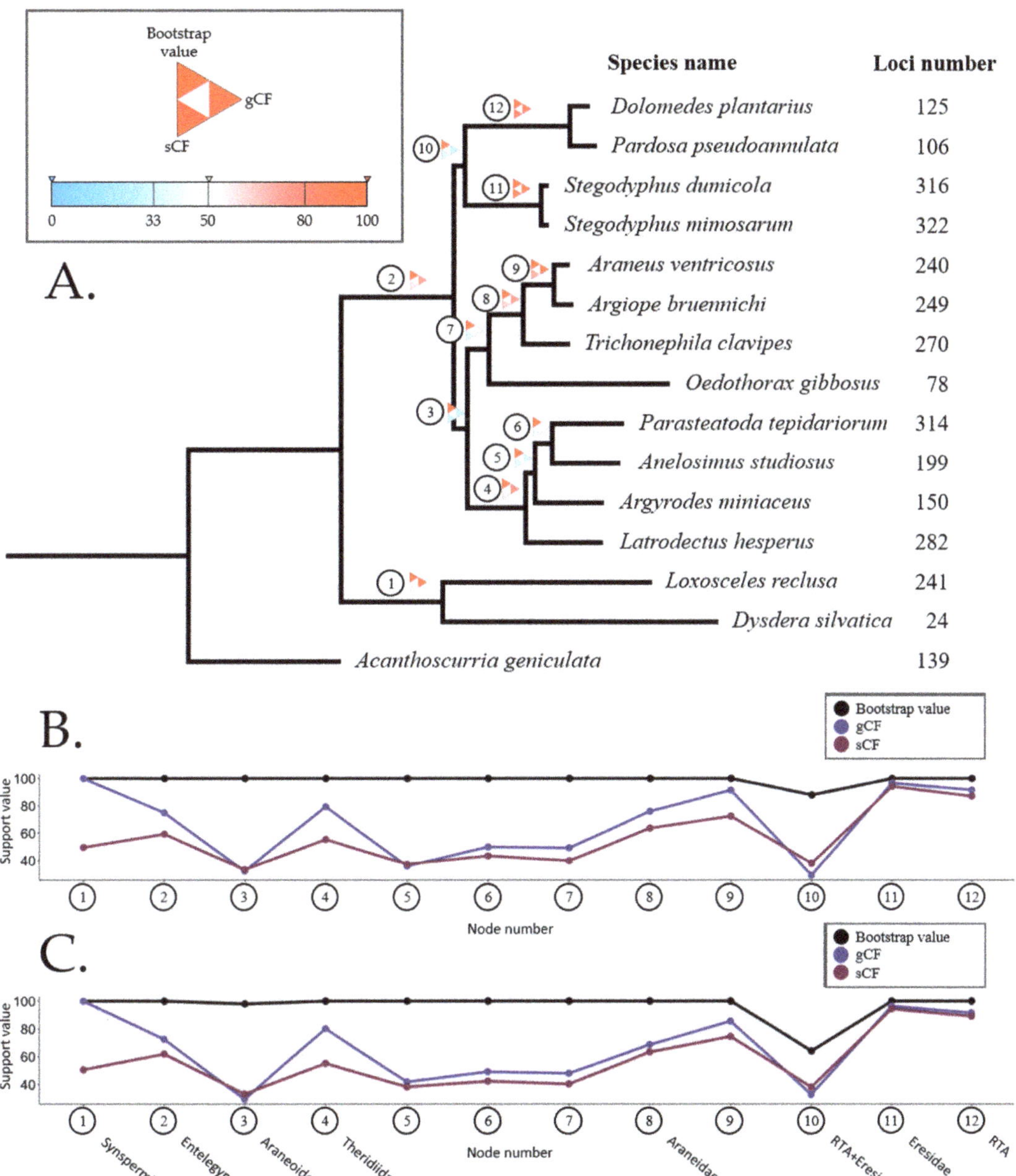

Figure 4. Phylogenetic tree of the data set that had the most captured loci (combined probe set filtered with 50% occupancy). (**A**) Tree topology: color of triangles represents three types of support values— bootstrap, gCF, and sCF—and dashed lines indicate the loci numbers used in tree reconstruction. (**B,C**) Support values for different nodes on reconstructed trees using combined probe set of 50% (**B**) and 75% (**C**) occupancies. Other trees, see Figures S1–S7.

4. Discussion

This study suggests that even when dealing with a monophyletic group (e.g., Araneae), an ancient evolutionary origin (e.g., ~380 MYA), the use of a specific probe set to test phylogenetic hypotheses within a sub-lineage could benefit via more lineage-specific loci, and potentially, more captured loci. A specific probe set is meant to enable the capture of a larger number of specific loci to facilitate phylogenomic analysis when combined with probes designed for higher taxonomic levels. The number of loci revealed by the Araneoidea probe set (7357 loci) was lower than that of the Araneae probe set (9103 loci). However, the loci captured in the Araneoidea species outperformed the probe sets designed for higher taxonomic levels (the Arachnid and Araneae probe sets, Figures 1 and 2). Incremental testing of occupancy from 10% to 100% revealed that the probe set designed specifically for Araneoidea presented a more gradual loss of retention than the other probe sets, including the combined probe set (Figure 3A,B). The higher retention rate made possible by the specific probe sets produced a larger number of orthologous loci that only occurred in the targeted clade. In tree reconstruction, the tree topologies were consistent across the eight data matrices in the basal nodes and the node related to Araneoidea (Figure 4). Note, however, that the nodal supports (node 10, Figure 4) in the *Stegodyphus* + RTA (retrolateral tibial apophysis) clade were unstable, thereby supporting our claim that a specially designed probe set is necessary for a sub-lineage (e.g., the RTA clade).

Our in silico results showed that the numbers of captured loci using the combined probe set generally outperformed other probe sets. Among the specifically designed taxon probe sets, the Araneoidea probe set captured a larger number of loci in five of the six genomes used to develop the probe set. However, both probe sets performed poorly in the RTA clade (Figures 1 and 2). We detected 490.4 ± 464.7 loci (range = 28 to 1083) in the Araneoidea probe set, and 951.4 ± 974.4 loci (range = 28 to 2493) in the combined probe set. Note that the newly assembled draft genome for *Argyrodes miniaceus* returned a relatively low number of loci (182 and 142 in the combined and Araneoideae probe sets, respectively). The other Araneoidea genomes included in this study, which assembled in lower qualities (Table 1), generated <270 loci, thereby demonstrating that the number of captured loci was biased toward the well-assembled genomes used in the design of the probes. The Araneae and Arachnid probe sets generated low numbers of loci in Araneoidea genomes (lower than the Araneoidea probe set, except *Parasteatoda tepidariorum* and *Argyrodes miniaceus*). These showed a taxon-specific trend that the probe sets designed for higher taxonomic levels tended to capture fewer, and nearly insufficient, loci for phylogenomic analyses. Together with the results obtained using the four probe sets, we found that the quality and completeness of the genomes could have a deterministic effect on the number of captured loci. Moreover, the taxonomic group played a role in the number of captured loci, i.e., if there were no representative genomes in a clade, a low number of captured loci would be observed (see the RTA clade in Figure 1).

We did not observe large variations in the tree topologies reconstructed using different data matrices (i.e., with loci captured from different probe sets). However, the nodal supports dropped in both traditional bootstrap statistics and in the concordance factors (gCF and sCF) when there were no representative genomes used for probe design (i.e., *Stegodyphus* + RTA clade, in our case) (Figure 4). Within Araneoidea, nodes 3 and 5 did not perform well in gCF and sCF; however, the results still met an acceptable level of >33, thereby indicating a possible downside of using existing probe sets to resolve these nodes. Bootstrap values tended to generate optimistically high support, as observed in other phylogenomic studies [55]. In the current study, we used 50% and 75% occupancies to generate data matrices for phylogenomic analysis, with the mean number of loci varying from 35.5 to 203.6 per genome in 50% occupancy matrices, and a mean of 18.4 to 80.1 per genome in 75% occupancy matrices. The number of captured loci in silico was significantly lower than would be expected in real-world, in-solution captured data. The mean number of captured loci was 589.3 in the Arachnid probe set [23] and 553.71 in the Araneae probe set [11]. Our in silico results, in rough estimation, only captured up to 1/3 of the loci compared with the in-solution captured results. We inferred that for in silico testing, we constrained the sequence identity to 80% for the capture of loci. From a practical perspective, in-solution capture could likely have required a lower degree of similarity to capture DNA fragments. As we aimed to relatively compare the numbers of the captured loci from different probe sets under the same in silico condition, we therefore expected to capture a larger number of loci when using our probe set for in-solution capture under laboratory conditions in the Araneoidea species.

5. Conclusions

This study designed specific probe sets using six genomes to facilitate the testing of phylogenetic hypotheses pertaining to Araneoidea. When using in silico capture, the data matrices generated using the combined and Araneoidea probes resolved most of the nodes in the sub-clades in Araneoidea, resulting in several hundred loci (relatively more loci compared with other non-targeted taxa). We expected that when conducting in-solution capture in a wet lab, it should be possible, using the estimated 1/3 in silico/in-solution ratio, to capture more than one thousand loci per genome. In our preliminary test using Argyrodinae as a targeted taxon, we captured 897.5 ± 62.9 loci per genome, which is about $4\times$ the number captured in our in silico *Argyrodes miniaceus* results (182 or 142 loci). However, there are disadvantages to using this newly designed probe set, e.g., (1) fewer applicability to other taxa such as the RTA clade, and (2) potentially higher costs when synthesizing this customized probe and the combined probe sets. Moreover, our combined approach showed that it broadens the application of the probe sets given there are representative genomes collected from a sub-lineage. However, these approaches should be tested in a wet lab to validate the applicability of the Araneoidea and combined probe sets.

Supplementary Materials: The following supporting information can be downloaded at https://www.mdpi.com/article/10.3390/d14030184/s1: all generated phylogenetic trees in this article, assembled genome statistics, and the raw data in Figures 1–4; Figure S1: Phylogenetic tree of the data set of the combined probe set filtered by 75% occupancy; Figure S2: Phylogenetic tree of the data set of the Arachnid probe set filtered by 50% occupancy; Figure S3: Phylogenetic tree of the data set of the Arachnid probe set filtered by 75% occupancy; Figure S4: Phylogenetic tree of the data set of the Araneae probe set filtered by 50% occupancy; Figure S5: Phylogenetic tree of the data set of the Araneae probe set filtered by 75% occupancy; Figure S6: Phylogenetic tree of the data set of the Araneoidea probe set filtered by 50% occupancy; Figure S7: Phylogenetic tree of the data set of the Araneoidea probe set filtered by 75% occupancy; Table S1: Statistics of genome and raw reads of Argyrodes miniaceus; Table S2: Results of probe set designed for Arachnid probe set in silico test on each genome; Table S3: Results of probe set designed for Araneae probe set in silico test on each genome; Table S4: Results of probe set designed for Araneoidea probe set in silico test on each genome; Table S5: Results of probe set designed for combined probe set in silico test on each genome; Table S6: Retention number of loci in data sets after filtering by different standards; Table S7: Loci number of each genome after filtering by occupancy.

Author Contributions: Conceptualization, Y.-C.S. and R.-C.C.; methodology, Y.-Y.L. and Y.-C.S.; software and hardware maintenance, J.-M.T. and Y.-C.W.; validation, C.-Y.W., Y.-F.C. and H.-Y.L.; formal analysis and data curation, Y.-Y.L., Y.-C.S. and C.-Y.W.; writing—original draft preparation, Y.-Y.L., Y.-F.C. and H.-Y.L.; writing—review and editing, Y.-C.S., R.-C.C. and N.W.; visualization, C.-Y.W. and Y.-Y.L.; funding acquisition, Y.-P.L.; supervision, project administration, and funding acquisition, Y.-C.S., R.-C.C. and N.W. All authors have read and agreed to the published version of the manuscript.

Funding: The funding sources are the Ministry of Science and Technology, Taiwan (110-2621-B-037 -001-MY3 and 107-2621-B-037-001-MY2 to Y.-C.S.; 108-2621-B-005-001 and 109-2621-B-005-003-MY2 to R.-C.C.), and the National Science and Technology Development Agency, Thailand (NSTDA: JRA-CO-2563-11148-TH to N.W.). Part of this research is from Y.-Y.L.'s thesis.

Institutional Review Board Statement: Our in silico study does not require Institutional Review Board approval as it does not involve human or vertebrate materials.

Data Availability Statement: The de novo assembled genome and the probe sets are available at GitHub: https://github.com/yiyenl/designing-probe-set-for-Araneoidea/blob/main/combine_probe.pl. Other visualized results are in the Supplementary Materials.

Acknowledgments: We thank all Ecology and Evolutionary Lab members at Kaohsiung Medical University, Taiwan.

Conflicts of Interest: The authors declare no conflict of interest.

References

1. Barrett, C.F.; Bacon, C.D.; Antonelli, A.; Cano, Á.; Hofmann, T. An Introduction to Plant Phylogenomics with a Focus on Palms. *Bot. J. Linn. Soc.* **2016**, *182*, 234–255. [CrossRef]
2. Pettengill, J.B.; Luo, Y.; Davis, S.; Chen, Y.; Gonzalez-Escalona, N.; Ottesen, A.; Rand, H.; Allard, M.W.; Strain, E. An Evaluation of Alternative Methods for Constructing Phylogenies from Whole Genome Sequence Data: A Case Study with *Salmonella*. *PeerJ* **2014**, *2*, e620. [CrossRef]
3. Brewer, M.S.; Cotoras, D.D.; Croucher, P.J.P.; Gillespie, R.G. New Sequencing Technologies, the Development of Genomics Tools, and Their Applications in Evolutionary Arachnology. *J. Arachnol.* **2014**, *42*, 1–15. [CrossRef]
4. Giribet, G. New Animal Phylogeny: Future Challenges for Animal Phylogeny in the Age of Phylogenomics. *Org. Divers. Evol.* **2016**, *16*, 419–426. [CrossRef]
5. Hirsch, C.D.; Evans, J.; Buell, C.R.; Hirsch, C.N. Reduced Representation Approaches to Interrogate Genome Diversity in Large Repetitive Plant Genomes. *Brief. Funct. Genom.* **2014**, *13*, 257–267. [CrossRef]
6. Ekblom, R.; Galindo, J. Applications of next Generation Sequencing in Molecular Ecology of Non-Model Organisms. *Heredity* **2011**, *107*, 1–15. [CrossRef]
7. McCormack, J.E.; Faircloth, B.C.; Crawford, N.G.; Gowaty, P.A.; Brumfield, R.T.; Glenn, T.C. Ultraconserved Elements Are Novel Phylogenomic Markers That Resolve Placental Mammal Phylogeny When Combined with Species-Tree Analysis. *Genome Res.* **2012**, *22*, 746–754. [CrossRef]
8. Mamanova, L.; Coffey, A.J.; Scott, C.E.; Kozarewa, I.; Turner, E.H.; Kumar, A.; Howard, E.; Shendure, J.; Turner, D.J. Target-Enrichment Strategies for next-Generation Sequencing. *Nat. Methods* **2010**, *7*, 111–118. [CrossRef]
9. Faircloth, B.C. Identifying Conserved Genomic Elements and Designing Universal Bait Sets to Enrich Them. *Methods Ecol. Evol.* **2017**, *8*, 1103–1112. [CrossRef]
10. Zhang, Y.M.; Williams, J.L.; Lucky, A. Understanding UCEs: A Comprehensive Primer on Using Ultraconserved Elements for Arthropod Phylogenomics. *Insect Syst. Divers.* **2019**, *3*, 3. [CrossRef]
11. Kulkarni, S.; Wood, H.; Lloyd, M.; Hormiga, G. Spider-Specific Probe Set for Ultraconserved Elements Offers New Perspectives on the Evolutionary History of Spiders (*Arachnida, Araneae*). *Mol. Ecol. Resour.* **2020**, *20*, 185–203. [CrossRef]
12. Ryu, T.; Seridi, L.; Ravasi, T. The Evolution of Ultraconserved Elements with Different Phylogenetic Origins. *BMC Evol. Biol.* **2012**, *12*, 236. [CrossRef]
13. Bejerano, G.; Pheasant, M.; Makunin, I.; Stephen, S.; Kent, W.J.; Mattick, J.S.; Haussler, D. Ultraconserved Elements in the Human Genome. *Science* **2004**, *304*, 1321–1325. [CrossRef]
14. Siepel, A.; Bejerano, G.; Pedersen, J.S.; Hinrichs, A.S.; Hou, M.; Rosenbloom, K.; Clawson, H.; Spieth, J.; Hillier, L.W.; Richards, S.; et al. Evolutionarily Conserved Elements in Vertebrate, Insect, Worm, and Yeast Genomes. *Genome Res.* **2005**, *15*, 1034–1050. [CrossRef]
15. Habic, A.; Mattick, J.S.; Calin, G.A.; Krese, R.; Konc, J.; Kunej, T. Genetic Variations of Ultraconserved Elements in the Human Genome. *OMICS A J. Integr. Biol.* **2019**, *23*, 549–559. [CrossRef]
16. Van Dam, M.H.; Henderson, J.B.; Esposito, L.; Trautwein, M. Genomic Characterization and Curation of UCEs Improves Species Tree Reconstruction. *Syst. Biol.* **2021**, *70*, 307–321. [CrossRef]
17. Hedin, M.; Derkarabetian, S.; Alfaro, A.; Ramírez, M.J.; Bond, J.E. Phylogenomic Analysis and Revised Classification of Atypoid Mygalomorph Spiders (*Araneae, Mygalomorphae*), with Notes on Arachnid Ultraconserved Element Loci. *PeerJ* **2019**, *7*, e6864. [CrossRef]
18. Faircloth, B.C.; McCormack, J.E.; Crawford, N.G.; Harvey, M.G.; Brumfield, R.T.; Glenn, T.C. Ultraconserved Elements Anchor Thousands of Genetic Markers Spanning Multiple Evolutionary Timescales. *Syst. Biol.* **2012**, *61*, 717–726. [CrossRef]
19. Thom, G.; Amaral, F.R.D.; Hickerson, M.J.; Aleixo, A.; Araujo-Silva, L.E.; Ribas, C.C.; Choueri, E.; Miyaki, C.Y. Phenotypic and Genetic Structure Support Gene Flow Generating Gene Tree Discordances in an Amazonian Floodplain Endemic Species. *Syst. Biol.* **2018**, *67*, 700–718. [CrossRef]
20. Winker, K.; Glenn, T.C.; Faircloth, B.C. Ultraconserved Elements (UCEs) Illuminate the Population Genomics of a Recent, High-Latitude Avian Speciation Event. *PeerJ* **2018**, *6*, e5735. [CrossRef]

21. Meiklejohn, K.A.; Faircloth, B.C.; Glenn, T.C.; Kimball, R.T.; Braun, E.L. Analysis of a Rapid Evolutionary Radiation Using Ultraconserved Elements: Evidence for a Bias in Some Multispecies Coalescent Methods. *Syst. Biol.* **2016**, *65*, 612–627. [CrossRef] [PubMed]

22. Bossert, S.; Murray, E.A.; Pauly, A.; Chernyshov, K.; Brady, S.G.; Danforth, B.N. Gene Tree Estimation Error with Ultraconserved Elements: An Empirical Study on *Pseudapis* Bees. *Syst. Biol.* **2021**, *70*, 803–821. [CrossRef] [PubMed]

23. Starrett, J.; Derkarabetian, S.; Hedin, M.; Bryson, R.W.; McCormack, J.E.; Faircloth, B.C. High Phylogenetic Utility of an Ultraconserved Element Probe Set Designed for Arachnida. *Mol. Ecol. Resour.* **2017**, *17*, 812–823. [CrossRef] [PubMed]

24. Gloor, D.; Nentwig, W.; Blick, T.; Kropf, C. World Spider Catalog. *Nat. Hist. Mus. Bern.* **2022**. [CrossRef]

25. Selden, P.A.; Shear, W.A.; Sutton, M.D. Fossil Evidence for the Origin of Spider Spinnerets, and a Proposed Arachnid Order. *Proc. Natl. Acad. Sci. USA* **2008**, *105*, 20781–20785. [CrossRef]

26. Garrison, N.L.; Rodriguez, J.; Agnarsson, I.; Coddington, J.A.; Griswold, C.E.; Hamilton, C.A.; Hedin, M.; Kocot, K.M.; Ledford, J.M.; Bond, J.E. Spider Phylogenomics: Untangling the Spider Tree of Life. *PeerJ* **2016**, *4*, e1719. [CrossRef]

27. Xu, X.; Su, Y.-C.; Ho, S.Y.W.; Kuntner, M.; Ono, H.; Liu, F.; Chang, C.-C.; Warrit, N.; Sivayyapram, V.; Aung, K.P.P.; et al. Phylogenomic Analysis of Ultraconserved Elements Resolves the Evolutionary and Biogeographic History of Segmented Trapdoor Spiders. *Syst. Biol.* **2021**, *70*, 1110–1122. [CrossRef]

28. Dimitrov, D.; Benavides, L.R.; Arnedo, M.A.; Giribet, G.; Griswold, C.E.; Scharff, N.; Hormiga, G. Rounding up the Usual Suspects: A Standard Target-Gene Approach for Resolving the Interfamilial Phylogenetic Relationships of Ecribellate Orb-Weaving Spiders with a New Family-Rank Classification (*Araneae, Araneoidea*). *Cladistics* **2017**, *33*, 221–250. [CrossRef]

29. Bolger, A.M.; Lohse, M.; Usadel, B. Trimmomatic: A Flexible Trimmer for Illumina Sequence Data. *Bioinformatics* **2014**, *30*, 2114–2120. [CrossRef]

30. Chikhi, R.; Medvedev, P. Informed and Automated K-Mer Size Selection for Genome Assembly. *Bioinformatics* **2014**, *30*, 31–37. [CrossRef]

31. Jackman, S.D.; Vandervalk, B.P.; Mohamadi, H.; Chu, J.; Yeo, S.; Hammond, S.A.; Jahesh, G.; Khan, H.; Coombe, L.; Warren, R.L.; et al. ABySS 2.0: Resource-Efficient Assembly of Large Genomes Using a Bloom Filter. *Genome Res.* **2017**, *27*, 768–777. [CrossRef] [PubMed]

32. Sanggaard, K.W.; Bechsgaard, J.S.; Fang, X.; Duan, J.; Dyrlund, T.F.; Gupta, V.; Jiang, X.; Cheng, L.; Fan, D.; Feng, Y.; et al. Spider Genomes Provide Insight into Composition and Evolution of Venom and Silk. *Nat. Commun.* **2014**, *5*, 3765. [CrossRef] [PubMed]

33. Purcell, J.; Pruitt, J.N. Are Personalities Genetically Determined? Inferences from Subsocial Spiders. *BMC Genom.* **2019**, *20*, 867. [CrossRef] [PubMed]

34. Kono, N.; Nakamura, H.; Ohtoshi, R.; Moran, D.A.P.; Shinohara, A.; Yoshida, Y.; Fujiwara, M.; Mori, M.; Tomita, M.; Arakawa, K. Orb-Weaving Spider Araneus Ventricosus Genome Elucidates the Spidroin Gene Catalogue. *Sci. Rep.* **2019**, *9*, 8380. [CrossRef] [PubMed]

35. Sheffer, M.M.; Hoppe, A.; Krehenwinkel, H.; Uhl, G.; Kuss, A.W.; Jensen, L.; Jensen, C.; Gillespie, R.G.; Hoff, K.J.; Prost, S. Chromosome-Level Reference Genome of the European Wasp Spider *Argiope Bruennichi*: A Resource for Studies on Range Expansion and Evolutionary Adaptation. *GigaScience* **2021**, *10*, giaa148. [CrossRef] [PubMed]

36. Wellcome Sanger Institute. 25 Genomes for 25 Years. Available online: https://www.sanger.ac.uk/collaboration/25-genomes-for-25-years/ (accessed on 16 December 2021).

37. Sánchez-Herrero, J.F.; Frías-López, C.; Escuer, P.; Hinojosa-Alvarez, S.; Arnedo, M.A.; Sánchez-Gracia, A.; Rozas, J. The Draft Genome Sequence of the Spider *Dysdera Silvatica* (*Araneae, Dysderidae*): A Valuable Resource for Functional and Evolutionary Genomic Studies in Chelicerates. *GigaScience* **2019**, *8*, giz099. [CrossRef]

38. i5K Consortium. The I5K Initiative: Advancing Arthropod Genomics for Knowledge, Human Health, Agriculture, and the Environment. *J. Hered.* **2013**, *104*, 595–600. [CrossRef]

39. Hendrickx, F.; De Corte, Z.; Sonet, G.; Van Belleghem, S.M.; Köstlbacher, S.; Vangestel, C. A Masculinizing Supergene Underlies an Exaggerated Male Reproductive Morph in a Spider. *Nat. Ecol. Evol.* **2022**, *6*, 195–206. [CrossRef]

40. Schwager, E.E.; Sharma, P.P.; Clarke, T.; Leite, D.J.; Wierschin, T.; Pechmann, M.; Akiyama-Oda, Y.; Esposito, L.; Bechsgaard, J.; Bilde, T.; et al. The House Spider Genome Reveals an Ancient Whole-Genome Duplication during Arachnid Evolution. *BMC Biol.* **2017**, *15*, 62. [CrossRef]

41. Yu, N.; Li, J.; Liu, M.; Huang, L.; Bao, H.; Yang, Z.; Zhang, Y.; Gao, H.; Wang, Z.; Yang, Y.; et al. Genome Sequencing and Neurotoxin Diversity of a Wandering Spider Pardosa Pseudoannulata (Pond Wolf Spider). *BioRxiv.* **2019**. [CrossRef]

42. Liu, S.; Aagaard, A.; Bechsgaard, J.; Bilde, T. DNA Methylation Patterns in the Social Spider, *Stegodyphus Dumicola*. *Genes* **2019**, *10*, 137. [CrossRef] [PubMed]

43. Babb, P.L.; Lahens, N.F.; Correa-Garhwal, S.M.; Nicholson, D.N.; Kim, E.J.; Hogenesch, J.B.; Kuntner, M.; Higgins, L.; Hayashi, C.Y.; Agnarsson, I.; et al. The Nephila Clavipes Genome Highlights the Diversity of Spider Silk Genes and Their Complex Expression. *Nat. Genet.* **2017**, *49*, 895–903. [CrossRef] [PubMed]

44. Faircloth, B.C. PHYLUCE is a software package for the analysis of conserved genomic loci. *Bioinformatics* **2016**, *32*, 786–788. [CrossRef] [PubMed]

45. Huang, W.; Li, L.; Myers, J.R.; Marth, G.T. ART: A next-Generation Sequencing Read Simulator. *Bioinformatics* **2012**, *28*, 593–594. [CrossRef]

46. Lunter, G.; Goodson, M. Stampy: A Statistical Algorithm for Sensitive and Fast Mapping of Illumina Sequence Reads. *Genome Res.* **2011**, *21*, 936–939. [CrossRef]
47. Li, H.; Handsaker, B.; Wysoker, A.; Fennell, T.; Ruan, J.; Homer, N.; Marth, G.; Abecasis, G.; Durbin, R.; 1000 Genome Project Data Processing Subgroup. The Sequence Alignment/Map Format and SAMtools. *Bioinformatics* **2009**, *25*, 2078–2079. [CrossRef]
48. Quinlan, A.R.; Hall, I.M. BEDTools: A Flexible Suite of Utilities for Comparing Genomic Features. *Bioinformatics* **2010**, *26*, 841–842. [CrossRef]
49. Hipp, R.D. SQLite Home Page. Available online: https://www.sqlite.org/index.html (accessed on 25 December 2020).
50. Harris, R.S. *Improved Pairwise Alignment of Genomic DNA*; The Pennsylvania State University: State College, PA, USA, 2007.
51. Katoh, K.; Standley, D.M. MAFFT Multiple Sequence Alignment Software Version 7: Improvements in Performance and Usability. *Mol. Biol. Evol.* **2013**, *30*, 772–780. [CrossRef]
52. Castresana, J. Selection of Conserved Blocks from Multiple Alignments for Their Use in Phylogenetic Analysis. *Mol. Biol. Evol.* **2000**, *17*, 540–552. [CrossRef]
53. Kalyaanamoorthy, S.; Minh, B.Q.; Wong, T.K.F.; von Haeseler, A.; Jermiin, L.S. ModelFinder: Fast Model Selection for Accurate Phylogenetic Estimates. *Nat. Methods* **2017**, *14*, 587–589. [CrossRef]
54. Minh, B.Q.; Schmidt, H.A.; Chernomor, O.; Schrempf, D.; Woodhams, M.D.; von Haeseler, A.; Lanfear, R. IQ-TREE 2: New Models and Efficient Methods for Phylogenetic Inference in the Genomic Era. *Mol. Biol. Evol.* **2020**, *37*, 1530–1534. [CrossRef] [PubMed]
55. Minh, B.Q.; Hahn, M.W.; Lanfear, R. New Methods to Calculate Concordance Factors for Phylogenomic Datasets. *Mol. Biol. Evol.* **2020**, *37*, 2727–2733. [CrossRef] [PubMed]
56. Wheeler, W.C.; Coddington, J.A.; Crowley, L.M.; Dimitrov, D.; Goloboff, P.A.; Griswold, C.E.; Hormiga, G.; Prendini, L.; Ramírez, M.J.; Sierwald, P.; et al. The Spider Tree of Life: Phylogeny of Araneae Based on Target-Gene Analyses from an Extensive Taxon Sampling. *Cladistics* **2017**, *33*, 574–616. [CrossRef]
57. Rambaut, A. FigTree v1.4.4. Available online: https://github.com/rambaut/figtree/releases (accessed on 25 October 2021).
58. Kallal, R.J.; Kulkarni, S.S.; Dimitrov, D.; Benavides, L.R.; Arnedo, M.A.; Giribet, G.; Hormiga, G. Converging on the Orb: Denser Taxon Sampling Elucidates Spider Phylogeny and New Analytical Methods Support Repeated Evolution of the Orb Web. *Cladistics* **2021**, *37*, 298–316. [CrossRef] [PubMed]

Article

Solenysa, a Cretaceous Relict Spider Group in East Asia

Jiahui Tian [1,2], Yongjia Zhan [1], Chengmin Shi [3], Hirotsugu Ono [4] and Lihong Tu [1,*]

[1] College of Life Sciences, Capital Normal University, Beijing 100048, China; tianjh0920@163.com (J.T.); zhanyj@cnu.edu.cn (Y.Z.)

[2] School of Ecology and Environment, Anhui Normal University, Wuhu 241002, China

[3] College of Plant Protection, Hebei Agricultural University, Baoding 071000, China; shichengmin@hebau.edu.cn

[4] Department of Zoology, National Museum of Nature and Science, Tsukuba-shi 305-0005, Japan; ono@kahaku.go.jp

* Correspondence: tulh@cnu.edu.cn; Tel.: +86-10-68907228

Abstract: A time scale of phylogenetic relationships contributes to a better understanding of the evolutionary history of organisms. Herein, we investigate the temporal divergence pattern that gave rise to the poor species diversity of the spider genus *Solenysa* in contrast with the other six major clades within linyphiids. We reconstructed a dated phylogeny of linyphiids based on multi-locus sequence data. We found that *Solenysa* diverged from other linyphiids early in the Cretaceous (79.29 mya), while its further diversification has been delayed until the middle Oligocene (28.62 mya). Its diversification trend is different from all of the other major lineages of linyphiids but is closely related with the Cenozoic ecosystem transition caused by global climate changes. Our results suggest that *Solenysa* is a Cretaceous relict group, which survived the mass extinction around the K-T boundary. Its low species diversity, extremely asymmetric with its sister group, is largely an evolutionary legacy of such a relict history, a long-time lag in its early evolutionary history that delayed its diversification. The limited distribution of *Solenysa* species might be related to their extreme dependence on highly humid environments.

Keywords: molecular phylogeny; divergence time; relict group; Linyphiidae

Citation: Tian, J.; Zhan, Y.; Shi, C.; Ono, H.; Tu, L. *Solenysa*, a Cretaceous Relict Spider Group in East Asia. *Diversity* **2022**, *14*, 120. https://doi.org/10.3390/d14020120

Academic Editors: Michael Wink and Martín J. Ramírez

Received: 21 November 2021
Accepted: 5 February 2022
Published: 8 February 2022

Publisher's Note: MDPI stays neutral with regard to jurisdictional claims in published maps and institutional affiliations.

1. Introduction

Molecular data has become an indispensable tool for the reconstruction of phylogenetic relationships among species and provides important insights on the evolutionary histories of many animal groups. It is common in systematics that a high-level molecular phylogeny may significantly conflict with the established taxonomic system based on morphological characters [1–4]. Understanding the evolutionary past that shaped the species diversity of lineages always attracts the interest of biologists [5–8]. Spiders are generalist predators, forming a successful terrestrial animal group, and their high species diversity is distributed unevenly across lineages [9], even extremely asymmetrically between sister groups [4,10]. Several hypotheses have been proposed to interpret the driving forces that promote spider diversification, such as co-diversification with insects [11–13], key innovations in silk structure and web architecture [10], repeated evolution of the respiratory system from book lungs to tracheae [14], and foraging changes from using capturing web to cursorial habits [15]. While these studies usually focus on the driving forces for fast diversifications that lead to a speciose clade, little attention having been exerted to the factors that might result in groups with poor species diversity. Herein, we investigated the evolutionary history of *Solenysa* spiders, one of the seven main clades within linyphiids, with poor species diversity compared to other clades [4].

Great conflicts exist between the molecular phylogeny of linyphiids and the classical taxonomic system. Linyphiidae is an ancient spider group and its earliest fossil record dates back to the early Cretaceous, about 125–135 mya [16]. As with many other spiders, linyphiids have experienced adaptive radiation, accompanying the fast radiation of insects in the Cretaceous. Currently, with more than 4700 recognized species, Linyphiidae represents the second largest group of the order Araneae [17]. Generally, linyphiids are conservative in somatic features but have complex genitalia with species-specific characters, which are used as criteria for species recognition. The classical taxonomic system of Linyphiidae consists of seven subfamilies [18]. However, this was not supported by molecular phylogenetic analyses [1,4,7]. Four of them, Linyphiinae, Erigoninae, Micronetinae, and Ipainae, were not monophyletic groups; the representatives of Mynogleninae and Dubiaraneinae fell into Linyphiinae; the Stemonyphantinae taxa were often clustered with pimoids, the sister group of linyphiids [1,4,19–21]. The subfamily Stemonyphantinae was newly revised by adding two ex-pimoid genera and another linyphiid genus in it [8]. However, the seven-clade topology of molecular phylogenies are robustly supported, and all of these seven major clades (clades A–F and S in [4]) are supported by some putative synapomorphic characters.

Spider diversity is distributed unevenly among the seven major clades within linyphiids, especially between sister groups [4]. The relationships among the seven major lineages are puzzling. The cladogenetic events of the seven-clade topology were correlated with successive transformations on the state of the epigynal plate that was defined by the location of the copulatory openings and tracings of epigynal tracts [4]. Generally, a set of epigynal characters forms an epigynal type that means a certain interaction pattern between male and female genitalia during copulation [22]. The series of state transformations of the epigynal plate coupled with those lineage divergence events make the seven-clade topology meaningful [4]. Nevertheless, the driving forces that shape the unbalanced diversification across lineages within linyphiids remain unresolved. Among them, the extremely asymmetric species diversity between *Solenysa* (15 spp.) and its sister group (2885 spp., [18]) provides us a model system to study the evolution of such an asymmetry.

Clade S in [4] is composed of a single genus, *Solenysa*, and is a unique lineage in linyphiids. All *Solenysa* species display a distinctive somatic appearance and special genital morphology. These make them easily distinguished from all other linyphiids [23,24]. However, the placement of *Solenysa* within linyphiids has long been controversial. Saaristo [25] placed *Solenysa* together with some micronetine genera into a new subfamily, Ipainae, largely based on the females having a movable epigynum. However, this treatment was disproved by the phylogenetic analyses either based on molecular data or morphological data [1,4,24]; the molecular phylogeny supported a sister relationship between *Solenysa* and clade B, which is a hodgepodge composed of all erigonines, some micronetines, and linyphiines. Although such a sister-relationship was supported by some putative synapomorphies, in comparison to its speciose and widespread sister-clade, the *Solenysa* clade appeared unusual in term of species diversity and limited distributions and is generally only known from type localities and small adjacent areas (Figure 1; [23,24,26–32]). The underlying evolutionary process that gave rise to such a biased diversity pattern between these sister clades remains to be explored.

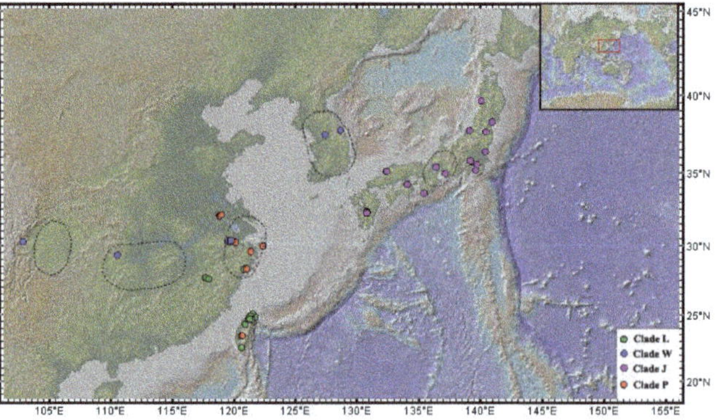

Figure 1. Distribution of *Solenysa* species. Circles in color represent species of different clades in Figure 2, green, clade L; blue, clade W; purple, clade J, red, clade P. Dash line circles indicate locations of some supposed Pleistocene ice age refugia in East Asia.

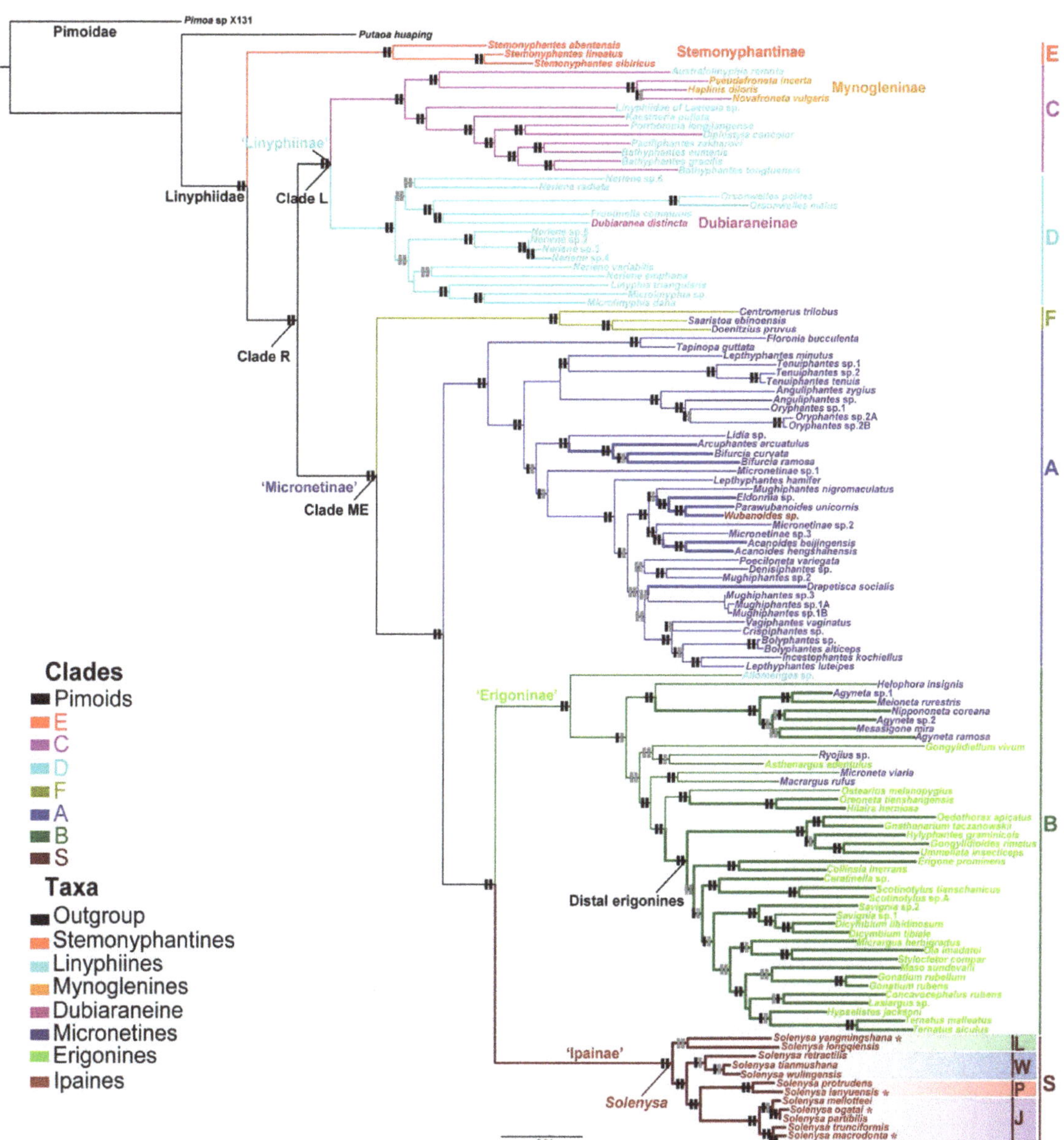

Figure 2. ML tree of linyphiid phylogeny with proportional branch lengths. Branches in color show the seven major clades within Linyphiidae. Taxa names in color indicate their current subfamilial placements. Thickened blue branches in clade A indicate taxa having a movable epigynum. Thickened green branches in clade B indicate taxa having a desmitracheate system. Thickened branches in clade S indicate all *Solenysa* taxa having both movable epigynum and an intermediate type tracheate system. Bars on branches indicate corresponding node supports of the main clades: the anterior show the maximum likelihood bootstrap (BS) and the posterior Bayesian posterior probability (PP), respectively. Bars in black indicate BS > 80%; PP > 0.95; bars in grey BS < 80%, PP < 0.95. Trees with all the node supports are included as Supplementary Material.

Studies on the defining features (synapomorphies) of a speciose group, especially in comparison to its sister group, may help us in search of potential drivers that promote fast diversification [9]. According to Wang et al. [4], the tracheal system in clade B repeatedly evolved from the haplotracheate to desmitracheate, in which the median pair tracheae extensively branched and extended into the prosoma [33]. Especially the distal erigonines clade, all species have a desmitracheate system forming the largest group in clade B [34], contributing more than half of the species diversity of Linyphiidae. While the tracheal system in *Solenysa* remains as an intermediate type, with the median pair unbranched in the opisthosoma but extended into the prosoma where they branch and extend into the legs (Tu, personal observation; [24]). Anterior extending tracheae would provide oxygen directly to the brain and legs [35,36], and extensively branched tracheae may help in reducing water loss [9,37]. These imply that the selective advantages of the desmitracheate system might trigger the fast diversification in clade B, while the diversification in the *Solenysa* clade remained slower, which resulted in the biased species diversity between them. Nevertheless, whether there are any other reasons is unclear.

In the present study, we aimed to explore the reason that caused the biased diversification between *Solenysa* and its sister group. Our results from phylogenetic analyses, molecular dating, and lineage-wise diversification tracing through time suggested that *Solenysa* represents a Cretaceous relict spider group in East Asia, and historical climate changes have played a pivotal role in shaping its evolutionary history.

2. Materials and Methods

2.1. Taxon Sampling

In the present study, we collected and sequenced four additional *Solenysa* species, *S. lanyuensis*, *S. yangmingshana*, *S. macrodonta*, and *S. ogatai* from their type localities on Taiwan Island and Japanese islands. The relevant sequence data for nine *Solenysa* species, *S. longqiensis*, *S. retractilis*, *S. tianmushana*, *S. wulingensis*, *S. protrudens*, *S. mellotteei*, *S. trunciformis*, *S. reflexilis*, and *S. partibilis*, have been generated in our earlier study [4]. Another two known species, *S. spiralis* from Sichuan, China, and *S. geumoensis* from the Korean Peninsula, were not sampled due to a lack of fresh materials for sequencing. Thus, 13 of the 15 known *Solenysa* species (86.7%) were included in the present study. To explore the phylogenetic position of *Solenysa* species within Linyphiidae, other linyphiid taxa of Wang et al. [4], except for the four unstable long-branch taxa and those repeated taxa, were compiled in our dataset. Pimoidae is the sister group of Linyphiidae and often used as outgroups for rooting. Given the recent revision of Pimoidae having the formerly pimoid genera *Putaoa* and *Weintrauboa* transferred to Linyphiidae, and redelimited Pimoidae as including only *Pimoa* and *Nanoa* [8], we also added representatives of *Pimoa* and *Putaoa* into the dataset. The final data set consisted of 127 taxa, including 13 *Solenysa* taxa, 113 other linyphiids, and one pimoid.

Collected information on the four newly sequenced *Solenysa* specimens is listed in Supporting Material Table S1. Specimens used for the molecular study were fixed in 95% ethanol and kept at −20 °C before DNA extraction. All newly collected specimens are deposited at the College of Life Sciences, Capital Normal University (CNU).

2.2. Laboratory Protocols for Molecular Data

Five loci used in [4], including two mitochondrial genes, cytochrome c oxidase subunit I (COI), and 16S rRNA (16S), and three nuclear genes, 18S rRNA (18S), 28S rRNA (28S), and histone H3 (H3), were sequenced for newly collected *Solenysa* specimens. Laboratory protocols and sequence curation follow those described in Wang et al. [4]. The primers and their annealing temperatures used for PCR amplification in the present study are provided in Table 1.

Table 1. Primer sequences, their sources, and reaction conditions used for PCR in this study.

Gene	Primer	Sequence (5′–3′)	Annealing Temperature	Voucher
COI	C1-J-1718(F)	GGA GGA TTT GGA AAT TGA TTA GTT CC	45 °C	Simon et al., 1994
	C1-N-219(R)	CCC GGT AAA ATT AAA ATA TAA ACT TC		Simon et al., 1994
16S	LR-N-13398 (F)	CGC CTG TTT AAC AAA AAC AT	42–45 °C	Arnedo et al., 2004
	LR-J-12864 (R)	CTC CGG TTT GAA CTC AGA TCA		Simon et al., 1994
18S	18Sa2.0 (F)	ATG GTT GCA AAG CTG AAA C	58 °C	Giribet et al., 1999
	5F (F)	GCG AAA GCA TTT GCC AAG AA		Giribet et al., 1999
	9R (R)	GAT CCT TCC GCA GGT TCA CCT AC		Giribet et al., 1999
28S	28Sa (F)	GAC CCG TCT TGA AAC ACG GA	48 °C	Whiting et al., 1997
	28Sb (R)	TCG AAG GAA CCA GCT ACT A		Whiting et al., 1997
H3	H3a (F)	ATG GCT CGT ACC AAG CAG AC(ACG) GC	48 °C	Colgan et al., 1998
	H3b (R)	ATA TCC TT(AG) GGC AT(AG) AT(AG) GTG AC		Colgan et al., 1998

2.3. Phylogenetic Analyses

Sequences for the five genes were aligned using MAFFT 7.490 [38] and were concatenated using Mesquite version 3.31 [39] in the order of 16S, 18S, 28S, COI, and H3. Maximum likelihood (ML) phylogenetic analysis was carried out using IQ-TREE v2.1.3 [40] with best-fit DNA substitution models selected using ModelFinder [41]. ML trees were inferred from (1) the concatenated super-matrix with a single overall the best-fit model and (2) the partitioned matrix defined by loci using PartitionFinder [42] with each partition applying its best-fit model. The GTR + F + R5 was selected as the best-fit model for the concatenated super-matrix. The models selected for partitions of the best scheme were: TIM2 + F + I + G4 for partition 16S, SYM + R4 for partition 18S + 28S, GTR + F + I + G4 for both partition COI and partition H3. Node supports were assessed through the ultrafast bootstrap method [43] with 1000 replicates, incurring the -bnni option to reduce the risk of overestimating branch supports. We also performed Bayesian phylogenetic inference based on the best partition scheme using MrBayes v3.2.6 [44]. We used the GTR model for the 16S data in place of the TIM2 model, as the latter is not implemented in MrBayes. The SYM model for the 18S + 28S data was converted from the GTR model by fixing the stationary state frequencies to be equal. MrBayes analyses were initiated with random starting trees employing four Markov chains (one cold and three hot). The Markov chains ran for 2×10^6 generations with trees and parameters being sampled every 100 generations. The "temperature" parameter was set to 0.2. The chains were converged and reached a stationary state after the iteration with the average standard deviation of split frequencies being smaller than 0.0074, and all values of potential scale reduction factor for all parameters being very close to 1.00. The majority-rule consensus tree was generated using the sample from the cold chain after the first 25% of the sample was discarded as burn-in. The topology of phylogeny inferred from the present study was statistically tested for robustness against alternative topologies with the approximately unbiased (AU) test [45].

2.4. Estimation of Divergence Times

The divergence times were estimated with a relaxed molecular clock approach implemented in BEAST2 version 2.6.6 [46]. The rate change was explicitly modeled using uncorrelated lognormal distribution across trees and a birth–death model was used for modeling speciation. The best-fit DNA substitution model was selected using bModelTest module [47] in BEAUti2. Two independent MCMC searches were run for 8×10^7 generations with trees and parameters being sampled every 1000 generations. The convergence of the MCMC chains was checked with Tracer version 1.7.1 [48]. The first 10% samples were discarded as burn-in.

Two calibration points were used: (1) the oldest linyphiid fossil from the Lower Cretaceous Lebanese amber and (2) the divergence between two endemic Hawaiian species *Orsonwelles polites* and *Orsonwelles malus*. The oldest linyphiid fossil was originally described as an undetermined linyphiid [16]. Several studies have used it as a calibration point based on different assumptions: as a stem linyphiid, a crown linyphiid, a crown clade containing all linyphiids except *Stemonyphantes* [7,8,19,20,49,50], or even as a stem araneoid [51]. Herein, following Arnedo and Hormiga [7], we assigned the age of the fossil to the most recent common ancestor (MRCA) of Linyphiidae and applied an exponential prior with a mean of 10.0 and offset = 125.0 for this calibration, which gave a 95% confident interval of 125–155 Myr. According to Hormiga et al. [52], the Hawaiian spiders *Orsonwelles malus* is endemic to Kaui Island (formed 5.1 million years ago), and *Orsonwelles polites* is endemic to the adjacent O'ahu Island (formed 2.6–3.7 million years ago). We assigned a normal prior for the divergence time between these two species with a mean of 3.0 Myr and standard deviation of 0.5 Myr, following Arnedo and Hormiga [7].

2.5. Lineages Tracing through Time

We use lineages through time (LTT) plot to gain insight into the history of diversification for clades S (*Solenysa*), A, and B. Samples of dated genealogies for each of these three clades were inferred using BEAST2. The times of the MRCA for the relevant clades came from the dated phylogeny. The LTT plots were generated using Tracer version 1.7.1 [48].

3. Results

3.1. DNA Sequence Data

A total of 590 sequences were obtained. Sequences for all five genes were acquired for 88 taxa (68.22%; 88/129), and at least four genes were acquired for the majority (96.06%; 122/127). Fragments from 16S, 18S, 28S, COI, and H3 were sequenced for the taxa sampled here are 91.34% (116/127), 97.64% (124/127), 81.89% (104/127), 95.28% (121/127), and 98.43% (125/127), respectively. After alignment, the concatenated matrix includes a total of 2678 sites. All newly acquired sequences have been deposited in GenBank. The accession numbers of all samples are listed in Table 2.

Table 2. Taxon list and sequence information. An asterisk (*) following species names indicate that they are newly sequenced for the present study.

Family	Subfamily	Species	16s	18s	28s	COI	H3	Voucher
Pimoidae		*Pimoa* sp. X131	AY230940	AY230893	AY231072	AY231025	AY230985	Arnedo et al., 2004
Linyphiidae	Micronetinae	*Acanoides beijingensis*	KJ027589	KJ027587	KJ027580	KJ027582	KJ027583	Wang et al., 2015
Linyphiidae	Micronetinae	*Acanoides hengshanensis*	KJ027585	KJ027588	KJ027584	KJ027586	KJ027581	Wang et al., 2015
Linyphiidae	Micronetinae	*Agyneta ramosa*	FJ838670	FJ838694		FJ838648	FJ838740	Arnedo et al., 2009
Linyphiidae	Micronetinae	*Agyneta* sp.1	KT003097	KT002904	KT003003	KT002707	KT002804	Wang et al., 2015
Linyphiidae	Micronetinae	*Agyneta* sp.2	KT003098	KT002905	KT003004	KT002708	KT002805	Wang et al., 2015
Linyphiidae	Micronetinae	*Allomengea* sp.	KT003099	KT002906		KT002709	KT002805	Wang et al., 2015
Linyphiidae	Micronetinae	*Anguliphantes* sp.		KT002907	KT003005	KT002710	KT002807	Wang et al., 2015
Linyphiidae	Micronetinae	*Anguliphantes zygius*	KT003100	KT002908	KT003006	KT002711	KT002808	Wang et al., 2015
Linyphiidae	Micronetinae	*Arcuphantes arcuatulus*			KT003007		KT002809	Wang et al., 2015
Linyphiidae	Erigoninae	*Asthenargus edentulus*	KT003101	KT002909	KT003008	KT002712	KT002810	Wang et al., 2015
Linyphiidae	Linyphiinae	*Australolinyphia remota*	FJ838671	FJ838695	FJ838718	FJ838649	FJ838741	Arnedo et al., 2009
Linyphiidae	Linyphiinae	*Bathyphantes eumenis*	KT003101	KT002910	KT003009	KT002713	KT002811	Wang et al., 2015
Linyphiidae	Linyphiinae	*Bathyphantes gracilis*	FJ838672	FJ838696		FJ838650	FJ838742	Arnedo et al., 2009
Linyphiidae	Linyphiinae	*Bathyphantes tongluensis*	KT003104	KT002912	KT003011	KT002715	KT002813	Wang et al., 2015
Linyphiidae	Micronetinae	*Bifurcia curvata*	KT003105	KT002913	KT003012	KT002716	KT002814	Wang et al., 2015
Linyphiidae	Micronetinae	*Bifurcia ramosa*	KT003106	KT002914	KT003013	KT002717	KT002815	Wang et al., 2015
Linyphiidae	Micronetinae	*Bolyphantes alticeps*	AY078660	AY078667	AY078678	AY078691	AY078700	Hormiga et al., 2003
Linyphiidae	Micronetinae	*Bolyphantes* sp.	KT003107	KT002915		KT002718	KT002816	Wang et al., 2015
Linyphiidae	Micronetinae	*Centromerus trilobus*	KT003108	KT002916	KT003014	KT002718	KT002817	Wang et al., 2015
Linyphiidae	Erigoninae	*Ceratinella* sp.	KT003109	KT002917	KT003015		KT002818	Wang et al., 2015
Linyphiidae	Erigoninae	*Collinsia inerrans*	KT003110	KT002918	KT003016	KT002720	KT002819	Wang et al., 2015
Linyphiidae	Erigoninae	*Concavocephalus rubens*	KT002919	KT003017		KT002721	KT002820	Wang et al., 2015
Linyphiidae	Micronetinae	*Crispiphantes* sp.	KT003111	KT002920	KT003018	KT002722	KT002821	Wang et al., 2015

Table 2. *Cont.*

Family	Subfamily	Species	16s	18s	28s	COI	H3	Voucher
Linyphiidae	Micronetinae	*Denisiphantes* sp.	KT003112	KT002921	KT003019	KT002723	KT002822	Wang et al., 2015
Linyphiidae	Erigoninae	*Dicymbium libidinosum*	KT003113	KT002922	KT003020	KT002724	KT002823	Wang et al., 2015
Linyphiidae	Erigoninae	*Dicymbium tibiale*	KT003114	KT002923	KT003021	KT002725	KT002824	Wang et al., 2015
Linyphiidae	Linyphiinae	*Diplostyla concolor*	FJ838673	FJ838697		FJ838651	FJ838743	Arnedo et al., 2009
Linyphiidae	Micronetinae	*Doenitzius pruvus*	KT003116	KT002925	KT003023	KT002727	KT002826	Wang et al., 2015
Linyphiidae	Micronetinae	*Drapetisca socialis*	FJ838674	FJ838698		FJ838652	FJ838744	Arnedo et al., 2009
Linyphiidae	Dubiaraneinae	*Dubiaranea distincta*	FJ838675	FJ838699	FJ838722	FJ838653	FJ838745	Arnedo et al., 2009
Linyphiidae	Micronetinae	*Eldonia* sp.	KT003117	KT002926	KT003024	KT002728	KT002827	Wang et al., 2015
Linyphiidae	Erigoninae	*Erigone prominens*		KT002927	KT003025	KT002729	KT002828	Wang et al., 2015
Linyphiidae	Micronetinae	*Floronia bucculenta*	FJ838676	FJ838700		FJ838654	FJ838746	Arnedo et al., 2009
Linyphiidae	Linyphiinae	*Frontinella communis*	FJ838677	FJ838701	FJ838724	FJ838655	FJ838747	Arnedo et al., 2009
Linyphiidae	Erigoninae	*Gnathonarium taczanowskii*	KT003119	KT002929	KT003027	KT002730	KT002830	Wang et al., 2015
Linyphiidae	Erigoninae	*Gonatium rubellum*	FJ838679	FJ838703		FJ838656	FJ838749	Arnedo et al., 2009
Linyphiidae	Erigoninae	*Gonatium rubens*	KT003120	KT002930	KT003028	KT002732	KT002831	Wang et al., 2015
Linyphiidae	Erigoninae	*Gongylidiellum vivum*	FJ838678	FJ838702	FJ838725		FJ838748	Arnedo et al., 2009
Linyphiidae	Erigoninae	*Gongylidioides rimatus*	KT003121	KT002931	KT003029	KT002733	KT002832	Wang et al., 2015
Linyphiidae	Mynogleninae	*Haplinis diloris*	FJ838680	FJ838704		FJ838657	FJ838750	Arnedo et al., 2009
Linyphiidae	Micronetinae	*Helophora insignis*	FJ838681	FJ838705		FJ838658	FJ838751	Arnedo et al., 2009
Linyphiidae	Erigoninae	*Hilaira herniosa*	KT003123	KT002933	KT003030	KT002735	KT002834	Wang et al., 2015
Linyphiidae	Erigoninae	*Hylyphantes graminicola*	KT003124	KT002934	KT003031	KT002736	KT002835	Wang et al., 2015
Linyphiidae	Erigoninae	*Hypselistes jacksoni*		KT002935	KT003032	KT002737	KT002836	Wang et al., 2015
Linyphiidae	Micronetinae	*Incestophantes kochiellus*	KT003125	KT002936	KT003033	KT002738	KT002836	Wang et al., 2015
Linyphiidae	Linyphiinae	*Kaestneria pullata*	KT003126	KT002937	KT003034	KT002739	KT002838	Wang et al., 2015
Linyphiidae	Erigoninae	*Lasiargus* sp.	KT003127	KT002938	KT003035	KT002740	KT002839	Wang et al., 2015
Linyphiidae	Micronetinae	*Lepthyphantes hamifer*	KT003128	KT002939	KT003036	KT002741	KT002840	Wang et al., 2015
Linyphiidae	Micronetinae	*Lepthyphantes luteipes*	KT003129	KT002940		KT002742	KT002841	Wang et al., 2015
Linyphiidae	Micronetinae	*Lepthyphantes minutus*	AY078663	AY078673	AY078681	AY078689	AY078705	Hormiga et al., 2003
Linyphiidae	Micronetinae	*Lidia* sp.	KT003130	KT002941	KT003037		KT002841	Wang et al., 2015
Linyphiidae	Linyphiinae	*Linyphia triangularis*	AY078664	AY078668	AY078682	AY078693	AY078702	Hormiga et al., 2003
Linyphiidae	Linyphiinae	*Laetesia* sp.	FJ838682	FJ838706		FJ838659	FJ838752	Arnedo et al., 2009
Linyphiidae	Micronetinae	*Macrargus rufus*	KT003133	KT002944	KT003040	KT002745	KT002845	Wang et al., 2015
Linyphiidae	Erigoninae	*Maso sundevalli*		KT002945	KT003041	KT002746	KT002846	Wang et al., 2015
Linyphiidae	Micronetinae	*Meioneta rurestris*	FJ838683	FJ838707		FJ838660	FJ838753	Arnedo et al., 2009
Linyphiidae	Micronetinae	*Mesasigone mira*	KT003134	KT002946		KT002746	KT002847	Wang et al., 2015
Linyphiidae	Erigoninae	*Micrargus herbigradus*	KT003135	KT002947	KT003042	KT002748	KT002848	Wang et al., 2015
Linyphiidae	Linyphiinae	*Microlinyphia dana*	AY078665	AY078677	AY078683	AY078690		Hormiga et al., 2003
Linyphiidae	Linyphiinae	*Microlinyphia* sp.	KT003136	KT002948	KT003043	KT002749	KT002849	Wang et al., 2015
Linyphiidae	Micronetinae	*Microneta viaria*	FJ838684	FJ838708		FJ838661	FJ838754	Arnedo et al., 2009
Linyphiidae	Micronetinae	*Micronetine* sp.1	KT003138	KT002950		KT002751	KT002851	Wang et al., 2015
Linyphiidae	Micronetinae	*Micronetine* sp.2	KT003139	KT002951	KT003045	KT002752	KT002851	Wang et al., 2015
Linyphiidae	Micronetinae	*Micronetine* sp.3	KT003140	KT002952	KT003046	KT002753	KT002853	Wang et al., 2015
Linyphiidae	Micronetinae	*Mughiphantes nigromaculatus*	KT003187	KT003001	KT003095	KT002802	KT002902	Wang et al., 2015
Linyphiidae	Micronetinae	*Mughiphantes* sp.1A	KT003141	KT002953	KT003047	KT002754	KT002854	Wang et al., 2015
Linyphiidae	Micronetinae	*Mughiphantes* sp.1B	KT003142	KT002954	KT003048	KT002755	KT002855	Wang et al., 2015
Linyphiidae	Micronetinae	*Mughiphantes* sp.2	KT003143	KT002955		KT002756	KT002855	Wang et al., 2015
Linyphiidae	Micronetinae	*Mughiphantes* sp.3	KT003144	KT002956	KT003049	KT002757	KT002857	Wang et al., 2015
Linyphiidae	Linyphiinae	*Neriene emphana*	KT003145	KT002957	KT003050	KT002758	KT002858	Wang et al., 2015
Linyphiidae	Linyphiinae	*Neriene radiata*	AY078710	AY078670	AY078684	AY078696	AY078709	Hormiga et al., 2003
Linyphiidae	Linyphiinae	*Neriene* sp.2	KT003146	KT002958	KT003051	KT002759	KT002859	Wang et al., 2015
Linyphiidae	Linyphiinae	*Neriene* sp.3	KT003147	KT002959	KT003052	KT002760	KT002859	Wang et al., 2015
Linyphiidae	Linyphiinae	*Neriene* sp.4	KT003148	KT002960	KT003053	KT002761	KT002861	Wang et al., 2015
Linyphiidae	Linyphiinae	*Neriene* sp.5	KT003149	KT002961	KT003054	KT002762	KT002862	Wang et al., 2015
Linyphiidae	Linyphiinae	*Neriene* sp.6	KT003150	KT002962	KT003055	KT002763	KT002863	Wang et al., 2015
Linyphiidae	Linyphiinae	*Neriene variabilis*	AY078711	AY078669	AY078685	AY078699	AY078706	Hormiga et al., 2003
Linyphiidae	Micronetinae	*Nippononeta coreana*	KT003151	KT002963	KT003056	KT002764	KT002864	Wang et al., 2015
Linyphiidae	Mynogleninae	*Novafroneta vulgaris*		FJ838710		FJ838663	FJ838756	Arnedo et al., 2009
Linyphiidae	Erigoninae	*Oedothorax apicatus*	KT003152	KT002964	KT003057	KT002765	KT002864	Wang et al., 2015
Linyphiidae	Erigoninae	*Oia imadatei*	KT003153	KT002965	KT003058	KT002765	KT002866	Wang et al., 2015
Linyphiidae	Erigoninae	*Oreoneta tienshangensis*	KT003154	KT002966	KT003059	KT002765	KT002867	Wang et al., 2015
Linyphiidae	Linyphiinae	*Orsonwelles malus*	AY078737	AY078716	AY078687	AY078697	AY078708	Hormiga et al., 2003
Linyphiidae	Linyphiinae	*Orsonwelles polites*	AY078725	AY078671	AY078686	AY078755	AY078701	Hormiga et al., 2003
Linyphiidae	Micronetinae	*Oryphantes* sp.1	KT003155	KT002967	KT003060	KT002768	KT002868	Wang et al., 2015
Linyphiidae	Micronetinae	*Oryphantes* sp.2A	KT003156	KT002968	KT003061	KT002769	KT002869	Wang et al., 2015
Linyphiidae	Micronetinae	*Oryphantes* sp.2B	KT003157	KT002969	KT003062	KT002801	KT002870	Wang et al., 2015
Linyphiidae	Erigoninae	*Ostearius melanopygius*	FJ838688	FJ838712			FJ838758	Arnedo et al., 2009
Linyphiidae	Linyphiinae	*Pacifiphantes zakharovi*	KT003159	KT002971	KT003064	KT002771	KT002872	Wang et al., 2015
Linyphiidae	Micronetinae	*Parawubanoides unicornis*	KT003160	KT002972	KT003065	KT002772	KT002873	Wang et al., 2015
Linyphiidae	Micronetinae	*Poeciloneta variegata*	KT003161	KT002973	KT003066	KT002772	KT002874	Wang et al., 2015
Linyphiidae	Linyphiinae	*Porrhomma longjiangense*	KT003162	KT002974	KT003067	KT002774	KT002875	Wang et al., 2015
Linyphiidae	Mynogleninae	*Pseudafroneta incerta*	FJ838690	FJ838714	FJ838737	FJ838666	FJ838760	Arnedo et al., 2009
Linyphiidae	Stemonyphantinae	*Putaoa huaping*	KT003163	KT002975	KT003068	KT002775	KT002876	Wang et al., 2015
Linyphiidae	Micronetinae	*Ryojius* sp.			KT003069	KT002776	KT002877	Wang et al., 2015
Linyphiidae	Micronetinae	*Saaristoa ebinoensis*	KT003164	KT002976	KT003070	KT002777	KT002878	Wang et al., 2015
Linyphiidae	Erigoninae	*Savignia* sp.1	KT003165	KT002977	KT003071	KT002778	KT002879	Wang et al., 2015
Linyphiidae	Erigoninae	*Savignia* sp.2	KT003166	KT002978	KT003072	KT002779	KT002880	Wang et al., 2015
Linyphiidae	Erigoninae	*Scotinotylus* sp.A	KT003167	KT002979	KT003073	KT002780	KT002881	Wang et al., 2015
Linyphiidae	Erigoninae	*Scotinotylus tianschanicus*	KT003188	KT003002	KT003096	KT002803	KT002903	Wang et al., 2015
Linyphiidae	Ipainae	*Solenysa lanyuensis* *	OL691622	OL691625	OL691629	OL693167	OL702838	CNU
Linyphiidae	Ipainae	*Solenysa longqiensis*	KT003169	KT002981	KT003075	KT002782	KT002883	Wang et al., 2015
Linyphiidae	Ipainae	*Solenysa macrodonta* *	OL691623	OL691627	OL691631	OL693169	OL702840	CNU
Linyphiidae	Ipainae	*Solenysa mellotteei*	KT003168	KT002980	KT003074	KT002780	KT002882	Wang et al., 2015

Table 2. *Cont.*

Family	Subfamily	Species	16s	18s	28s	COI	H3	Voucher
Linyphiidae	Ipainae	*Solenysa ogatai* *		OL691626	OL691630	OL693168	OL702839	CNU
Linyphiidae	Ipainae	*Solenysa partibilis*	KT003170	KT002983	KT003077	KT002784	KT002885	Wang et al., 2015
Linyphiidae	Ipainae	*Solenysa protrudens*	KT003171	KT002984	KT003078	KT002785	KT002886	Wang et al., 2015
Linyphiidae	Ipainae	*Solenysa reflexilis*	KT003172	KT002985	KT003079	KT002786	KT002887	Wang et al., 2015
Linyphiidae	Ipainae	*Solenysa retractilis*	KT003174	KT002987	KT003081	KT002788	KT002889	Wang et al., 2015
Linyphiidae	Ipainae	*Solenysa tianmushana*	KT003175	KT002988	KT003082	KT002788	KT002890	Wang et al., 2015
Linyphiidae	Ipainae	*Solenysa trunciformis*		KT002982	KT003076	KT002783	KT002884	Wang et al., 2015
Linyphiidae	Ipainae	*Solenysa wulingensis*	KT003176	KT002989	KT003083	KT002790		Wang et al., 2015
Linyphiidae	Ipainae	*Solenysa yangmingshana* *		OL691624	OL691628	OL693166	OL702837	CNU
Linyphiidae	Stemonyphantinae	*Stemonyphantes abantensis*	KT003177	KT002990	KT003084	KT002791	KT002891	Wang et al., 2015
Linyphiidae	Stemonyphantinae	*Stemonyphantes lineatus*	FJ838691	FJ838715		FJ838667	FJ838761	Arnedo et al., 2009
Linyphiidae	Stemonyphantinae	*Stemonyphantes sibiricus*	FJ838692			FJ838668	FJ838762	Arnedo et al., 2009
Linyphiidae	Erigoninae	*Styloctetor compar*	KT003178	KT002991	KT003085	KT002792	KT002892	Wang et al., 2015
Linyphiidae	Micronetinae	*Tapinopa guttata*	KT003179	KT002992	KT003086		KT002893	Wang et al., 2015
Linyphiidae	Micronetinae	*Tenuiphantes* sp.1		KT002993	KT003087	KT002793	KT002894	Wang et al., 2015
Linyphiidae	Micronetinae	*Tenuiphantes* sp.2	KT003180	KT002994	KT003087	KT002794	KT002895	Wang et al., 2015
Linyphiidae	Micronetinae	*Tenuiphantes tenuis*	FJ838693	FJ838716		FJ838669	FJ838763	Arnedo et al., 2009
Linyphiidae	Erigoninae	*Ternatus malleatus*	KT003181	KT002995	KT003089	KT002795	KT002896	Wang et al., 2015
Linyphiidae	Erigoninae	*Ternatus siculus*	KT003182	KT002996	KT003090	KT002795	KT002897	Wang et al., 2015
Linyphiidae	Erigoninae	*Ummeliata insecticeps*	KT003184	KT002998	KT003092	KT002798	KT002899	Wang et al., 2015
Linyphiidae	Micronetinae	*Vagiphantes vaginatus*	KT003185	KT002999	KT003093	KT002799	KT002900	Wang et al., 2015
Linyphiidae	Ipainae	*Wubanoides* sp.	KT003186	KT003000	KT003094	KT002800	KT002901	Wang et al., 2015

3.2. Phylogeny of Solenysa

The ML trees inferred from the concatenated super-matrix and from the best partition scheme and the Bayesian tree are highly congruent as far as the major clades we concern in the present study. All analyses recovered seven major clades equivalent to those found in previous studies [4]. Given the ML estimate obtained from the partitioned analysis was significantly better than from the concatenated analysis, (log-likelihoods, −45494.13 vs. −46688.08), we reported the result from the partitioned analysis in Figure 2 (see Supplementary Material Figures S1 and S2 for node supports in ML tree and Bayesian tree). All linyphiid taxa formed a monophyletic lineage sister to the pimoid clade and were grouped into seven strongly supported (bootstrap support/Bayesian posterior = 100/1.00), named as clades A, B, C, D, E, F, and S (*sensu* Wang et al. [4]). The monophyly of *Solenysa* (clade S, 100/1.00) and its sister relationship with the species-rich clade B (100/1.00) were robustly supported (100/1.00). Relationships among the seven major lineages remain the same as those of the analyses of Wang et al. [4]. One major conflict between the phylogeny we inferred here and the phylogeny reported by [8] involved relationship between *Putaoa hauaping* and the *Stemonyphantes* species (clade E). In our results, *Putaoa hauaping* and the *Stemonyphantes* species failed to form a monophyletic clade. However, our phylogeny was strongly supported by the AU test (*p* = 0.99).

The thirteen *Solenysa* taxa were further divided into four clades (Figure 2): the *longqiensis* clade (clade L) including *S. longqiensis* and *S. yangmingshana* (68/0.81); the *wulingensis* clade (clade W) including *S. wulingensis*, *S. retractilis*, and *S. tianmushana* (99/1.00); the *mellotteei* clade (clade J) including six Japanese species, *S. mellotteei*, *S. reflexilis*, *S. trunciformis*, *S. macrodonta*, *S. partibili*, and *S. ogatai* (100/1.00); and the *protrudens* clade (clade P), including *S. protrudens* and *S. lanyuensis* (100/1.00). Compositions of the four clades were largely congruent with the four clades recognized in phylogenetic analysis based on morphological data, each of which was supported by several synapomorphies [12]. However, the relationships among these clades in our results were: the *longqiensis* clade was the most basal lineage sister to all other *Solenysa* clades (100/1.00); the *mellotteei* clade was sister to the *protrudens* clade (100/1.00), and the clade (*mellotteei* + *protrudens*) was sister to the *wulingensis* clade (95/1.00).

Mapping the characters of the epigynum with an extensible base and the desmitracheate system onto the phylogenetic framework show that the movable epigynum independently evolved multiple times in clades A and S. The *Solenysa* clade was distantly related with the taxon of Ipainae (*Wubanoides* sp.) and those micronetines also having a movable epigynum. The desmitracheate system independently evolved in clade B multiple times.

Distributions of the four *Solenysa* clades have different patterns (Figure 1). All species of Clade J are limited to the Japanese Archipelago; the species of clades L and P are known from the southeast coast of China and Taiwan Island; those of clade W scatter in southern China, as well as the Korean Peninsula (not sampled here). All *Solenysa* species have a disjunct distribution, and all the materials studied here were collected from leaf litter with high ambient humidity.

3.3. Divergence Times Estimation

Divergence times of the relevant nodes within linyphiids estimated using a relaxed molecular clock method in BEAST are provided in Table 3. The dated phylogeny is shown in Figure 3, with branches of the other six major clades collapsed and the complete chronogram is presented in Supplementary Material Figure S3.

Table 3. Divergence times of relevant nodes within linyphiids estimated using a relaxed molecular clock method in BEAST.

Node	Mean	95% HPD Lower	95% HPD Upper
1	128.80	125.14	136.35
2	116.55	104.48	128.18
3	101.67	88.88	113.61
4	90.85	72.69	108.83
5	89.22	77.24	100.88
6	79.29	67.69	90.67
7	28.62	20.09	37.98
8	22.06	13.71	31.27
9	21.99	15.74	28.85
10	18.49	12.75	24.66
11	12.34	6.29	18.98
12	3.46	1.09	6.50
13	6.35	3.10	10.10
14	6.59	4.03	9.39
15	3.46	1.80	5.27
16	2.52	1.12	4.07
17	2.56	0.99	4.33
18	0.83	0.02	1.94

TMRCA of the oldest linyphiids fossil: Exponential prior (Mean = 10, Offset = 125). TMRCA of *Orsonwelles polites* and *Orsonwelles malus*: Normal prior (Mean = 3.0 mya, SD = 0.5). Time in million years ago (mya). Nodes labeled as in Figure 3. HPD, highest posterior density; TMRCA, time of most recent common ancestor.

The chronogram of linyphiids suggests that all the seven major lineages survived the K-T boundary. The MRCA of all linyphiids (node 1) can be traced back to the early Cretaceous (128.80 mya, 95% HPD, 125.14–136.36 mya). Clade S and clade B (node 6) diverged around 79.29 (67.69–90.67) mya in the Cretaceous. Further diversifications in the linyphiid lineages largely took place after the K-T boundary, except for the *Solenysa* clade. The earliest split of *Solenysa* species (node 7) can only be traced back to 28.62 (20.09–37.98) mya in the middle Oligocene. In the following 10 Myr until the early Miocene, all of the four *Solenysa* clades (nodes 8–10) have appeared. The speciation of extant species (nodes 11–17) took place largely in the middle Miocene and the Pliocene, ranging between 12.34 (6.29–18.98) mya to 2.56 (0.99–4.33) mya, before the onset of the Quaternary climatic oscillations (2.6 mya–present [53,54]). Two exceptions involved the divergence of Japanese species, between *S. partibilis* and *S. ogatai* (node 18), which occurred much recently, around 0.83 (0.02–1.94) mya and between *S. macrodonta* and *S. reflexilis* (node 16), around 2.52 (1.12–4.07) mya during the Quaternary glaciations. In contrast, diversification in clades A and B crossed the K-T boundary and continued during the whole Cenozoic.

3.4. Temporal Patterns of Linyphiids Diversification

Analysis of lineage accumulation over time using LTT plots suggested that species of clade A and clade B began to accumulate long before the K-T boundary, while the lineage number in clade S (*Solenysa*) did not change until the Oligocene, long after the K-T boundary (Figure 4). Both clade S and clade B experienced obvious increases in the accumulation of lineages during the Oligocene. Besides, clade S experienced a unique fast lineage-accumulating phase during the last 10 Myr.

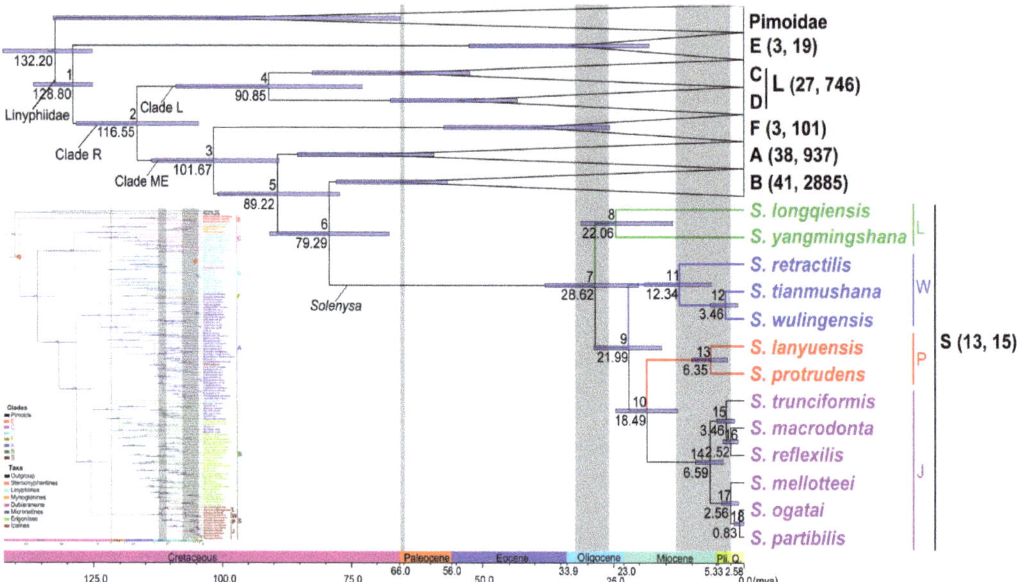

Figure 3. Chronogram of linyphiids. Branches and taxon names in color show the four clades within *Solenysa*. Branches of other six major clades collapsed. In the complete chronogram shown in microimage, two red dots indicate calibration points. Numbers in parentheses after clade names refer to the sampling numbers of the clades and species numbers represented, respectively. Numbers above branches label the divergence nodes of the seven major clades within linyphiids and internal nodes within *Solenysa*. Values below branches show divergence time of nodes, and node bars show confidence intervals. Three grey belts refer to K-T boundary, early and middle Oligocene, and middle Miocene and Pliocene, respectively.

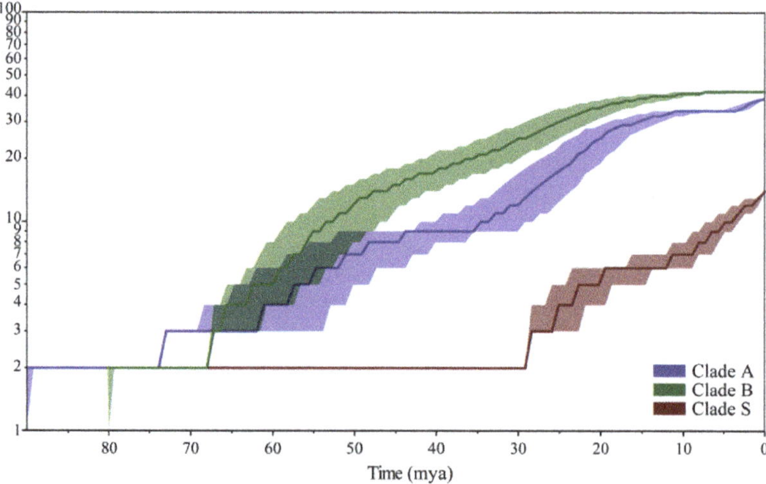

Figure 4. Temporal patterns of linyphiid lineage diversifications. a, stepped line in color indicate species number of lineages vary through time for *Solenysa* clade (brown), clade A (purple), and clade B (green). The shadowed areas in the same color show 95% confidence intervals.

4. Discussion

For a long time, a temporal framework was lacking for linyphiids, the second-largest group of spiders (but see [7,8]). Through phylogenetic reconstruction and molecular dating based on sequence data of five genes for all major groups of linyphiids, we brought a time scale to this important spider group and gained some illuminating insights into the evolutionary history that gave rise to the poor species diversity of *Solenysa* spiders contrasted with its sister group. We further use LTT plots to demonstrate the history of *Solenysa* diversification.

Solenysa spiders originated in the first radiation of linyphiids and missed the second burst of speciation in this group. The chronogram shows that the MRCA of linyphiids can be traced back to the early Cretaceous (Figure 3); it might even be traced to the Jurassic [20]. All the seven major clades, including the *Solenysa* clade, emerged during the Cretaceous. Being generalist predators in natural ecosystems, linyphiid spiders weave horizonal sheet webs to catch prey [1]. Their web-building level varies a lot across the seven major clades (Figure 2): most taxa of clade C + D, traditionally grouped in Linyphiinae, build aerial webs at various levels of vegetation, especially often at the crown level, while the taxa of clade ME mainly formed by species of Micronetinae and Erigoninae build their webs much closer to the ground. Generally, those micronetines with dorsal spots on their abdomen usually build aerial webs at the leaf-litter surface in forests, and those of clade B, especially those erigonines having an abdomen in grey to dark black without dorsal spots, build substrate-webs close to or even on the ground. *Solenysa* spiders, as members of clade ME, inhabit the litter layer in forests and have a grayish abdomen without dorsal spots; they build their webs close to the ground (see the figure in [31]). These suggest that linyphiids had their first diversification, as in many other spiders, accompanied by the co-adaptative radiation of insects and angiosperms in the Cretaceous [1,11,12,55], and diversification among the seven clades were accompanied by the divergence of their web-building levels. Furthermore, the subsequent diversification within these major clades mainly flourished in the Cenozoic in general. This indicates that the species diversity of all seven clades was significantly affected by the mass extinction of the K-T boundary (65 mya, [56]); their diversifications in the Cenozoic was, in fact, the second radiation in linyphiid evolution. Unlike other major clades, diversification in the *Solenysa* clade has a long-time lag of more than 50 Myr in their early history. The single lineage crossed the K-T boundary, with the earliest split occurring 28.62 mya in the middle Oligocene, much later than its sister group clade B, and their sister group clade A (Figure 4), although we expanded the sampling of *Solenysa* taxa in the present study.

The low species diversity of *Solenysa* spiders might result from both the lower diversification rate than that of their sister group clade B and the long-time lag in their evolutionary history. Among all the major clades of linyphiids, only in clade B did the desmitracheate system independently evolve multiple times (Figure 2). Although its selective advantages, either the high efficiency of anteriorly extending tracheae in providing oxygen directly to the brain and legs [35,36] or the assistance of extensively branched tracheae on water retention [9,37], have never been physiologically tested in web-weaver spiders (but see the morphological test in [57]), such a tracheal system repeatedly evolved in clade B, as well as in several litter-dweller spider groups [14,58,59], which implies its selective advantage for these spiders. Furthermore, the great species diversity of clade B, especially of the distal erigonines clade, suggest that the desmitracheate system might be a key innovation that triggered the fast diversification in clade B. Accordingly, without driving by the selection advantages of the desmitracheate system, it is not a surprise that the diversification rate of *Solenysa* clade is not as fast as that of clade B. Nevertheless, this is not the only reason attributed to the extreme asymmetry in the species diversity between the *Solenysa* clade and clade B. The long time lag across the K-T boundary to the middle Oligocene makes the *Solenysa* clade have no origination of any further extant groups for more than half of the common historical time shared with clades A and B (Figure 4).

Being highly dependent on high-humidity environments may be a major limit impact on *Solenysa* spiders. Unlike book lungs, tracheal respiration in spiders does not depend on hemolymph [36,60], and intensive branched tracheae are helpful in saving water [9,37]. Nevertheless, the intermediate-type tracheal system in *Solenysa* has the unbranched median pair tracheae extending into the prosoma [24]. Furthermore, their living habits are usually in areas with high ambient humidity. Although building sheet-webs at litter level, *Solenysa* spiders failed to leave forests for more open and less humid ecosystems, such as grasslands, or even as pioneers to colonize the ecological bare grounds as those erigonines of their sister group. Accordingly, we infer that their extreme dependence on high environmental humidity is the main constraint for their distribution.

Our results show the temporal patterns of linyphiid lineage diversification were closely related to global climate changes (Figures 3 and 4). Although the Paleocene Earth was commonly considered ice-free and the global climate was warm and humid [61,62], evidence has shown that the climate in Asia was dry during most of the Paleogene. There was an arid belt that existed from the western-most part to the eastern coasts, and arid and semi-arid conditions dominated in large areas of China [63]. This led to the formation of temperate grasslands and savannah ecosystems on most land at the expense of forest decline [61,62,64]. Such ecosystem transformations might have acted as a driving force that promoted the rapid diversification of micronetines (clade A) and erigonines (clade B) but might be a major constraint on the survival of *Solenysa* species. Nevertheless, such a dry belt retreated northwestward from the Eocene to the Oligocene [63]. Until the late Oligocene (28–24 mya), the southeast part of China, including the southeast coasts and Taiwan Island, became humid. This might have triggered the diversification of extant *Solenysa* spiders (28.62 mya, Figure 3). In the following Miocene (24–5.3 mya) and the Pliocene (5.3–2.6 mya), the humid belt further expanded northwards, the whole eastern part of China transformed into humid conditions. During this time, clade S experienced a unique fast lineage-accumulating phase. Therefore, such climate changes might have released the constraint from arid conditions and promoted the diversification of *Solenysa* in the Neogene.

Solenysa spiders display typical characters of a relict group, with low species diversity and narrow distribution [65–67]. The chronogram of linyphiids shows that most *Solenysa* species have emerged before the onset of the Quaternary glaciations (2.6 mya, [53,54]). However, all of the 15 known species have a disjunct distribution, most of them being only known from type localities that fall near the supposed Pleistocene glacial refugia in East Asia (Figure 1; [68–72]). This suggests that the refugia have played an important role in maintaining these *Solenysa* species during the glacial period. The dramatically cold climate during the glacial period would generally incur contractions of the distribution range [65,66,73–75]. Such an interpretation may partially explain the current disjunct distribution patterns of *Solenysa* species. However, it also implies that *Solenysa* species have failed to expand their distributional ranges as the climate became warmer in the post-glacial periods. Generally, linyphiids, as well as most other spiders, are capable of dispersal by balloon [60]. Nevertheless, long-distance dispersal by ballooning means spiders staying a long time in the air without a water supply. The survival of *Solenysa* spiders depends extremely on highly humid environments that not only constrain their distribution but also limit their capacity for long-time ballooning for dispersal, especially when suitable habits are segmentized. Accordingly, the current distribution pattern of *Solenysa* spiders might be shaped by both the locations of those refugia they survived during the Quaternary glaciations and their weak dispersal capacity.

5. Conclusions

Our results suggest that *Solenysa* is a Cretaceous relict group, having survived the mass extinctions at the K-T boundary and the ecosystem transition caused by the global climate changes in the Cenozoic. Its diversification was shaped by the climatic oscillations in the Cenozoic. The low species diversity of the *Solenysa* clade, in contrast to its sister group, is largely due to the long time lag in its early evolutionary history. Given *Solenysa* represents a strongly supported major clade in linyphiid multi-locus phylogeny and is supported by several synapomorphies, it warrants a subfamilial status in Linyphiidae.

Supplementary Materials: The following supporting information can be downloaded at: https://www.mdpi.com/article/10.3390/d14020120/s1, Table S1: Collecting information of the *Solenysa* spider sequenced in the present study; Figure S1: Bayesian tree with all nodes supports; Figure S2: ML tree with all nodes supports; Figure S3: The complete chronogram.

Author Contributions: Conceptualisation, J.T., C.S., and L.T.; methodology, J.T., Y.Z., C.S., and L.T.; acquisition, J.T., H.O., L.T.; formal analysis, J.T., Y.Z., C.S., and L.T.; investigation, all authors; writing—original draft preparation, J.T., C.S., and L.T.; writing—review and editing, all authors; funding acquisition, L.T., and C.S. All authors have read and agreed to the published version of the manuscript.

Funding: Please add: This research was funded by the National Natural Sciences Foundation of China (grant Nos. 31572244, 31872188 and 31772435).

Institutional Review Board Statement: Not applicable.

Informed Consent Statement: Not applicable.

Data Availability Statement: The new genetic sequences are available on GenBank (ac-cession codes: OL693166 to OL693169 (COI), OL691622 to OL691623 (16S), OL691624 to OL691627 (18S), OL691628 to OL691631 (28S), OL702837 to OL702840 (H3)) and also as a supplement to this paper.

Acknowledgments: We thank J. Fu and the anonymous reviewers for their critical comments on the early versions of this manuscript. We also thank F. Wang, F. Zhang, S. Tian., Z. Zhao, and R. Li for their kind assistance in field work, and A. Andoh and A. Tanikawa for kindly providing *Solenysa* material collected from Japan.

Conflicts of Interest: The authors declare no conflict of interest.

References

1. Arnedo, M.A.; Hormiga, G.; Scharff, N. Higher-level phylogenetics of linyphiid spiders (Araneae, Linyphiidae) based on morphological and molecular evidence. *Cladistics* **2009**, *25*, 231–262. [CrossRef] [PubMed]
2. Dabert, M.; Witalinski, W.; Kaźmierski, A.; Olszanowski, Z.; Dabert, J. Molecular phylogeny of acariform mites (Acari, Arachnida): Strong conflict between phylogenetic signal and long-branch attraction artifacts. *Mol. Phylogenetics Evol.* **2010**, *56*, 222–241. [CrossRef] [PubMed]
3. Sharma, P.P.; Kaluziak, S.T.; Perez-Porro, A.; González, V.L.; Hormiga, G.; Wheeler, W.C.; Giribet, G. Phylogenomic interrogation of Arachnida reveals systemic conflicts in phylogenetic signal. *Mol. Biol. Evol.* **2014**, *31*, 2963–2984. [CrossRef] [PubMed]
4. Wang, F.; Ballesteros, J.A.; Hormiga, G.; Chesters, D.; Zhan, Y.; Sun, N.; Zhu, C.; Chen, W.; Tu, L. Resolving the phylogeny of a speciose spider group, the family Linyphiidae (Araneae). *Mol. Phylogenetics Evol.* **2015**, *91*, 135–149. [CrossRef] [PubMed]
5. Bininda-Emonds, O.R.P.; Cardillo, M.; Jones, K.E.; MacPhee, R.D.E.; Beck, R.M.D.; Grenyer, R.; Price, S.A.; Vos, R.A.; Gittleman, J.L.; Purvis, A. The delayed rise of present-day mammals. *Nature* **2007**, *446*, 507–512. [CrossRef]
6. Near, T.J.; Dornburg, A.; Kuhn, K.L.; Eastman, J.T.; Pennington, J.N.; Patarnello, T.; Zane, L.; Fernández, D.A.; Jones, C.D. Ancient climate change, antifreeze, and the evolutionary diversification of Antarctic fishes. *Proc. Natl. Acad. Sci. USA* **2012**, *109*, 3434–3439. [CrossRef]
7. Arnedo, M.A.; Hormiga, G. Repeated colonization, adaptive radiation and convergent evolution in the sheet-weaving spiders (Linyphiidae) of the south Pacific Archipelago of Juan Fernandez. *Cladistics* **2020**, *37*, 317–342. [CrossRef]
8. Hormiga, G.; Kulkarni, S.; Moreira, T.D.S.; Dimitrov, D. Molecular phylogeny of pimoid spiders and the limits of Linyphiidae, with a reassessment of male palpal homologies (Araneae, Pimoidae). *Zootaxa* **2021**, *5026*, 71–101. [CrossRef]
9. Dimitrov, D.; Hormiga, G. Spider Diversification Through Space and Time. *Annu. Rev. Entomol.* **2021**, *66*, 225–241. [CrossRef]
10. Bond, J.E.; Opell, B.D. Testing adaptive radiation and key innovation hypotheses in spiders. *Evolution* **1998**, *52*, 403–414. [CrossRef]
11. Penney, D. Does the fossil record of spiders track that of their principal prey, the insects? *Trans. R. Soc. Edinburgh Earth Sci.* **2003**, *94*, 275–281. [CrossRef]
12. Penney, D.; Ortuño, V.M. Oldest true orb-weaving spider (Araneae: Araneidae). *Biol. Lett.* **2006**, *2*, 447–450. [CrossRef] [PubMed]
13. Selden, P.A.; Penney, D. Fossil spiders. *Biol. Rev.* **2010**, *85*, 171–206. [CrossRef]
14. Lopardo, L.; Michalik, P.; Hormiga, G. Take a deep breath . . . The evolution of the respiratory system of symphytognathoid spiders (Araneae, Araneoidea). *Org. Divers. Evol.* **2021**, 1–33. [CrossRef]
15. Garrison, N.L.; Rodriguez, J.; Agnarsson, I.; Coddington, J.A.; Griswold, C.E.; Hamilton, C.A.; Hedin, M.; Kocot, K.M.; Ledford, J.M.; Bond, J.E. Spider phylogenomics: Untangling the Spider Tree of Life. *PeerJ* **2016**, *4*, e1719. [CrossRef] [PubMed]

16. Penney, D.; Selden, P.A. The oldest linyphiid spider, in Lower Cretaceous Lebanese amber (Araneae, Linyphiidae, Linyphiinae). *J. Arachnol.* **2002**, *30*, 487–493. [CrossRef]

17. World Spider Catalog. Version 22.5. Natural History Museum Bern. Available online: http://wsc.nmbe.ch (accessed on 20 November 2021). [CrossRef]

18. Tanasevitch, A.V. Linyphiid Spiders of the World. Available online: http://www.andtan.newmail.ru/list (accessed on 20 November 2021).

19. Dimitrov, D.; Lopardo, L.; Giribet, G.; Arnedo, M.; Álvarez-Padilla, F.; Hormiga, G. Tangled in a sparse spider web: Single origin of orb weavers and their spinning work unravelled by denser taxonomic sampling. *Proc. R. Soc. B Biol. Sci.* **2011**, *279*, 1341–1350. [CrossRef]

20. Dimitrov, D.; Benavides, L.R.; Arnedo, M.; Giribet, G.; Griswold, C.E.; Scharff, N.; Hormiga, G. Rounding up the usual suspects: A standard target-gene approach for resolving the interfamilial phylogenetic relationships of ecribellate orb-weaving spiders with a new family-rank classification (Araneae, Araneoidea). *Cladistics* **2017**, *33*, 221–250. [CrossRef]

21. Wheeler, W.C.; Coddington, J.A.; Crowley, L.M.; Dimitrov, D.; Goloboff, P.A.; Griswold, C.E.; Hormiga, G.; Prendini, L.; Ramírez, M.J.; Sierwald, P.; et al. The spider tree of life: Phylogeny of Araneae based on target-gene analyses from an extensive taxon sampling. *Cladistics* **2017**, *33*, 574–616. [CrossRef]

22. Eberhard, W.G. Evolution of genitalia: Theories, evidence, and new directions. *Genetica* **2010**, *138*, 5–18. [CrossRef]

23. Tu, L.; Li, S. A review of the linyphiid spider genus *Solenysa* (Araneae, Linyphiidae). *J. Arachnol.* **2006**, *34*, 87–97. [CrossRef]

24. Tu, L.; Hormiga, G. Phylogenetic analysis and revision of the linyphiid spider genus *Solenysa* (Araneae: Linyphiidae: Erigoninae). *Zool. J. Linn. Soc.* **2011**, *161*, 484–530. [CrossRef]

25. Saaristo, M.I. A new subfamily of linyphiid spiders based on a new genus created for the keyserlingi-group of the genus *Lepthyphantes* (Aranei: Linyphiidae). *Arthropoda Sel.* **2007**, *16*, 33–42.

26. Namkung, J. Two unrecorded species of linyphiid spiders from Korea. *Korean Arachnol.* **1986**, *2*, 11–18.

27. Li, S.; Song, D. On two new species of soil linyphiid spiders from China (Araneae: Linyphiidae: Erigoninae). *Acta Arachnol. Sin.* **1992**, *1*, 6–9.

28. Gao, J.C.; Zhu, C.D.; Sha, Y.H. Two new species of the genus *Solenysa* from China (Araneae: Linyphiidae: Erigoninae). *Acta Arachnol. Sin.* **1993**, *2*, 65–68.

29. Tu, L.; Ono, H.; Li, S. Two new species of *Solenysa* Simon, 1894 (Araneae: Linyphiidae) from Japan. *Zootaxa* **2007**, *1426*, 57–62. [CrossRef]

30. Ono, H. Notes on Japanese spiders of the genera *Paikiniana* and *Solenysa* (Araneae, Linyphiidae). *Bull. Natl. Mus. Nat. Sci. Ser. A* **2011**, *37*, 121–129.

31. Tu, L.; Wang, F.; Ono, H. A review of *Solenysa* spiders from Japan (Araneae, Linyphiidae), with a comment on the type species *S. mellotteei* Simon, 1894. *ZooKeys* **2015**, *481*, 39–56. [CrossRef]

32. Tian, J.; Tu, L. A new species of the spider genus *Solenysa* from China (Araneae, Linyphiidae). *Zootaxa* **2018**, *4531*, 142–146. [CrossRef]

33. Blest, A.D. The tracheal arrangement and the classification of linyphiid spiders. *J. Zool.* **1976**, *180*, 185–194. [CrossRef]

34. Hormiga, G. Higher level phylogenetics of erigonine spiders (Araneae, Linyphiidae, Erigoninae). *Smithson. Contrib. Zool.* **2000**, 1–160. [CrossRef]

35. Schmitz, A. Metabolic rates during rest and activity in differently tracheated spiders (Arachnida, Araneae): *Pardosa lugubris* (Lycosidae) and *Marpissa muscosa* (Salticidae). *J. Comp. Physiol. B* **2004**, *174*, 519–526. [CrossRef]

36. *Spider Ecophysiology*; Springer Science and Business Media LLC: Berlin/Heidelberg, Germany, 2013; pp. 1–529.

37. Levi, H.W. Adaptations of respiratory systems of spiders. *Evolution* **1967**, *21*, 571–583. [CrossRef] [PubMed]

38. Katoh, K.; Rozewicki, J.; Yamada, K.D. MAFFT online service: Multiple sequence alignment, interactive sequence choice and visualization. *Briefings Bioinform.* **2019**, *20*, 1160–1166. [CrossRef]

39. Maddison, W.P.; Maddison, D.R. Mesquite: A Modular System for Evolutionary Analysis. Version 3.51. Available online: http://www.mesquiteproject.org (accessed on 20 November 2021).

40. Nguyen, L.-T.; Schmidt, H.A.; Von Haeseler, A.; Minh, B.Q. IQ-TREE: A fast and effective stochastic algorithm for estimating maximum-likelihood phylogenies. *Mol. Biol. Evol.* **2015**, *32*, 268–274. [CrossRef]

41. Kalyaanamoorthy, S.; Minh, B.Q.; Wong, T.K.F.; Von Haeseler, A.; Jermiin, L.S. ModelFinder: Fast model selection for accurate phylogenetic estimates. *Nat. Methods* **2017**, *14*, 587–589. [CrossRef]

42. Lanfear, R.; Calcott, B.; Ho, S.Y.W.; Guindon, S. PartitionFinder: Combined selection of partitioning schemes and substitution models for phylogenetic analyses. *Mol. Biol. Evol.* **2012**, *29*, 1695–1701. [CrossRef]

43. Hoang, D.T.; Chernomor, O.; Von Haeseler, A.; Minh, B.Q.; Vinh, L.S. UFBoot2: Improving the ultrafast bootstrap approximation. *Mol. Biol. Evol.* **2018**, *35*, 518–522. [CrossRef]

44. Ronquist, F.; Teslenko, M.; van der Mark, P.; Ayres, D.L.; Darling, A.; Höhna, S.; Larget, B.; Liu, L.; Suchard, M.A.; Huelsenbeck, J.P. MrBayes 3.2: Efficient Bayesian phylogenetic inference and model choice across a large model space. *Syst. Biol.* **2012**, *61*, 539–542. [CrossRef]

45. Shimodaira, H. An approximately unbiased test of phylogenetic tree selection. *Syst. Biol.* **2002**, *51*, 492–508. [CrossRef]

46. Bouckaert, R.; Heled, J.; Kühnert, D.; Vaughan, T.; Wu, C.-H.; Xie, D.; Suchard, M.A.; Rambaut, A.; Drummond, A.J. BEAST 2: A software platform for bayesian evolutionary analysis. *PLoS Comput. Biol.* **2014**, *10*, e1003537. [CrossRef] [PubMed]

47. Bouckaert, R.R.; Drummond, A.J. bModelTest: Bayesian phylogenetic site model averaging and model comparison. *BMC Evol. Biol.* **2017**, *17*, 1–11. [CrossRef] [PubMed]
48. Rambaut, A.; Drummond, A.J.; Xie, D.; Baele, G.; Suchard, M.A. Posterior summarization in Bayesian phylogenetics using Tracer 1.7. *Syst. Biol.* **2018**, *67*, 901–904. [CrossRef] [PubMed]
49. Fernández, R.; Kallal, R.J.; Dimitrov, D.; Ballesteros, J.A.; Arnedo, M.A.; Giribet, G.; Hormiga, G. Phylogenomics, diversification dynamics, and comparative transcriptomics across the Spider Tree of Life. *Curr. Biol.* **2018**, *28*, 1489–1497.e5. [CrossRef]
50. Scharff, N.; Coddington, J.A.; Blackledge, T.A.; Agnarsson, I.; Framenau, V.; Szűts, T.; Hayashi, C.Y.; Dimitrov, D. Phylogeny of the orb-weaving spider family Araneidae (Araneae: Araneoidea). *Cladistics* **2020**, *36*, 1–21. [CrossRef]
51. Magalhaes, I.L.F.; Azevedo, G.H.F.; Michalik, P.; Ramírez, M.J. The fossil record of spiders revisited: Implications for calibrating trees and evidence for a major faunal turnover since the Mesozoic. *Biol. Rev.* **2020**, *95*, 184–217. [CrossRef]
52. Hormiga, G.; Arnedo, M.; Gillespie, R.G. Speciation on a conveyor belt: Sequential colonization of the Hawaiian Islands by *Orsonwelles* spiders (Araneae, Linyphiidae). *Syst. Biol.* **2003**, *52*, 70–88. [CrossRef]
53. Hewitt, G.M. The genetic legacy of the Quaternary ice ages. *Nature* **2000**, *405*, 907–913. [CrossRef]
54. Hewitt, G.M. Genetic consequences of climatic oscillations in the Quaternary. *Philos. Trans. R. Soc. B Biol. Sci.* **2004**, *359*, 183–195. [CrossRef]
55. Blackledge, T.A.; Scharff, N.; Coddington, J.A.; Szüts, T.; Wenzel, J.W.; Hayashi, C.Y.; Agnarsson, I. Reconstructing web evolution and spider diversification in the molecular era. *Proc. Natl. Acad. Sci. USA* **2009**, *106*, 5229–5234. [CrossRef] [PubMed]
56. Nayar, A. Asteroid impact may have gassed Earth. *Nature* **2009**, 2009–2012. [CrossRef]
57. Opell, B.D. The influence of web monitoring tactics on the tracheal systems of spiders in the family Uloboridae (Arachnida, Araneida). *Zoomorphology* **1987**, *107*, 255–259. [CrossRef]
58. Lopardo, L.; Hormiga, G. On the synaphrid spider *Cepheia longiseta* (Simon 1881) (Araneae, Synaphridae). *Am. Museum Novit.* **2007**, *20052*, 18. [CrossRef]
59. Lopardo, L.; Hormiga, G. Out of the twilight zone: Phylogeny and evolutionary morphology of the orb-weaving spider family Mysmenidae, with a focus on spinneret spigot morphology in symphytognathoids (Araneae, Araneoidea). *Zool. J. Linn. Soc.* **2015**, *173*, 527–786. [CrossRef]
60. Foelix, R.F. *Biology of Spiders*, 3rd ed.; Oxford University Press: New York, NY, USA, 2011.
61. Zachos, J.C.; Dickens, G.R.; Zeebe, R.E. An early Cenozoic perspective on greenhouse warming and carbon-cycle dynamics. *Nature* **2008**, *451*, 279–283. [CrossRef]
62. Strömberg, C.A. Evolution of grasses and grassland ecosystems. *Annu. Rev. Earth Planet. Sci.* **2011**, *39*, 517–544. [CrossRef]
63. Guo, Z.T.; Sun, B.; Zhang, Z.S.; Peng, S.Z.; Xiao, G.Q.; Ge, J.Y.; Hao, Q.Z.; Qiao, Y.S.; Liang, M.Y.; Liu, J.F.; et al. A major reorganization of Asian climate by the Early Miocene. *Clim. Past* **2008**, *4*, 153–174. [CrossRef]
64. Law, C.J. Evolutionary shifts in extant mustelid (Mustelidae: Carnivora) cranial shape, body size and body shape coincide with the Mid-Miocene Climate Transition. *Biol. Lett.* **2019**, *15*, 20190155. [CrossRef]
65. Hampe, A.; Jump, A.S. Climate Relicts: Past, Present, Future. *Annu. Rev. Ecol. Evol. Syst.* **2011**, *42*, 313–333. [CrossRef]
66. Grandcolas, P.; Nattier, R.; Trewick, S. Relict species: A relict concept? *Trends Ecol. Evol.* **2014**, *29*, 655–663. [CrossRef] [PubMed]
67. Grandcolas, P.; Trewick, S. *Biodiversity Conservation and Phylogenetic Systematics, Topics in Biodiversity and Conservation*; Springer Nature: Basel, Switzerland, 2016.
68. Fu, J.; Weadick, C.J.; Zeng, X.; Wang, Y.; Liu, Z.; Zheng, Y.; Li, C.; Hu, Y. Phylogeographic analysis of the *Bufo gargarizans* species complex: A revisit. *Mol. Phylogenetics Evol.* **2005**, *37*, 202–213. [CrossRef] [PubMed]
69. Borzée, A.; Santos, J.; Sanchez-Ramirez, S.; Bae, Y.; Heo, K.; Jang, Y.; Jowers, M.J. Phylogeographic and population insights of the Asian common toad (*Bufo gargarizans*) in Korea and China: Population isolation and expansions as response to the ice ages. *PeerJ* **2017**, *5*, e4044. [CrossRef] [PubMed]
70. Song, W.; Cao, L.-J.; Li, B.-Y.; Gong, Y.-J.; Hoffmann, A.A.; Wei, S.-J. Multiple refugia from penultimate glaciations in East Asia demonstrated by phylogeography and ecological modelling of an insect pest. *BMC Evol. Biol.* **2018**, *18*, 152. [CrossRef]
71. Du, Z.; Ishikawa, T.; Liu, H.; Kamitani, S.; Tadauchi, O.; Cai, W.; Li, H. Phylogeography of the assassin bug *Sphedanolestes impressicollis* in East Asia inferred from mitochondrial and nuclear gene sequences. *Int. J. Mol. Sci.* **2019**, *20*, 1234. [CrossRef]
72. Kurata, S.; Sakaguchi, S.; Hirota, S.K.; Kurashima, O.; Suyama, Y.; Nishida, S.; Ito, M. Refugia within refugium of *Geranium yesoense* (Geraniaceae) in Japan were driven by recolonization into the southern interglacial refugium. *Biol. J. Linn. Soc.* **2021**, *132*, 552–572. [CrossRef]
73. Habel, J.C.; Assmann, T. *Relict Species: Phylogeography and Conservation Biology*; Springer eBooks: Berlin/Heidelberg, Germany, 2010.
74. Shi, C.-M.; Ji, Y.-J.; Liu, L.; Wang, L.; Zhang, D.-X. Impact of climate changes from Middle Miocene onwards on evolutionary diversification in Eurasia: Insights from the mesobuthid scorpions. *Mol. Ecol.* **2013**, *22*, 1700–1716. [CrossRef]
75. Mammola, S.; Hormiga, G.; Arnedo, M.A.; Isaia, M. Unexpected diversity in the relictual European spiders of the genus *Pimoa* (Araneae: Pimoidae). *Invertebr. Syst.* **2016**, *30*, 566–587. [CrossRef]

Article

Old Brains in Alcohol: The Usability of Legacy Collection Material to Study the Spider Neuroarchitecture

F. Andres Rivera-Quiroz [1,2,*] and Jeremy Abraham Miller [1,*]

1. Understanding Evolution Research Group, Naturalis Biodiversity Center, Darwinweg 2, 2333 CR Leiden, The Netherlands
2. Institute for Biology Leiden (IBL), Leiden University, Sylviusweg 72, 2333 BE Leiden, The Netherlands
* Correspondence: andres.riveraquiroz@naturalis.nl (F.A.R.-Q.); jeremy.miller@naturalis.nl (J.A.M.)

Abstract: Natural history collections include rare and significant taxa that might otherwise be unavailable for comparative studies. However, curators must balance the needs of current and long-term research. Methods of data extraction that minimize the impact on specimens are therefore favored. Micro-CT has the potential to expose new character systems based on internal anatomy to taxonomic and phylogenetic analysis without dissection or thin sectioning for histology. However, commonly applied micro-CT protocols involve critical point drying, which permanently changes the specimen. Here, we apply a minimally destructive method of specimen preparation for micro-CT investigation of spider neuroanatomy suitable for application to legacy specimens in natural history collections. We used two groups of female spiders of the common species *Araneus diadematus*—freshly captured (n = 11) vs. legacy material between 70 and 90 years old (n = 10)—to qualitatively and quantitatively assess the viability of micro-CT scanning and the impact of aging on their neuroarchitecture. We statistically compared the volumes of the supraesophageal ganglion (syncerebrum) and used 2D geometric morphometrics to analyze variations in the gross shape of the brain. We found no significant differences in the brain shape or the brain volume relative to the cephalothorax size. Nonetheless, a significant difference was observed in the spider size. We considered such differences to be explained by environmental factors rather than preservation artifacts. Comparison between legacy and freshly collected specimens indicates that museum specimens do not degrade over time in a way that might bias the study results, as long as the basic preservation conditions are consistently maintained, and where lapses in preservation have occurred, these can be identified. This, together with the relatively low-impact nature of the micro-CT protocol applied here, could facilitate the use of old, rare, and valuable material from collections in studies of internal morphology.

Keywords: Arachnida; Arthropoda; tissue; X-rays; micro-CT; cerebrum; nervous system; neuroanatomy; imaging

Citation: Rivera-Quiroz, F.A.; Miller, J.A. Old Brains in Alcohol: The Usability of Legacy Collection Material to Study the Spider Neuroarchitecture. *Diversity* 2021, 13, 601. https://doi.org/10.3390/d13110601

Academic Editors: Michael Wink and Matjaž Kuntner

Received: 7 October 2021
Accepted: 16 November 2021
Published: 21 November 2021

1. Introduction

Specimens in natural history collections represent one of our largest and most complete archives of biodiversity records. They provide verifiable evidence of the existence of species in space and time, as well as molecular and morphological information, and cues about biodiversity dynamics, ecological interactions, and even physiological processes [1–6]. In a changing world, this historical baseline will become increasingly precious. High-value specimens in natural history collections include many rare species known from only one or a few specimens collected long ago, type specimens that are the essential vouchers for taxonomic research, and specimens from now extinct populations or species. The long-term preservation of such collections for future study is vital [7,8]. Curators must be responsive to the needs of current research without curtailing future research. Methods that extract data with the minimum possible impact on specimens are therefore preferable. Minimally destructive methods have been developed to facilitate DNA extraction from high-value specimens [7–9]. With some exceptions for specialized collections, natural history collection conditions are generally not adequate to completely prevent the degradation of specimen DNA [10–13]. However, the advent of new molecular technologies such as target-capture methods have substantiated the use of museum specimens as a source of DNA sequences from material collected decades or even centuries ago [2]. Similarly, cutting-edge imaging methods such as micro-CT scans have allowed the observation and characterization of internal anatomy while producing a minimal impact on the specimen by comparison with other methods for documenting internal anatomy (e.g., histology or dissection) [3,14]. Still, the effect of preservatives on the volume and shape of the structures may pose a challenge to the correct visualization, interpretation, and quantification of a specimen's anatomy [15].

Micro-CT has gained traction in recent years as a way to observe the internal and external anatomy and reconstruct 3D models of a variety organs and systems in different invertebrate taxa [3,14,16–24]. Its ability to visualize and reconstruct models of internal and external anatomical features, without the need for dissection or thin slicing, makes this approach ideal for the digitization of both common and rare material. Ideal contrast in scans can be achieved through a staining process (critical for soft tissues) from which several protocols are available; these vary in their impact on specimen preservation and visualization [14,17,25]. Remarkably, in spiders, its use has allowed the documentation of a variety of sexual, muscular, and nervous organs and systems. This has permitted a quick and reliable way to generate 3D reconstructions and volume measurements that have the potential to become a powerful tool in studies as diverse as systematics, sexual selection, character evolution, and development, among many others [25–30]. Most of these studies on spiders have relied on the use of freshly collected material following, in many cases, specific fixation protocols. However, biological collections offer a massive library of taxa that can likely grant access to rare and relevant species while also broadening our potential taxonomic sampling. Furthermore, widely used protocols involve critical point drying of specimens, which can lead to distortions of the internal anatomy and interrupts ethanol preservation, permanently changing the specimen in ways that curtail other common research applications [25].

Here, we used a recently developed micro-CT protocol for spiders based on phosphotungstic acid (PTA) [25] to document and compare two groups (freshly collected vs. collection material) of the common spider *Araneus diadematus* Clerck, 1757. We found that given the correct preservation of specimens, gross neuroarchitectural features (as well as other anatomical traits) can be visualized, measured, and reconstructed, even after decades of 70% Et-OH storage.

2. Materials and Methods

Two groups of adult female *Araneus diadematus* were used. Group 1 ($n = 11$) was composed of spiders freshly collected at the Singelpark, Leiden, the Netherlands in October 2020. Group 2 ($n = 10$) was formed by legacy specimens archived in the collection of the Naturalis Biodiversity Center and collected within the Netherlands between the years 1929 and 1948 (Table 1). Fresh spiders were directly placed and stored in 70% Et-OH; collection specimens were stored in 70% Et-OH as well. All the specimens had their legs and palps removed and were stained with a contrast enhancing solution (1% PTA-70% Et-OH) for several days following Rivera-Quiroz and Miller [25]. After staining, all spiders were washed three times with 96% Et-OH and transferred into 1.5 mL eppendorf tubes filled with 96% Et-OH for scanning. Micro-CT scanning was performed with a Zeiss X-radia 520 versa. The software Avizo 2021.2 (Thermo Fisher Scientific, Waltham, MA, USA) was used to visualize the scannings and make 3D reconstructions of the supraesophageal ganglion. Labeling of the brain structures was performed by topological correspondence to the anatomy of *Araneus diadematus* and *Argiope lobata* (as *Epeira* in the original publication) [31], *Argiope trifasciata* [32,33], and other araneomorph spiders [27,29,34].

Table 1. Data of the freshly collected and legacy specimens used in our study.

Specimen Code	Coll. Date	Collector	Coll. Country	Province	Region	Locality	Coord Lat	Coord Lon
AD_01	7-Oct-2020	A. Rivera-Quiroz, D, Arguedas	NL	Zuid Holland	Leiden	Singel Park	52.16088	4.504739
AD_02	7-Oct-2020	A. Rivera-Quiroz, D, Arguedas	NL	Zuid Holland	Leiden	Singel Park	52.16088	4.504739
AD_03	7-Oct-2020	A. Rivera-Quiroz, D, Arguedas	NL	Zuid Holland	Leiden	Singel Park	52.16088	4.504739
AD_04	7-Oct-2020	A. Rivera-Quiroz, D, Arguedas	NL	Zuid Holland	Leiden	Singel Park	52.16088	4.504739
AD_05	7-Oct-2020	A. Rivera-Quiroz, D, Arguedas	NL	Zuid Holland	Leiden	Singel Park	52.16088	4.504739
AD_06	7-Oct-2020	A. Rivera-Quiroz, D, Arguedas	NL	Zuid Holland	Leiden	Singel Park	52.16088	4.504739
AD_07	7-Oct-2020	A. Rivera-Quiroz, D, Arguedas	NL	Zuid Holland	Leiden	Singel Park	52.16088	4.504739
AD_09	7-Oct-2020	A. Rivera-Quiroz, D, Arguedas	NL	Zuid Holland	Leiden	Singel Park	52.16088	4.504739
AD_10	7-Oct-2020	A. Rivera-Quiroz, D, Arguedas	NL	Zuid Holland	Leiden	Singel Park	52.16088	4.504739
AD_11	7-Oct-2020	A. Rivera-Quiroz, D, Arguedas	NL	Zuid Holland	Leiden	Singel Park	52.16088	4.504739
AD_12	7-Oct-2020	A. Rivera-Quiroz, D, Arguedas	NL	Zuid Holland	Leiden	Singel Park	52.16088	4.504739
AD_13	22-Apr-1940	B. de Jong	NL	ND	ND	ND	ND	ND
AD_14	24-Aug-1941	B. de Jong	NL	Noord Holland	Amsterdam	Kruislaan	ND	ND
AD_15	9-Oct-1929	W. B. Begerinck, B. de Jong	NL	Noord Holland	ND	ND	ND	ND
AD_16	9-Oct-1929	W. B. Begerinck, B. de Jong	NL	Noord Holland	ND	ND	ND	ND
AD_17	18-Oct-1946	B. de Jong	NL	Noord Holland	Hilversum	Lage Vuursche	ND	ND
AD_18	18-Oct-1946	B. de Jong	NL	Noord Holland	Hilversum	Lage Vuursche	ND	ND
AD_19	22-Apr-1940	B. de Jong	NL	ND	ND	ND	ND	ND
AD_20	18-Oct-1946	B. de Jong	NL	Noord Holland	Hilversum	Lage Vuursche	ND	ND
AD_22	18-Oct-1948	B. de Jong	NL	Noord Holland	ND	ND	ND	ND
AD_23	3-Nov-1947	B. de Jong	NL	Overijssel	Wanneperveen	ND	ND	ND

ND = no data.

For the volumetric analysis, we obtained the synganglion volumes from the 3D reconstruction, and measurements of cephalothorax and brain length and width in Avizo. The size–volume ratio was obtained by dividing the cephalothorax and brain measurements by the total synganglion volume in mm^3. Box plots were generated in R using the package GGplot2 [35]. Colored boxes indicate the 25 and 75 percentiles or interquartile range (IQR), the thick horizontal lines indicate the group's median and error lines (whiskers) indicate the maximum and minimum distribution for the group (calculated as minimum value $- 1.5 \times$ IQR and maximum value $+ 1.5 \times$ IQR). Individual values are represented with a dot; values that fall outside the error lines indicate potential outliers. Samples were tested for normality following a graphical approach (based on the boxplots) and a Shapiro–Wilk normality test performed in R [36]. In case Shapiro–Wilk suggested a non-normal distribution, outliers were removed and samples were tested a second time. Welch two sample *t*-test analyses were performed to test for significant differences between the normally distributed groups.

Due to the asymmetric nature of the optic nerves and other incidental abnormalities we found in the brain (see discussion), there was no satisfactory view that included right and left sides of the brain and optic lobes using the slice function in Avizo. Therefore, we used a simplified reconstruction of the whole supraesophageal ganglion on an axial view to perform the shape analysis. A scalebar of 200 micrometers was added to each image in Avizo and was used to calibrate the size before setting the landmarks. The software tpsUtil64 [37] was used to build a TPS file from the Axial views of the brain reconstructions.

A total of fourteen landmarks were per image were manually laid using tpsDig264 [38]. Landmarks were placed on the most prominent features of the optic lobes in the anterior portion of the brain in the following order: Landmark 1 is placed at the intersection of the secondary eyes optic nerve and the left brain's cellular cortex (CC) on the medial side. Landmarks 2 and 14 are placed on either side of the first visual neuropile of the left secondary eye tract. Landmarks 11, 10, and 12 mirror 1, 2, and 14 on the right side. Landmark 13 is placed on the cleavage of the right and left optic lobes. The posterior part lacks clear morphological features; therefore, to keep a consistent placement, landmarks were evenly distributed every 45 degrees in a semicircular fashion taking the middle axis as the starting point (see Figure A1). The program MorphoJ v.1.07a [39] was used to perform the geometric morphometric analysis. A preliminary procrustes fit was performed aligning the coordinates by principal axes. Additionally, as part of the preliminaries, a wire frame was created to help visualize the mean shape and two classifiers (group and size) were created. Group classifier indicates whether the specimens were freshly collected (*New*) or were legacy material (*Old*). A principal component analysis was performed to observe the distribution of our data (see Theska [40]), after which we performed a regression analysis to predict association between the procrustes values (dependent variable) and brain width (independent variable). Finally, we used a discriminant function analysis for further comparison of the shape variation between both groups.

3. Results

3.1. Size and Volumetric Analyses

Our analyses showed an unequivocal difference in sizes between the freshly collected samples and the legacy material. The mean cephalothorax length was of 5.82 ± 0.363 mm for the *New* group while the *Old* group was smaller with 4.03 ± 0.647 mm. Similarly, the mean carapace width was 4.59 ± 0.370 mm and 3.46 ± 0.484 mm, respectively. The Shapiro–Wilk test showed all our samples to be normally distributed (Table 2). The *t*-test results show a significant difference between the groups in carapace length (t (14.823) $= -7.2144$, $p = 3.215 \times 10^{-6}$) and carapace width (t (16.795) $= -6.024$, $p = 1.44 \times 10^{-5}$). Interestingly, even the largest specimen from the legacy material is far below the mean size of the freshly collected spiders. We found a significant difference in the volume of the supraesophageal ganglion, with the *New* group averaging 0.16 ± 0.048 mm^3 and the *Old* spiders 0.11 ± 0.022 mm^3 (t (18.989) $= -3.8665$, $p = 0.00104$). The Shapiro–Wilk test showed brain size data to be normally distributed, except for volume of the *New* samples, which achieved normality after removal of one outlier (Table 2). By three metrics (carapace length, carapace width, brain volume), the fresh spiders were larger than the legacy specimens. However, we found no significant difference in the width of the supraesophageal ganglion, 0.61 ± 0.078 mm in the *New* group vs. 0.61 ± 0.084 mm in the *Old* group (t (19.841) $= 0.48921$, $p = 0.6301$. Moreover, when comparing the brain volume–cephalothorax width ratio as a way of scaling for differences in overall size, the groups were not significantly different: 0.036 ± 0.009 mm in the *New* group vs. 0.032 ± 0.006 mm in the *Old* group (t (19.714) $= -0.89161$, $p = 0.3834$) (Figure 1).

Table 2. Results of normality test.

	p-Values		[Outliers Removed]	
	New	*Old*	*New*	*Old*
Carapace width	0.854	0.2008	–	–
Carapace length	0.8829	0.1361	–	–
Brain width	0.3331	0.2035	–	–
Brain volume	0.009821	0.3665	0.688 [1]	–
BR vol. vs. Car. W	0.1129	0.3808	–	–

All subsets we compared were tested for normality using a Shapiro–Wilk test. The *p*-values in the first two columns are were calculated from the original sample ($p > 0.05$ indicates a normal distribution). Values in the second two columns were obtained after the removal of outliers; the number of discarded outliers is shown in brackets.

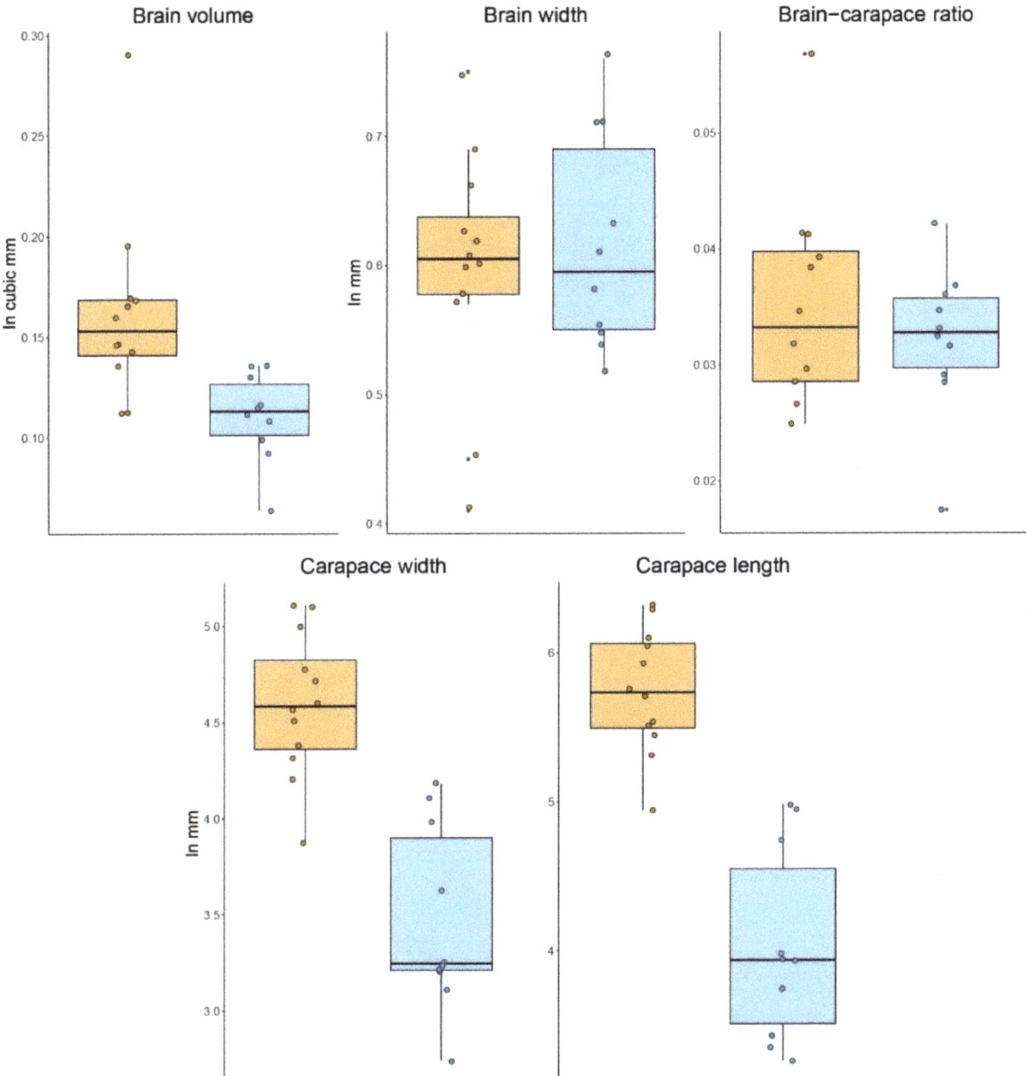

Figure 1. Volumetric and size comparisons. Boxplots comparing the volume and size variation in both groups. Fresh material (*New*) is shown in dark yellow and legacy material (*Old*) in blue. Note that although there is a considerable variation in absolute brain volume and carapace size, the brain width and brain–carapace ratio are not significantly different between the groups.

3.2. Shape Variation

The PCA (Figure 2b) shows that the brain shapes of both groups largely overlap. PC1 is related to changes in the width of the brain, the length of its middle axis, and the length of the optic lobes. The shape variation accounted for by PC2 is principally related to the aperture and width of the optic lobes (see Figures 2c and A2a). These first two PCs together explain almost 60% of the shape variation, having both a tendency towards a more regular and symmetric shape. PC3 and PC4, on the other hand, cover a more asymmetrical variation of the anterior half of the brain (see Figure A2b). PC3 shows a broader left optic lobe, with the two optic lobes being closer together in the center. Finally,

PC4 indicates a skewedness of the optic lobes towards the right side, a deeper cleavage and therefore a shorter median axis of the brain and a slight enlargement on the posterior left side of the supraesophageal ganglion. The regression analysis shows an association between brain shape (procrustes coordinates) and brain width (Figure 2d) ($p = 0.3633$) that is not significantly different between the Old and New groups. Similarly, the discriminant analysis (Figure A3) ($p = 0.7229$) shows a clear overlap between both groups and suggests no significant difference between them.

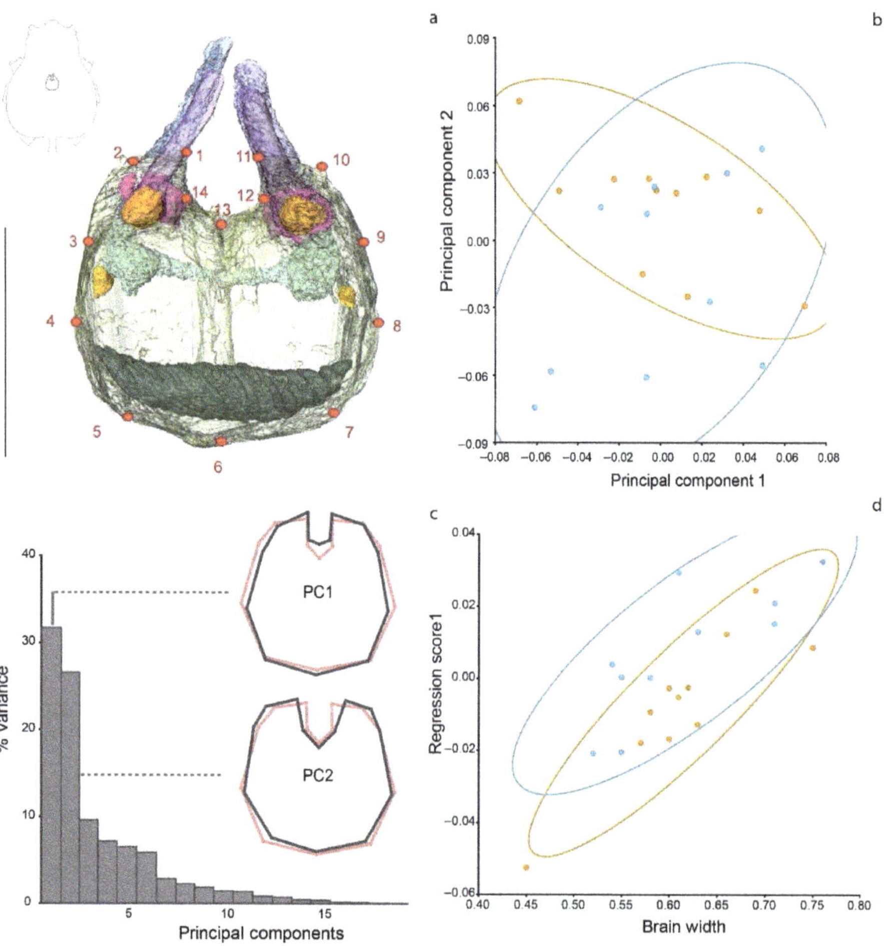

Figure 2. Shape analysis of the supraesophageal ganglion. (**a**) Volume rendering of the spider brain (dorsal view) showing the placement of the 14 landmarks we measured; inset line drawing shows the brain in relation to the cephalothorax in dorsal view. (**b**) Scatter plot of the PCA scores for PC1 and PC2. The distribution suggests no differences in the brain shape of the fresh and legacy specimens. (**c**) Bar plot showing the variance coverage of each PC. PC1 and PC2, the most meaningful PC (together explain almost 60% of the variance) are illustrated. The red outlines indicate the mean shape, and the grey outlines show the shape variation associated to each PC. (**d**) Regression analysis of the association between brain shape (procrustes coordinates) and brain width. The overlapping distributions of both groups show no statistical significance in the shape variation between *Old* and *New* spider groups. (**b,d**) fresh material (*New*) is shown in dark yellow and collection material (*Old*) in blue. Scale bar = 0.5 mm.

3.3. Qualitative Assessment

The internal anatomy and, most importantly, the central nervous system (CNS) architecture was recognizable and usable in both groups. We were able to visually identify the different parts of the visual systems (e.g., optic nerves, optic neuropiles and central body) and other parts of the CNS (e.g., ganglia, blood vessels) in most of our specimens, regardless of the time spent in preservation (Figure 3). Nevertheless, not all specimens were equally useful; two legacy specimens showed signs of tissue deterioration (probably caused by desiccation), and others had their synganglion displaced and deformed by the digestive diverticula (Figure 4). Still, in all of those cases, the supraesophageal ganglion could be identified, segmented, and compared from an axial view together with the rest of the samples.

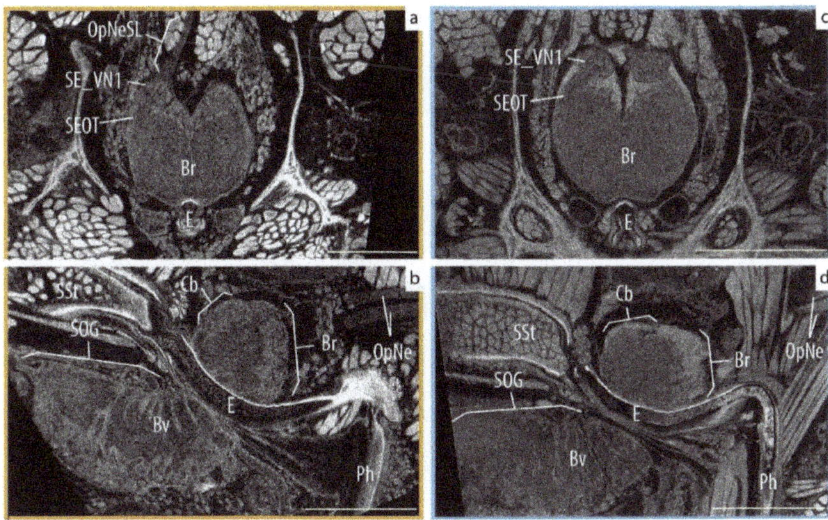

Figure 3. Comparison of *Araneus diadematus* neuroanatomy. *New* material was collected in October 2020; the *Old* specimen was captured in April 1940. (**a**,**c**) Axial slice of the brain focusing on the secondary eye pathway. (**b**,**d**) Sagittal slice of the whole CNS. Abbreviations: Br, brain; Bv, blood vessels; CB, central body (arcuate body); E, esophagus; OpNe, optic nerves; OpNeSL, optic nerves of the left secondary eyes; Ph, pharinx; SE_VN1, visual neuropiles 1 (lamina) of the secondary eye pathway; SEOT, secondary optic track; SOG, sub-esophagic ganglion; SSt, sucking stomach. Fresh material (*New*) is marked in dark yellow and legacy material (*Old*) in blue. Scale bars = 0.5 mm.

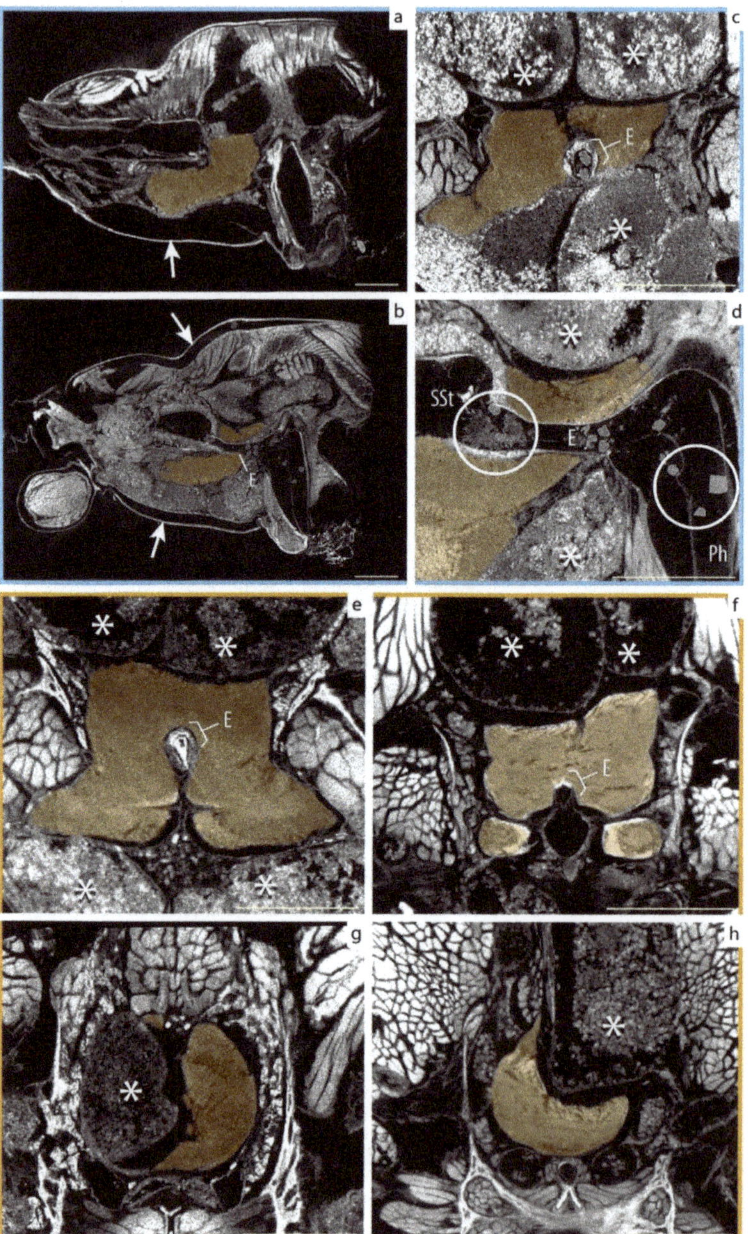

Figure 4. Artifacts and incidental brain morphologies. Two factors to take in account when assessing the individual variation: (1) incorrect preservation of specimens; evidence of shrinkage caused by desiccation (arrows in (**a**,**b**)); and formation of crystals in the pharynx (Ph), esophagus (E), and sucking stomach (SSt) (marked with circles in (**d**)). (2) Naturally occurring abnormalities, for example, irregular growth of the digestive diverticula (* in (**c**–**h**)) pushing and deforming the central nervous system ganglia (in orange) but being more evident on the supraesophageal ganglion (see (**g**,**h**)). Fresh material (*New*) is marked in dark yellow and legacy material (*Old*) in blue. Scale bars = 0.5 mm.

4. Discussion

The brain and visual system in the order Araneae have been mostly studied only in a handful of species (e.g., *Cupiennius salei* [41–43]) with broad comparative anatomy studies being rare. Still, these few multi-taxon surveys [31–33] have shown spider neuroanatomy to be a potentially rich and interesting taxonomic character system. Internal anatomy is inherently difficult to study due to the necessity of performing invasive histological techniques and dissections in order to document it. Nevertheless, the last decade has seen a growing number of studies that rely on the utilization of X-ray micro-CT scanning for reconstructing and quantifying various internal organs and tissues in spiders [25–29,44–46]. Still, we consider that the conscientious use of legacy material can be an important route to fill the taxonomic sampling gaps and more easily and promptly understand the evolution of the spider central nervous system.

Our results show that this approach is feasible. Specimens preserved in 70% Et-OH for 70–90 years did not show signs of structural degradation in their internal anatomy, except in cases where proper storage had not been constantly maintained (see Figure 4). Furthermore, the tissue staining by PTA achieved identical results in fresh and archive material, allowing us to visually identify the main brain structures (Figure 3) without the need of critical point or HDMS drying. Thus, our investigation did not interrupt the consistent proper preservation of legacy specimens and provides a template for applying this method to a broader sample of legacy specimens. Our specimens, despite the individual variation seen in the synganglion's 3D model reconstructions, did not show a significant difference in their brain morphology according to our geometric morphometric analyses (Figures 2, A2 and A3). Thus, we conclude that that long-stored specimens can be used to accurately examine spider neuroanatomy, opening the door to the use of collection material for studies of internal anatomy.

Nonetheless, the analysis of the cephalothorax size and brain volume did show a significant difference between both groups. We found an interesting pattern where archived specimens are consistently smaller than the fresh spiders. It was suggested to us that we consider whether this could be attributed to shrinking due to EtOH preservation, but this would require hard and soft parts to shrink proportionately, and also suggests that spider collections (or alcohol preserved specimens in general) might be shrinking over time, neither of which seems realistic. Therefore, we formulated three other non-mutually exclusive explanations to our observations: **(1) Seasonal effects.** Although we tried to minimize as many variables as possible, the *Old* material was sampled in a range of times and places within the Netherlands. These spiders were captured between 1929 and 1948, with most of them being captured between August and November, and two collected in April (although out of season, both specimens were adults). On the other hand, all the *New* spiders were collected in a single sampling event, from a single population in October (near the end of the mating season [47]). **(2) Environmental variables.** Our *New* group was consistently bigger than the legacy spiders, with the largest individuals from the *Old* group not even reaching the mean size of the fresh material. Therefore, we consider that access to food might had played an important role in the growth of the fresh specimens. The food level has been shown to have a strong positive impact on the body size development on the cellar spider *Pholcus phalangioides*, with females investing primarily in weight and size gain when food was abundant [48]. Likewise, the brain volume has shown be very plastic and at least partially dependent on the food supply during development of the jumping spider *Marpissa muscosa* [28]. **(3) Urban Evolution.** Differential adaptations of urban and rural populations have been documented for song patterns in birds [49,50], and colorations and morphologies in snails [51]. Therefore, changes in the body size over decades in response to changes in urban environments could be a viable explanation for our observations. The addition of natural history collections material can be an important resource for assessing the anthropogenic impact in urban-living species [52]. However, a bigger sampling effort from different populations and times, out of the scope of this study, would be required to test these hypotheses.

Finally, we did find some incidental factors that degraded the morphology of the brain and other internal organs (Figure 4). The most conspicuous was tissue shrinking induced by the accidental desiccation of two specimens. In both cases, the specimen's condition was not visible externally but only became evident during the 3D data analysis. The aberrant morphologies were easy to identify and diagnose due to the evident shrinking in muscles and other organs, and the formation of crystals (Figure 4d) in the digestive tract. The other factor was a naturally occurring deformation of the supra- and subesophageal ganglia caused by the abnormal growth of the digestive diverticula (Figure 4c–h). This condition was observed in both groups. In all of those cases, the gross morphology of the supraesophageal ganglion could be identified, segmented, and compared together with the rest of the samples. By contrast, detailed assessment of the brain substructures in the dried samples was futile.

Broad comparative anatomy studies are necessary to elucidate the plasticity and evolution of the brain in a phylogenetic context. We hope that our results encourage the responsible use of legacy material to fill critical taxonomic gaps and illustrate the evolution of the CNS and other internal organs and systems whose comparative anatomy has so far largely remained a mystery.

Author Contributions: Conceptualization, F.A.R.-Q. and J.A.M.; formal analysis, F.A.R.-Q.; funding acquisition, F.A.R.-Q. and J.A.M.; investigation, F.A.R.-Q.; methodology, F.A.R.-Q.; supervision, J.A.M.; visualization, F.A.R.-Q.; writing—original draft, F.A.R.-Q.; writing—review and editing, J.A.M. All authors have read and agreed to the published version of the manuscript.

Funding: This research was funded by a Martin & Temminck Fellowship to FARQ from Naturalis Biodiversity Center.

Institutional Review Board Statement: Not applicable.

Data Availability Statement: The data presented in this study are available on request from the corresponding author. The data are not publicly available due to the size of the original scan files.

Acknowledgments: Thanks to Rob Langelaan and Dirk van der Marel for their valuable assistance obtaining micro-CT scans and for their suggestions on the protocol. Martin Rücklin facilitated the use of Avizo software. Davinia Arguedas helped find and collect the fresh *Araneus diadematus* specimens used in this study. Thanks to Vincent Merckx and the Naturalis Understanding Evolution research group for the support necessary to obtain the scans. Thanks to the two anonymous reviewers for their useful comments and suggestions. Thanks to Naturalis Biodiversity Center for providing support for FARQ through a Martin & Temminck Fellowship. Thanks to the reviewers and the editor for their constructive comments on the manuscript.

Conflicts of Interest: The authors declare no conflict of interest.

Appendix A

Supplementary figures providing additional details of the analyses presented.

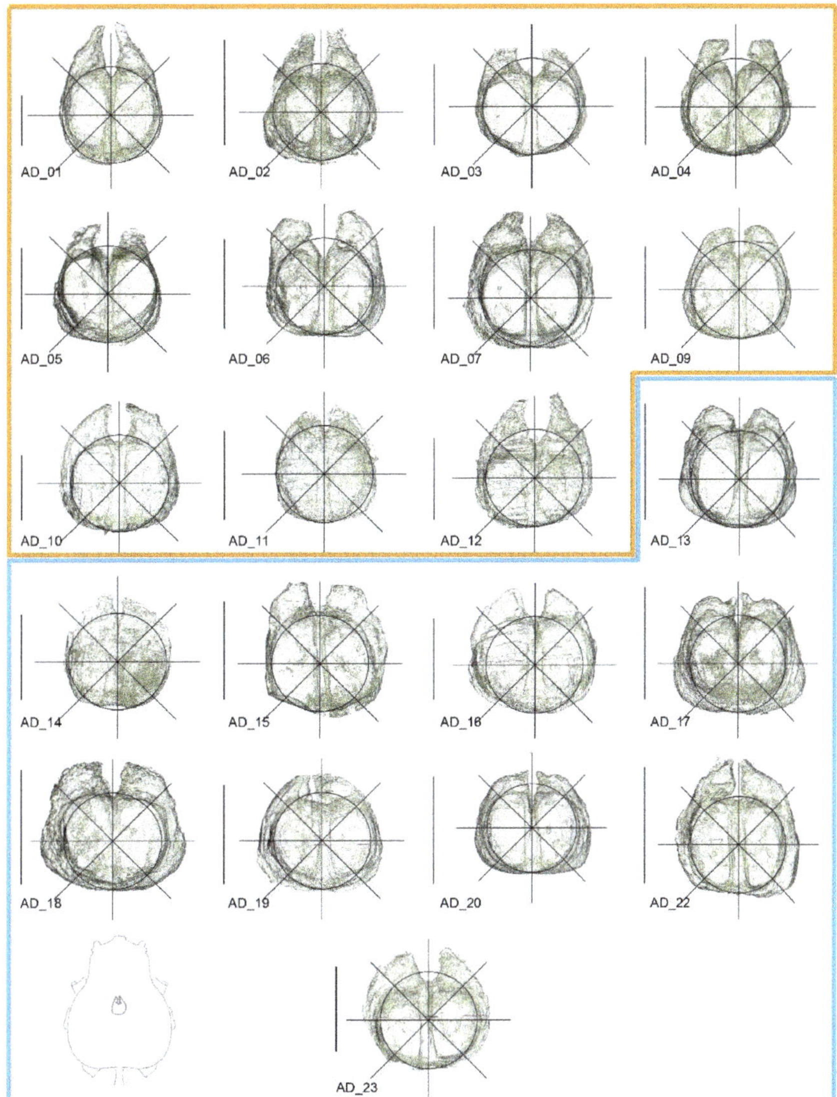

Figure A1. Volume reconstructions of *Araneus diadematus* brains in dorsal view. Fresh material is marked in dark yellow and legacy material in blue. Landmarks were placed on the crossing of the lines and the edge of the reconstruction following the template in Figure 2. Line drawing in the lower left corner shows the brain in relation to the cephalothorax in dorsal view. Scale bars = 0.5 mm.

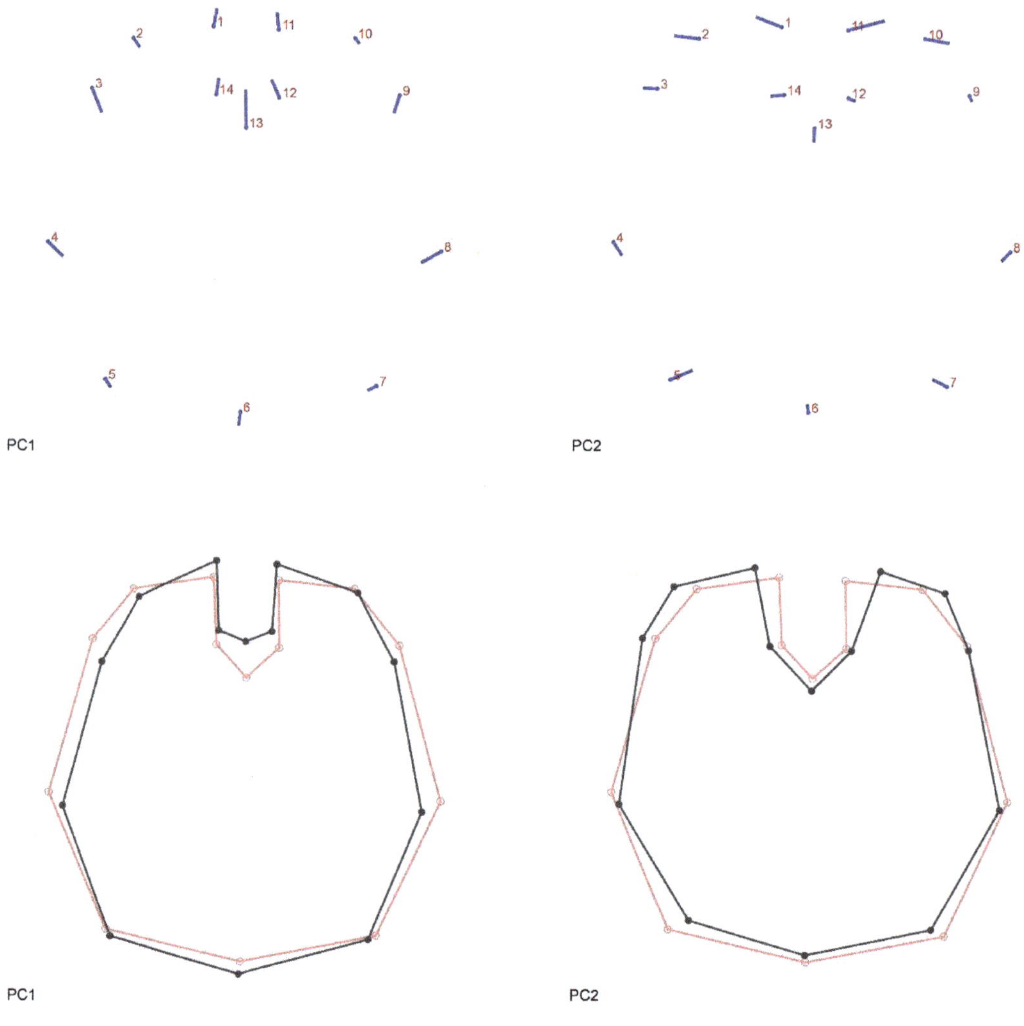

(a)

Figure A2. *Cont.*

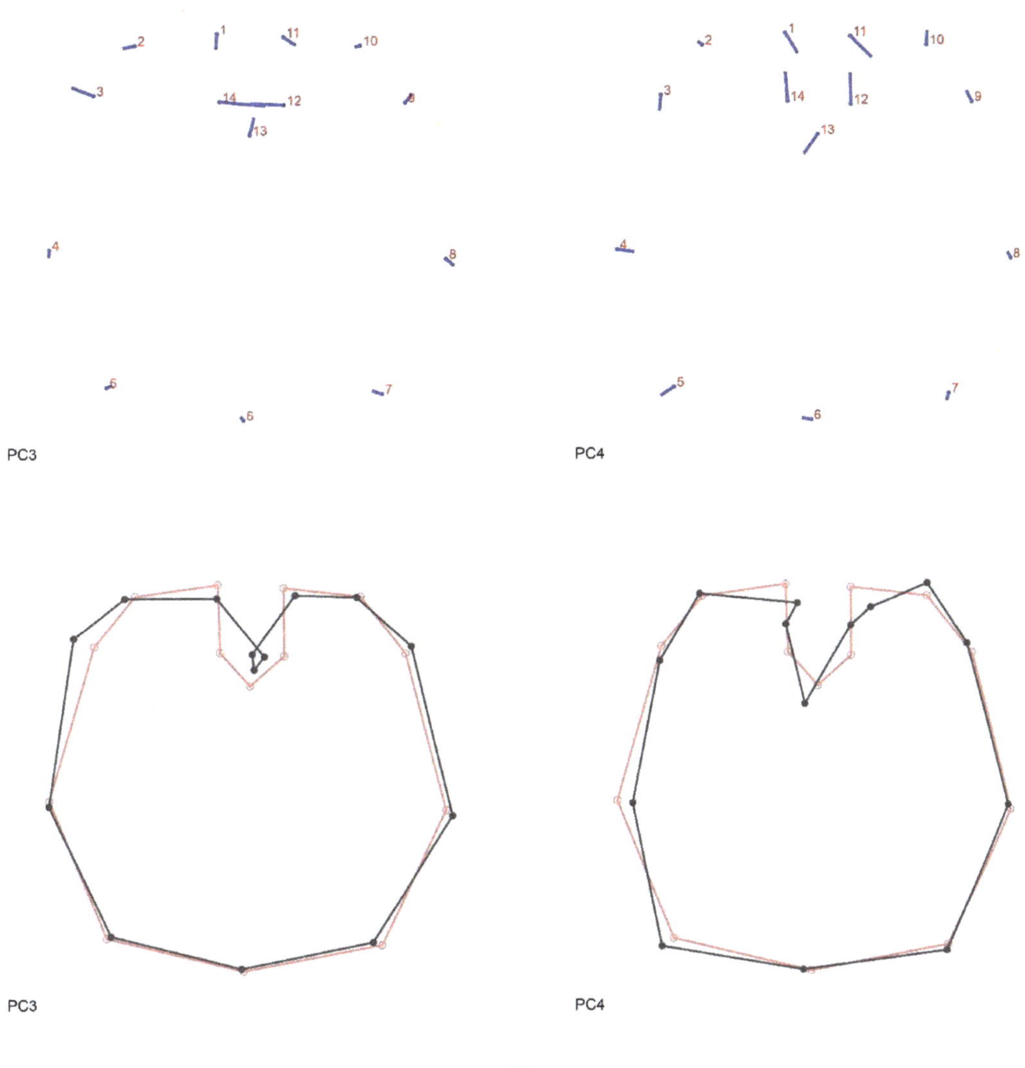

(b)

Figure A2. Principal component analysis shape changes. (**a**) PC1 (31.709%) and PC2 (26.582%); (**b**) PC3 (9.610%) and PC4 (7.106%). Lollipop graphs indicate the direction and magnitude of change per landmark associated with each PC. Wireframes indicate the mean shape (red) and the shape variation (dark grey).

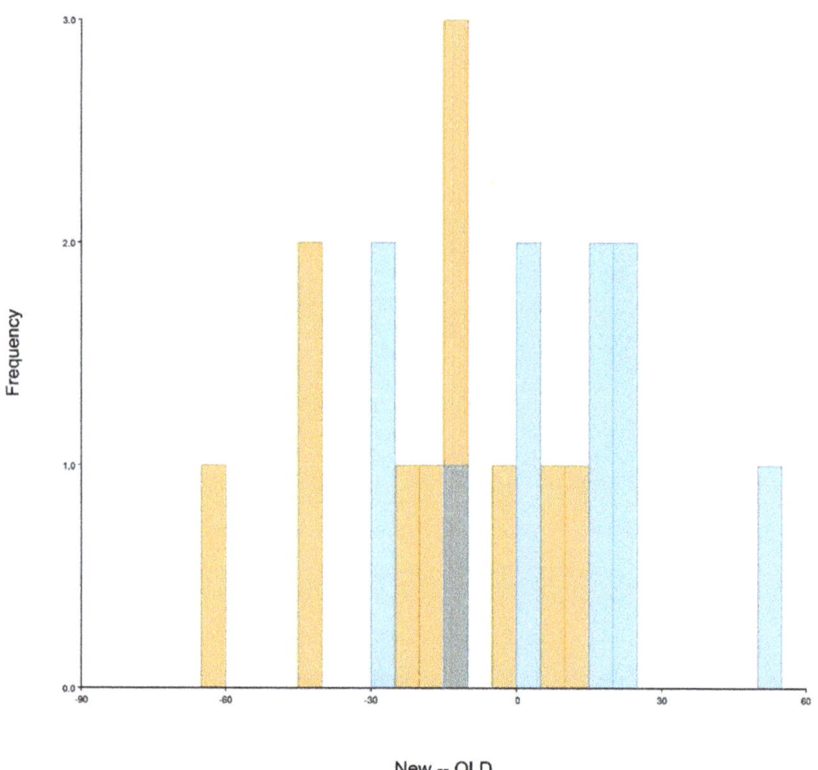

Figure A3. Discriminant analysis of brain shape variation for the *New* and *Old* spider groups. Discriminant scores (frequencies) after a "leave-one-out" cross-validation are shown using histogram bars with the default 24 bins. Overlap between both groups of specimens and $p = 0.7229$ suggest no significant shape difference between the groups. Fresh material (*New*) is marked in dark yellow and legacy material (*Old*) in blue.

References

1. Meineke, E.K.; Davies, T.J.; Daru, B.H.; Davis, C.C. Biological collections for understanding biodiversity in the Anthropocene. *Philos. Trans. R. Soc. B Biol. Sci.* **2019**, *374*, 20170386. [CrossRef]
2. Derkarabetian, S.; Benavides, L.R.; Giribet, G. Sequence capture phylogenomics of historical ethanol-preserved museum specimens: Unlocking the rest of the vault. *Mol. Ecol. Resour.* **2019**, *19*, 1531–1544. [CrossRef] [PubMed]
3. Fernández, R.; Kvist, S.; Lenihan, J.; Giribet, G.; Ziegler, A. Sine systemate chaos? A versatile tool for earthworm taxonomy: Non-destructive imaging of freshly fixed and museum specimens using micro-computed tomography. *PLoS ONE* **2014**, *9*, e96617. [CrossRef] [PubMed]
4. Beaman, R.S.; Cellinese, N. Mass digitization of scientific collections: New opportunities to transform the use of biological specimens and underwrite biodiversity science. *Zookeys* **2012**, *209*, 7–17. [CrossRef] [PubMed]
5. Nelson, G.; Ellis, S. The history and impact of digitization and digital data mobilization on biodiversity research. *Philos. Trans. R. Soc. B Biol. Sci.* **2019**, *374*, 20170391. [CrossRef]
6. Short, A.E.Z.; Dikow, T.; Moreau, C.S. Entomological Collections in the Age of Big Data. *Annu. Rev. Entomol.* **2018**, *63*, 513–530. [CrossRef]
7. Hofreiter, M. Nondestructive DNA extraction from museum specimens. In *Ancient DNA: Methods and Protocols, Methods in Molecular Biology*; Shapiro, B., Hofreiter, M., Eds.; Humana Press: New York, NY, USA, 2012; Volume 840, pp. 93–100. [CrossRef]
8. Gilbert, M.T.P.; Moore, W.; Melchior, L.; Worebey, M. DNA extraction from dry museum beetles without conferring external morphological damage. *PLoS ONE* **2007**, *2*, e272. [CrossRef]
9. Castalanelli, M.A.; Severtson, D.L.; Brumley, C.J.; Szito, A.; Foottit, R.G.; Grimm, M.; Munyard, K.; Groth, D.M. A rapid non-destructive DNA extraction method for insects and other arthropods. *J. Asia-Pac. Entomol.* **2010**, *13*, 243–248. [CrossRef]

10. Miller, J.A.; Beentjes, K.K.; van Helsdingen, P.; Ijland, S. Which specimens from a museum collection will yield DNA barcodes? A time series study of spiders in alcohol. *Zookeys* **2013**, *365*, 245–261. [CrossRef]
11. Vink, C.J.; Thomas, S.M.; Paquin, P.; Hayashi, C.Y.; Hedin, M. The effects of preservatives and temperatures on arachnid DNA. *Invertebr. Syst.* **2005**, *19*, 99–104. [CrossRef]
12. Marquina, D.; Buczek, M.; Ronquist, F.; Lukasik, P. The effect of ethanol concentration on the morphological and molecular preservation of insects for biodiversity studies. *PeerJ* **2021**, *9*, e10799. [CrossRef] [PubMed]
13. King, J.R.; Porter, S.D. Recommendations on the use of alcohols for preservation of ant specimens (Hymenoptera, Formicidae). *Insectes Soc.* **2004**, *51*, 197–202. [CrossRef]
14. Sombke, A.; Lipke, E.; Michalik, P.; Uhl, G.; Harzsch, S. Potential and limitations of X-ray micro-computed tomography in arthropod neuroanatomy: A methodological and comparative survey. *J. Comp. Neurol.* **2015**, *523*, 1281–1295. [CrossRef] [PubMed]
15. Weisbecker, V. Distortion in formalin-fixed brains: Using geometric morphometrics to quantify the worst-case scenario in mice. *Brain Struct. Funct.* **2012**, *217*, 677–685. [CrossRef]
16. Ziegler, A.; Menze, B.H. Accelerated acquisition, visualization, and analysis of zoo-anatomical data. In *Computation for Humanity*; Zander, J., Mosterman, J., Eds.; CRC Press: Boca Raton, FL, USA, 2013; pp. 233–261.
17. Keklikoglou, K.; Faulwetter, S.; Chatzinikolaou, E.; Wils, P.; Brecko, J.; Kvaček, J.; Metscher, B.; Arvanitidis, C. Micro-computed tomography for natural history specimens: A handbook of best practice protocols. *Eur. J. Taxon.* **2019**, *522*, 1–55. [CrossRef]
18. Faulwetter, S.; Dailianis, T.; Vasileiadou, A.; Arvanitidis, C. Investigation of contrast enhancing techniques for the application of Micro-CT in marine biodiversity studies. *Microsc.Anal.* **2013**, *27*, S4–S7.
19. Sakurai, Y.; Ikeda, Y. Development of a contrast-enhanced micro computed tomography protocol for the oval squid (*Sepioteuthis lessoniana*) brain. *Microsc. Res. Tech.* **2019**, *82*, 1941–1952. [CrossRef]
20. Swart, P.; Wicklein, M.; Sykes, D.; Ahmed, F.; Krapp, H.G. A quantitative comparison of micro-CT preparations in Dipteran flies. *Sci. Rep.* **2016**, *6*, 39380. [CrossRef]
21. Alba-Tercedor, J.; Hunter, W.B.; Alba-Alejandre, I. Using micro-computed tomography to reveal the anatomy of adult *Diaphorina citri* Kuwayama (Insecta: Hemiptera, Liviidae) and how it pierces and feeds within a citrus leaf. *Sci. Rep.* **2021**, *11*, 1358. [CrossRef]
22. Alba-Alejandre, I.; Hunter, W.B.; Alba-Tercedor, J. Micro-CT study of male genitalia and reproductive system of the Asian citrus psyllid, *Diaphorina citri* Kuwayama, 1908 (Insecta: Hemiptera, Liviidae). *PLoS ONE* **2018**, *13*, e0202234. [CrossRef]
23. Castejón, D.; Alba-Tercedor, J.; Rotllant, G.; Ribes, E.; Durfort, M.; Guerao, G. Micro-computed tomography and histology to explore internal morphology in decapod larvae. *Sci. Rep.* **2018**, *8*, 14399. [CrossRef] [PubMed]
24. van de Kamp, T.; Schwermann, A.H.; dos Santos Rolo, T.; Lösel, P.D.; Engler, T.; Etter, W.; Faragó, T.; Göttlicher, J.; Heuveline, V.; Kopmann, A.; et al. Parasitoid biology preserved in mineralized fossils. *Nat. Commun.* **2018**, *9*, 3325. [CrossRef] [PubMed]
25. Rivera-Quiroz, F.A.; Miller, J.A. Micro-CT visualization of the CNS: Performance of different contrast-enhancing techniques for documenting the spider brain. *J. Comp. Neurol.* **2021**. under review.
26. Steinhoff, P.O.M.; Sombke, A.; Liedtke, J.; Schneider, J.M.; Harzsch, S.; Uhl, G. The synganglion of the jumping spider *Marpissa muscosa* (Arachnida: Salticidae): Insights from histology, immunohistochemistry and microCT analysis. *Arthropod Struct. Dev.* **2017**, *46*, 156–170. [CrossRef]
27. Stafstrom, J.A.; Michalik, P.; Hebets, E.A. Sensory system plasticity in a visually specialized, nocturnal spider. *Sci. Rep.* **2017**, *7*, 46627. [CrossRef]
28. Steinhoff, P.O.M.; Liedtke, J.; Sombke, A.; Schneider, J.M.; Uhl, G. Early environmental conditions affect the volume of higher-order brain centers in a jumping spider. *J. Zool.* **2018**, *304*, 182–192. [CrossRef]
29. Steinhoff, P.O.M.; Uhl, G.; Harzsch, S.; Sombke, A. Visual pathways in the brain of the jumping spider *Marpissa muscosa*. *J. Comp. Neurol.* **2020**, *528*, 1883–1902. [CrossRef]
30. Lin, S.-W.; Lopardo, L.; Uhl, G. Diversification through sexual selection on gustatorial courtship traits in dwarf spiders. *bioRxiv* **2021**. [CrossRef]
31. Saint-Remy, G. *Contribution a L'étude du Cerveau Chez les Arthropodes Trachéates*; Faculté des Sciences de Paris: Paris, France, 1890.
32. Long, S.M. *Spider Brain Morphology & Behavior*; University of Massachusetts: Amherst, MA, USA, 2016.
33. Long, S.M. Variations on a theme: Morphological variation in the secondary eye visual pathway across the order of Araneae. *J. Comp. Neurol.* **2021**, *529*, 259–280. [CrossRef]
34. Hill, D.E. The Structure of the Central Nervous System of Jumping Spiders of the Genus *Phidippus* (Araneae: Salticidae). Master's Thesis, Oregon State University, Corvallis, OR, USA, 2006.
35. Wickham, H. *ggplot2 Elegant Graphics for Data Analysis*, 2nd ed.; Springer International Publishing: Cham, Switzerland, 2016. [CrossRef]
36. Wickham, H.; François, R.; Henry, L.; Müller, K. dplyr: A Grammar of Data Manipulation. 2021. Available online: https://cran.r-project.org/package=dplyr (accessed on 27 September 2021).
37. Rohlf, F.J. *tpsUtil, v1.81*; Department of Anthropology, and Ecology & Evolution, Stony Brook University: New York, NY, USA, 2021.
38. Rohlf, F.J. *tpsDig2, v2.32*; Department of Anthropology, and Ecology & Evolution, Stony Brook University: New York, NY, USA, 2021.
39. Klingenberg, C.P. MorphoJ: An integrated software package for geometric morphometrics. *Mol. Ecol. Resour.* **2011**, *11*, 353–357. [CrossRef]

40. Theska, T.; Sieriebriennikov, B.; Wighard, S.S.; Werner, M.S.; Sommer, R.J. Geometric morphometrics of microscopic animals as exemplified by model nematodes. *Nat. Protoc.* **2020**, *15*, 2611–2644. [CrossRef]

41. Loesel, R.; Seyfarth, E.A.; Bräunig, P.; Agricola, H.J. Neuroarchitecture of the arcuate body in the brain of the spider *Cupiennius salei* (Araneae, Chelicerata) revealed by allatostatin-, proctolin-, and CCAP-immunocytochemistry and its evolutionary implications. *Arthropod Struct. Dev.* **2011**, *40*, 210–220. [CrossRef]

42. Babu, K.S.; Barth, F.G. Neuroanatomy of the central nervous system of the wandering spider, *Cupiennius salei* (Arachnida, Araneida). *Zoomorphology* **1984**, *104*, 344–359. [CrossRef]

43. Schmid, A.; Becherer, C. Distribution of histamine in the CNS of different spiders. *Microsc. Res. Tech.* **1999**, *44*, 81–93. [CrossRef]

44. Lipke, E.; Hammel, J.U.; Michalik, P. First evidence of neurons in the male copulatory organ of a spider (Arachnida, Araneae). *Biol. Lett.* **2015**, *11*, 20150465. [CrossRef] [PubMed]

45. Dederichs, T.M.; Müller, C.H.G.; Sentenská, L.; Lipke, E.; Uhl, G.; Michalik, P. The innervation of the male copulatory organ of spiders (Araneae)—A comparative analysis. *Front. Zool.* **2019**, *16*, 39. [CrossRef] [PubMed]

46. Rivera-Quiroz, F.A.; Petcharad, B.; Miller, J.A. First records and three new species of the family Symphytognathidae (Arachnida, Araneae) from Thailand, and the circumscription of the genus *Crassignatha* Wunderlich, 1995. *Zookeys* **2021**, *1012*, 21–53. [CrossRef]

47. iNaturalist: Cross Orbweaver (*Araneus diadematus*). Available online: https://www.inaturalist.org/taxa/52628-Araneus-diadematus (accessed on 27 September 2021).

48. Uhl, G.; Schmitt, S.; Schäfer, M.A.; Blanckenhorn, W. Food and sex-specific growth strategies in a spider. *Evol. Ecol. Res.* **2004**, *6*, 523–540. [CrossRef]

49. Dowling, J.L.; Luther, D.A.; Marra, P.P. Comparative effects of urban development and anthropogenic noise on bird songs. *Behav. Ecol.* **2012**, *23*, 201–209. [CrossRef]

50. Narango, D.L.; Rodewald, A.D. Urban-associated drivers of song variation along a rural-urban gradient. *Behav. Ecol.* **2016**, *27*, 608–616. [CrossRef]

51. Kerstes, N.A.G.; Breeschoten, T.; Kalkman, V.J.; Schilthuizen, M. Snail shell colour evolution in urban heat islands detected via citizen science. *Commun. Biol.* **2019**, *2*, 264. [CrossRef] [PubMed]

52. Shultz, A.J.; Adams, B.J.; Bell, K.C.; Ludt, W.B.; Pauly, G.B.; Vendetti, J.E. Natural history collections are critical resources for contemporary and future studies of urban evolution. *Evol. Appl.* **2021**, *14*, 233–247. [CrossRef] [PubMed]

Article

Incorporating Topological and Age Uncertainty into Event-Based Biogeography of Sand Spiders Supports Paleo-Islands in Galapagos and Ancient Connections among Neotropical Dry Forests

Ivan L. F. Magalhaes [1,2,*], Adalberto J. Santos [2] and Martín J. Ramírez [1]

[1] División Aracnología, Museo Argentino de Ciencias Naturales "Bernardino Rivadavia"—CONICET, Av. Ángel Gallardo 470, C1405DJR Buenos Aires, Argentina; ramirez@macn.gov.ar

[2] Departamento de Zoologia, Instituto de Ciências Biológicas, Universidade Federal de Minas Gerais, Av. Antônio Carlos 6627, 31270-901 Belo Horizonte, MG, Brazil; oxyopes@yahoo.com

* Correspondence: magalhaes@macn.gov.ar

Abstract: Event-based biogeographic methods, such as dispersal-extinction-cladogenesis, have become increasingly popular for attempting to reconstruct the biogeographic history of organisms. Such methods employ distributional data of sampled species and a dated phylogenetic tree to estimate ancestral distribution ranges. Because the input tree is often a single consensus tree, uncertainty in topology and age estimates are rarely accounted for, even when they may affect the outcome of biogeographic estimates. Even when such uncertainties are taken into account for estimates of ancestral ranges, they are usually ignored when researchers compare competing biogeographic hypotheses. We explore the effect of incorporating this uncertainty in a biogeographic analysis of the 21 species of sand spiders (Sicariidae: *Sicarius*) from Neotropical xeric biomes, based on a total-evidence phylogeny including a complete sampling of the genus. Using a custom R script, we account for uncertainty in ages and topology by estimating ancestral ranges over a sample of trees from the posterior distribution of a Bayesian analysis, and for uncertainty in biogeographic estimates by using stochastic maps. This approach allows for counting biogeographic events such as dispersal among areas, counting lineages through time per area, and testing biogeographic hypotheses, while not overestimating the confidence in a single topology. Including uncertainty in ages indicates that *Sicarius* dispersed to the Galapagos Islands when the archipelago was formed by paleo-islands that are now submerged; model comparison strongly favors a scenario where dispersal took place before the current islands emerged. We also investigated past connections among currently disjunct Neotropical dry forests; failing to account for topological uncertainty underestimates possible connections among the Caatinga and Andean dry forests in favor of connections among Caatinga and Caribbean + Mesoamerican dry forests. Additionally, we find that biogeographic models including a founder-event speciation parameter ("+J") are more prone to suffer from the overconfidence effects of estimating ancestral ranges using a single topology. This effect is alleviated by incorporating topological and age uncertainty while estimating stochastic maps, increasing the similarity in the inference of biogeographic events between models with or without a founder-event speciation parameter. We argue that incorporating phylogenetic uncertainty in biogeographic hypothesis-testing is valuable and should be a commonplace approach in the presence of rogue taxa or wide confidence intervals in age estimates, and especially when using models including founder-event speciation.

Keywords: BioGeoBEARS; Caatinga; dispersal; Galapagos; Neotropical; speciation; spiders; tropical dry forests; vicariance

Citation: Magalhaes, I.L.F.; Santos, A.J.; Ramírez, M.J. Incorporating Topological and Age Uncertainty into Event-Based Biogeography of Sand Spiders Supports Paleo-Islands in Galapagos and Ancient Connections among Neotropical Dry Forests. *Diversity* **2021**, *13*, 418. https://doi.org/10.3390/d13090418

Academic Editor: Matjaž Kuntner

Received: 17 July 2021
Accepted: 28 August 2021
Published: 31 August 2021

Publisher's Note: MDPI stays neutral with regard to jurisdictional claims in published maps and institutional affiliations.

1. Introduction

The reconstruction of the biogeographic history of organisms is one of the main aims of systematic biology. Based on a known phylogeny, researchers may attempt to glimpse into the ancestral area where a particular clade originated [1], infer the number and direction of dispersal events [2], or estimate the number of vicariant events along the evolutionary history of the group [3]. More often than not, the geographic distribution of a group is known only from its extant species, occasionally accompanied by a few fossil forms. This represents but a fraction of the diversity of a group throughout its history, and we cannot directly assess the distribution range of unknown extinct species. Thus, biogeographers interested in such questions must resort to methods that attempt to estimate past distribution ranges from the data available from known species.

Initially, such estimates of ancestral geographical ranges relied on algorithms such as Fitch or Camin-Sokal parsimony optimization [1,4]. This approach treats distribution ranges as discrete characters and optimizes them as such along the phylogenies, but has some drawbacks. For instance, only tips of the phylogeny may be 'polymorphic' and occur in more than one area simultaneously, and important biogeographic processes such as vicariance are not modeled at all. This changed with the advent of event-based biogeography methods, pioneered by dispersal-vicariance analysis (DIVA) [3]. Such methods attempt to explain the current distribution of organisms by modeling biogeographic events such as dispersal (range expansion into a new area), range contractions (extinction of a species in a particular area) and allopatry (allopatric speciation leading to each descendant inheriting only part of the range of the ancestor species). DIVA brought a fundamental advance with respect to previous methods: changes in a lineage's geographic distribution may happen as anagenetic events (dispersal or extinction along branches) or cladogenetic events (allopatry at nodes). Furthermore, the states in such estimates are geographic ranges, which may consist of one or more areas; thus, both tips and ancestors may be 'polymorphic'. By using a parsimony framework, DIVA assigns costs to dispersal and extinction events, and treats allopatry as the null expectation for explaining biogeographic history. A similar rationale was used later to elaborate a likelihood-based approach to ancestral range estimation named dispersal-extinction-cladogenesis analysis (DEC) [5,6]. DIVA's parsimony-based optimization is agnostic to branch lengths. On the other hand, DEC models anagenetic events as a time-continuous function that takes branch lengths into account; thus, longer branches are more likely to contain anagenetic events such as dispersal or extinction. More importantly, DEC can accommodate dated trees, and thus prior information on geological history can be utilized for explicit testing of biogeographic hypothesis [7]. Modifications of DIVA and DEC have been implemented and further elaborated in software packages for biogeography, such as RASP [8] and BioGeoBEARS [9,10]. The latter has become particularly popular due to its flexible implementation of biogeographic models allowing for different types of cladogenetic events, such as allopatry, subset sympatry, and founder-event speciation [10,11]; in addition, all models are implemented in a likelihood framework, allowing direct model comparison. It also allows incorporating prior information, such as dispersal probabilities among areas, to put biogeographic hypotheses to test in an explicit manner (e.g., [12]).

A unifying feature of all methods discussed above is that they rely on the knowledge of the geographic distribution of each of the taxa, and of the phylogenetic tree describing their interrelationships. As in any comparative method, estimates of ancestral ranges are only as reliable as the primary data underlying it. The knowledge of the geographic distribution of tips may be affected by the Linnean and Wallacean shortfalls [13], i.e., gaps in the data about existing species and their distributions. However, it is arguable that estimates of ancestral ranges are usually carried out by systematists that specialize in a particular taxon, who strive to include all or most known species and distribution records in their sampling, so as to mitigate any negative effects of such shortfalls. On the other hand, the imperfect knowledge of the underlying phylogenetic tree can be potentially more problematic. In recent years, the accumulation of genomic-scale studies has shown that the phylogenetic relationships of some clades are elusive even with massive quantities of data due to e.g., incomplete lineage sorting and/or very short internodes (e.g., [14,15]), and some portions of such trees are shrouded in topological uncertainty. There is uncertainty not only regarding the topology, but also in the estimates of divergence times. This uncertainty in age estimates is also resistant to the accumulation of genetic markers (see [16]). Rates of molecular evolution rarely conform to a strict molecular clock, and branch lengths estimated from molecular sequences are a product of substitution rate and time, such that each of these two parameters are individually non-identifiable [17]. In addition, estimation of dated trees builds upon several assumptions, some of which may not be met by empirical datasets [18], and require fossil or other external calibrations, whose selection and justification is not free of difficulties [19,20]. In this scenario, analyses that depend on phylogenetic trees might benefit from accounting for any uncertainties regarding their topologies and/or ages [21]. In this paper, we refer to "phylogenetic uncertainty" as any uncertainty regarding the topology and/or node ages.

Efforts have been made to incorporate such phylogenetic uncertainty into ancestral range estimation. Nylander et al. [22] used multiple trees from the posterior distribution of a Bayesian analysis to estimate ancestral ranges, and thus infer the biogeographic history of a clade of birds whose phylogeny had proven difficult to solve; they called this approach Bayes-DIVA. Similar pipelines have been incorporated into RASP (S-DIVA; [23]) and BioGeoBEARS (using the *run_bears_optim_on_multiple_trees* function). These functions summarize estimates of ancestral ranges of several trees on a target tree, such as a majority-rule consensus (as output by e.g., MrBayes) or a maximum clade credibility tree (MCC tree; as output by e.g., BEAST) and are routinely employed by researchers interested in incorporating phylogenetic uncertainty (e.g., [24,25]). However, the results are summarized in a single tree, and the underlying variability in the estimates is difficult to grasp. A more elaborate approach is the joint estimation of the phylogeny and ancestral ranges (e.g., [26,27]), although it can be computationally intensive in complex systems. Furthermore, the individual underlying trees usually are not taken into account when comparing competing biogeographic hypotheses regarding particular events. We argue that it is important to understand the effect of analyzing multiple trees during hypothesis-testing. This is especially relevant in time-stratified analyses, where different time periods can have different biogeographic possibilities (e.g., different availability of areas or dispersal probabilities), as confidence intervals of age estimates may cross boundaries of time slices.

In addition to phylogenetic uncertainty, there is the uncertainty associated with stochastic time-continuous models such as DEC. Because of this, estimates for ancestral nodes might include several possible states, each with its own probability. This hampers counting biogeographic events or estimating their ages. Dupin et al. [2] solved this by introducing biogeographic stochastic mapping (BSM), a method to count biogeographic events while accounting for uncertainty in ancestral range estimation. Several replicates are run, and in each one the states at each node are resolved by taking into account the probabilities of each state. This allows, for instance, counting the number of dispersal events among areas, or estimating the time when such transitions took place. This approach is elegant, but it is usually employed on estimates based on a single tree, and thus incorporates uncertainty on ancestral range estimates conditional on a single topology. If there is topological uncertainty leading to different ancestral range estimates, this approach will underestimate the uncertainty in the biogeographic reconstruction (Figure 1). Furthermore, using a single tree as input ignores the confidence intervals in age estimates, which are usually large. We argue that it would be productive to run stochastic maps over a sample of trees to incorporate both phylogenetic and stochastic uncertainty simultaneously. This approach has been successfully implemented to estimate dispersal rates among areas through time in some studies (e.g., [27–29]).

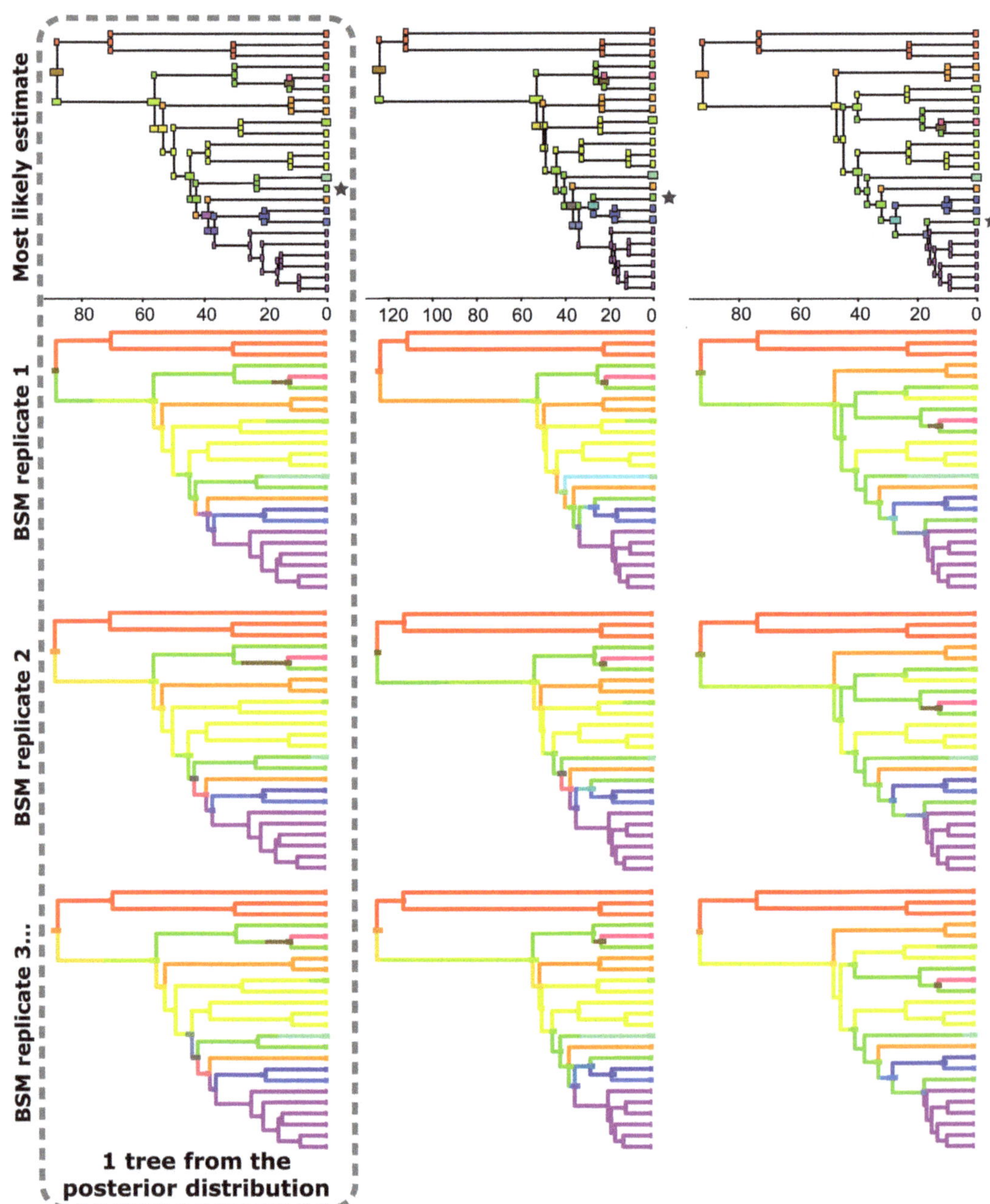

Figure 1. Biogeographic stochastic maps are insufficient to account for uncertainty in ancestral range estimates. Each column represents a single tree from the posterior distribution of the Bayesian analysis of our dataset, with the most likely estimated states in the top row and different stochastic maps in the bottom rows. While estimates in each stochastic map are different, the most remarkable differences are found among trees with different topologies. Note the tip marked with a grey star, which is a rogue taxon (*Sicarius andinus*). As it shifts its position across the different trees of the stationary phase of the chain, it leads to substantially different ancestral range estimates in each tree.

To illustrate our argument, we explore the effect of uncertainty in the biogeographic history of Neotropical sand spiders (*Sicarius*). These spiders represent an ideal system for this test because they are moderately diverse (21 species) and all species have been included in a dated total-evidence phylogeny [30]. The genus has a disjunct distribution, and each species is restricted to one or two arid areas surrounded by mesic habitats; phylogeographic and phylogenetic patterns suggest they are very poor dispersers [31–33]. Thus, their areas of distribution are clearly delimited and could be interpreted as "islands" of dry biomes inserted in a matrix of unsuitable humid habitats. Because of their distribution and moderate diversity, most of the biogeographic transitions should be straightforward to interpret, thus allowing us to easily measure the effects of phylogenetic uncertainty in biogeographic estimates.

Specifically, we focus on two particularly pressing questions. The first is the timing of arrival of *Sicarius* in the Galapagos archipelago. The Galapagos are currently inhabited by a single species of sand spider, *Sicarius utriformis* (Butler), which is sister to *S. peruensis* (Keyserling) from the Peruvian coastal deserts [30,33]. Although the oldest emerged islands are ~3.5 million years (Myr) old [34], geological evidence suggests that the archipelago existed for at least 14.5 Myr, when it was composed of paleo-islands that are now submerged [35,36]. The age of divergence between *S. utriformis* and *S. peruensis* has a 95% confidence interval between 1.2 and 22.2 Myr (median 9.7; [30]), and thus it is possible that this pair of species split before the current islands were formed, but during the time the archipelago was formed by paleo-islands. This makes this system ideal to test the effect of the uncertainty of age estimates in biogeographic inference.

The second question is the estimation of the ancestral distribution range of a clade endemic to the Brazilian Caatinga. This area is one of the largest and most diverse tropical dry forests in the world [37], and six *Sicarius* species inhabit the region [33]. These six species form a well-supported monophyletic group [30], suggesting a single dispersal into this area. In this case, from which area did the ancestor of this clade come? The American tropical dry forests and other xeric biomes such as deserts and scrublands currently have a disjunct distribution, but the similarity in their biota suggests they have been connected in the past (see [37]). Based on plant distributions, some argued that the Caatinga could have been connected to dry forests in Bolivia and Argentina, or to dry forests in the Caribbean coast of northern South America [38,39]; these connections would have taken place by expansion of dry forests over areas that now are covered by mesic biomes. Connections between the Caatinga and the Caribbean dry forests imply a northern route passing through present-day Amazon (a rainforest); alternatively, the Caatinga could have been connected with southern formations, such as the Monte or the Chiquitano dry forests, passing through present-day Cerrado (a savannah). We thus aim to identify the most likely route for the occupation of the Caatinga by sand spiders. However, this is hampered by the fact that one species, *Sicarius andinus* Magalhaes et al. from the Peruvian Andes, is a rogue taxon in the phylogeny and does not have a well-resolved phylogenetic position [30]. Different positions of this species may yield different biogeographic estimates for the Caatinga clade, and thus we must take this into account.

In this paper, we test the effect of taking phylogenetic uncertainty into account to address the two biogeographic questions mentioned above. We combine processing of several trees of the stationary phase of the Markov chains of a Bayesian analysis with biogeographic stochastic maps for each tree, so that both phylogenetic and biogeographic uncertainties are considered simultaneously. Specifically, we test (1) whether data from sand spiders support dispersal to Galapagos in the last 3.5 Myr (age of oldest emerged island), in the last 14.5 Myr (age of oldest known submerged paleo-island) or an unconstrained model (representing the possibility of older, yet-undetected paleo-islands), and (2) whether the Caatinga was connected to northern (Caribbean, Mesoamerican or Andean dry forests) or southern (Chiquitano dry forests or Monte) biomes, as well as the age of such connections. We anticipate that analyzing a sample of trees, instead of a single target tree, provides invaluable insights for testing competing biogeographic hypotheses. Finally, we provide scripts for R [40] to replicate the analyses described below, in the hope they will be useful for further studies.

2. Material and Methods

2.1. Summaries of Biogeographic Inferences in Face of Uncertainty

We prepared an R script to (1) sample *n* trees randomly from BEAST output files and prune them to the taxa of interest, (2) estimate ancestral ranges using BioGeoBEARS for each of these trees, (3) run biogeographic stochastic maps for each of these estimates, and finally (4) parse the results and summarize them. The summaries include: (1) maximum likelihood parameter estimates, log-likelihoods and AICc scores for ancestral range estimates of each tree, (2) tables containing the transitions between geographic ranges for each stochastic map of each of the *n* sampled trees, along with the age of such transitions; (3) a graphic of lineages through time per area averaged over all trees and stochastic maps; (4) tables with all possible geographic ranges in both rows and columns, and average counts of how many times a transition between a particular pair of ranges took place; such tables are broken down by each type of transition (dispersal, extinction, allopatry, etc.) and summarized in a table containing all types of transitions; and (5) a summary of the most common biogeographic transitions (as the mean number of transitions per BSM replicate). The script (Online Supplementary File S1) uses functions from the packages phytools [41], ape [42], sjmisc [43] and BioGeoBEARS [9] and has been written as to be easily adapted to most datasets. The necessary input files are the same as those used in BioGeoBEARS, except that users may provide multiple trees instead of a single target tree.

2.2. Model Selection

BioGeoBEARS implements three different biogeographic models that differ mainly in the events that may take place during cladogenesis when the ancestor has a widespread range (i.e., its range consists of two or more areas; see [10] for a summary). DEC has an identical implementation to the model described in [6] and allows narrow vicariance (one descendant inherits exactly a single area, the other inherits the rest of the range) or subset sympatry (one descendant inherits exactly a single area, while the other inherits the whole distribution range of the ancestor). DIVA-like allows the same cladogenetic events as the original implementation of DIVA [3]: narrow allopatry (as in DEC) and wide allopatry (each descendant may inherit two or more areas from the ancestor). BAYAREA-like does not allow changes in distribution ranges during cladogenesis, and each descendant inherits exactly the same range as the ancestor. Each of these three models can be modified by the addition of a "jump-dispersal" free parameter ("+J") that allows for founder-event speciation during cladogenesis, i.e., one of the descendants occupies a single area that is not part of the range of the ancestor, while the other descendant inherits the same range as the ancestor [11]. Thus, models including founder-event speciation are unique in that they allow dispersal to take place during cladogenetic events. Please note that founder-event speciation is also called jump dispersal, which leads some researchers to interpret it as long-distance dispersal; it should be noted that this notion is incorrect, as this parameter merely models dispersal taking place instantaneously during cladogenetic events (hence the "jump"). It is affected by dispersal matrices in the same way as anagenetic dispersal is, and thus the distances among areas are not more important for jump dispersal than they are to other parameters.

Biogeographic history may be estimated under each of these six models and the fit of the data to each of them can be compared by using the Akaike information criterion (AIC; [44]) and Akaike weights (AICw; [45]). This procedure has been used to guide the selection of models that best explain the data and for testing biogeographic hypotheses [10,11]. We here estimate the fit of the data to six different models (DEC, DIVA-like, BAYAREA-like, DEC + J, DIVA-like + J, BAYAREA-like + J) for initial exploration of the behavior of the estimates. The script used for model selection can be found as Online Supplementary File S5.

We wanted to investigate whether the choice of a particular biogeographic model interacts in any way with the decision to run analyses over a sample of posterior trees. Thus, we ran the aforementioned script under two models with (DIVA-like + J and BAYAREA-like + J) and without (DIVA-like and DEC) founder-event speciation. The choice of these four models was made to include those that are a better fit to the data (DIVA-like and the +J variant), a model that is frequently used in empirical studies (DEC) and a model excluding the possibility of allopatry, and thus more similar to simple parsimony reconstructions (BAYAREA-like + J). Each of these four models was used in both unconstrained and time-stratified analysis (see below), and using either the MCC tree or 100 posterior trees as source. In this latter case, to make these analyses directly comparable, the same sample of 100 trees was used for all runs. The combination of four biogeographic models, three time-stratification scenarios and two sources of trees resulted in 24 different runs (see Online Supplementary Figure S10). For each individual tree, 100 biogeographic stochastic maps were estimated to count biogeographic events and estimate the number of lineages through time by area (see below).

2.3. Phylogeny and Distribution Data

To reconstruct the biogeographic history of *Sicarius*, we used a recently published phylogeny estimated using morphology and DNA sequences and dated using a combination of fossil calibrations and substitution rates for the histone H3, subunit A gene [30]. To understand the effect of incorporating uncertainty into biogeographic inference, analyses were run both on the maximum clade credibility (MCC) tree, and on a sample of 100 trees randomly drawn from the posterior distribution of the analysis, after removing the first 10% samples as burn-in (see above). *Hexophthalma* is the African sister group of *Sicarius* and was included in the analyses; the remaining terminals (*Loxosceles* and non-sicariid outgroups) were pruned from the trees. Species geographic ranges and trees can be found as Online Supplementary Files S2–S4.

The distribution of each species has been fully mapped in recent taxonomic publications on the genus including ca. 1800 adult specimens from natural history collections and recent field expeditions [33,46]. We classified species distribution in ten areas: southern Africa deserts and xeric scrublands (F), Argentinean Monte (O), Atacama Desert and neighboring Chilean xeric scrublands (T), Sechura desert in the Peruvian coast (S), Andean dry forests (D), Chiquitano dry forests in Bolivia (C), Mesoamerican dry forests (M), dry forests in the Caribbean coast of Colombia (B), Caatinga dry forest in Brazil (I), and the Galapagos Islands (G) (Figure 2). Most of these areas are clearly delimited by geographic barriers such as oceans or mountain ranges, or correspond to well-recognized ecoregions or phytogeographic units [37,47].

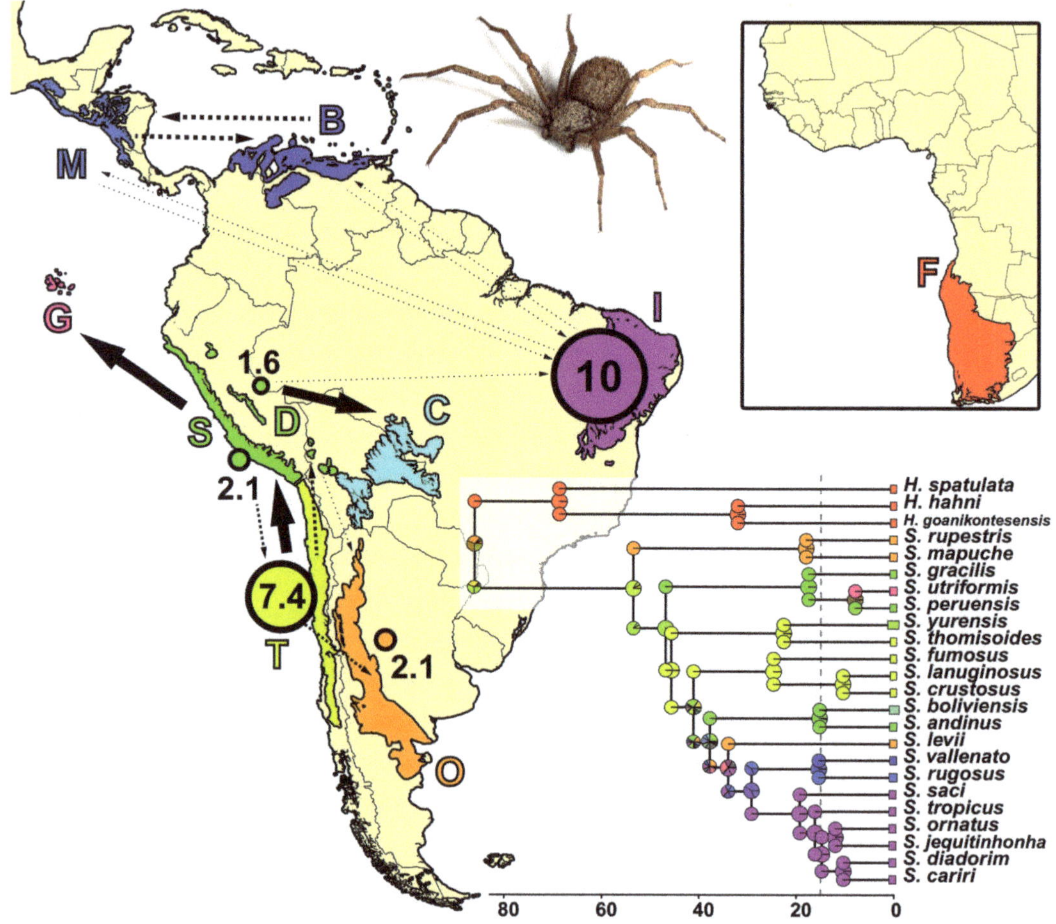

Figure 2. Map depicting the areas inhabited by *Sicarius* and *Hexophthalma*, and the ancestral range estimates under DIVA-like using the maximum clade credibility tree. Solid arrows among areas represent one dispersal event between areas that is robust to topological and biogeographical uncertainty. Dashed lines represent inferred dispersal events that are sensitive to uncertainty; arrow widths are proportional to the frequency at which such dispersals are inferred. Numbered circles indicate the inferred number of within-area speciation in each of the areas. Area abbreviations: B = dry forests in the Caribbean coast of Colombia; F = southern Africa deserts and xeric scrublands; C = Chiquitano dry forests in Bolivia; D =Andean dry forests; G = Galapagos Islands; I = Caatinga dry forest in Brazil; M = Mesoamerican dry forests; O = Argentinean Monte; S = Sechura desert in the Peruvian coast; T = Atacama Desert and neighboring Chilean xeric scrublands. Africa not to scale.

2.4. BioGeoBEARS Parameters and Time Stratification

Sand spiders are poor dispersers [30] and most species are restricted to a single area, with only two species occurring in two areas [33]. For this reason, and to speed up calculations, we restricted the maximum range size to include three areas. Likelihood calculations were carried out with *optimx* [48]. We compared ancestral ranges estimates under three scenarios: (1) unconstrained, allowing dispersal to the islands in any time, and (2 & 3) two different time-stratified scenarios, each with two time slices, where the Galapagos Islands were only available for occupation in the more recent slice. The boundary between the two slices was set to (2) 15 Myr, as geological evidence points to submerged islands that are at least 14.5 Myr old [36], or (3) 3.5 Myr, representing the approximate age of the oldest emerged island [34]. Files for implementing the time-stratified model are available as Online Supplementary Files S6–S8.

3. Results

3.1. Model Selection and Estimates of Ancestral Ranges

Regarding runs on MCC trees, overall DEC and DIVA-like resulted in similar ancestral range estimates among them, as did all models including a +J parameter. Parameter estimates and fit of data under different models are summarized in Online Supplementary Table S1. AIC values indicate that DIVA-like (log-likelihood: −51.29, AIC: 107.16) is the favored model among those not including a founder-speciation free parameter, while DIVA-like + J (log-likelihood: −44.24, AIC: 95.69) is favored among models including this parameter. Because it has been demonstrated that models including founder-event speciation are prone to over-fitting [5]; but see [49], we here show results of the ancestral range estimates for the MCC tree under DIVA-like (Figure 2); results under DIVA-like + J can be seen in Online Supplementary Figure S11.

We briefly investigated the effect of taking topological uncertainty into account during model selection by comparing the AICc values of the estimates of each of the 100 trees for the models DIVA-like, DIVA-like + J and DEC. A histogram of such values displays some overlap among values of the different models (Online Supplementary Figure S12). However, for each individual tree the relationship of the preferred model is maintained (i.e., DIVA-like + J is preferred to DIVA-like, which is preferred to DEC), and is identical to the order found by running the analysis using the MCC tree. Thus, taking uncertainty into account has not affected the results of model selection.

3.2. Lineages through Time by Area

We estimated the number of lineages occupying each individual area through time. For this, the tree has been divided in 1 Myr slices, and we counted the number of species in each area in each slice; widespread species are counted once in each area of their distribution range. The number of lineages in each area increases with time (Figure 3) as a combination of within-area speciation and new dispersals into the area. When using a single tree, changes in diversity associated with cladogenetic events (e.g., within-area speciation or founder-event speciation) appear as abrupt increases in the plot (Figure 3a,b). This effect is stronger in the model including founder-event speciation (Figure 3a), but disappears when topological and age uncertainty is taken into account (Figure 3c,d). It should be noted that now extinct lineages do not contribute to the plots, so the graphs only portray minimal (not actual) regional species richness over time (see Discussion).

The results (Figure 3) indicate that the ancestor of *Hexophthalma* + *Sicarius* lived in a range composed of (1) southern African deserts and scrublands, and (2) either the Atacama desert, Sechura desert, or Argentinean Monte. These latter three have similar probabilities of being part of the ancestral range of sand spiders due to uncertainty in topology, divergence times, and biogeographic estimates. There has been a slow and steady increase in diversity in these temperate desert areas for the last ~100–80 Myr. On the other hand, lineages only started occupying tropical dry forests later, around 50 Myr. The Caatinga has become the most species-rich region relatively rapidly due to within-area speciation, mainly during the Miocene.

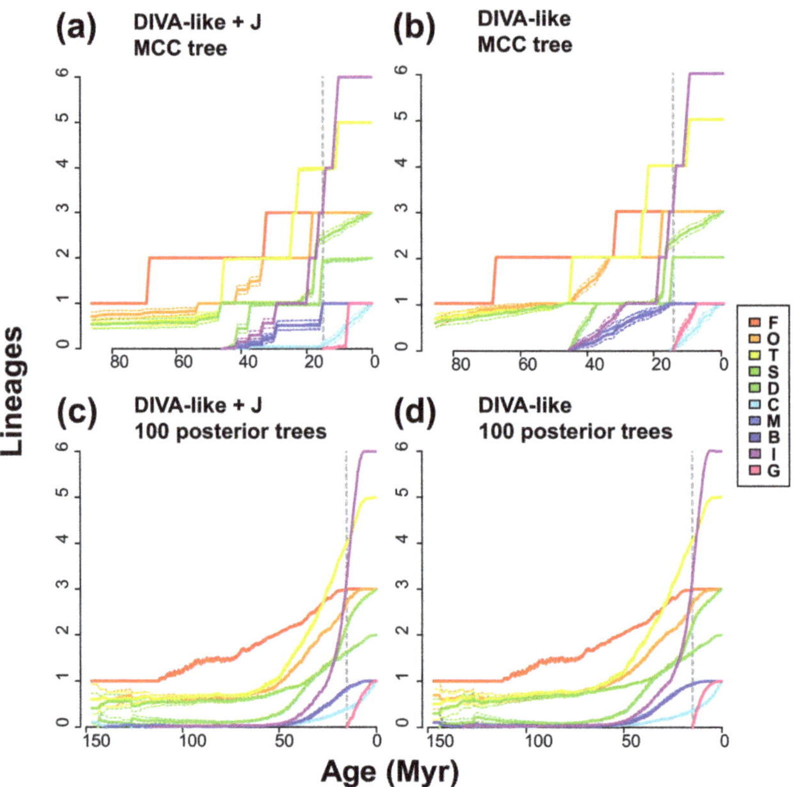

Figure 3. Number of lineages occupying each area through time. Solid lines are the average of 100 biogeographic stochastic maps, and dashed lines are the 95% confidence interval. See Figure 2 for area abbreviations. The dashed vertical line indicates the boundary between time slices in the time-stratified analysis. Runs on a single maximum clade credibility tree (**a,b**) show abrupt changes in the number of lineages that are related to cladogenetic events, whose age does not vary because there is no uncertainty associated to the tree. This effect is more pronounced in the model with founder-event speciation (**a**) because it relies more on cladogenetic events for estimating ancestral ranges. Using several trees (each with 100 stochastic maps) to account for topological and age uncertainty removes this effect and smooths the curves (**c,d**), reducing differences between models with and without founder-event speciation.

3.3. Comparing Time-Stratified vs. Unconstrained Models in the Face of Age Uncertainty

We compared the fit of the data to unconstrained models (occupation of Galapagos possible at any time) to time-stratified models (occupation of Galapagos possible only in the last 15 Myr, or in the last 3.5 Myr). First, we investigated the inferred age of dispersal to the islands in the unconstrained analysis. In the analyses using a MCC tree as input, dispersal to the Galapagos has been inferred to occur more recently than 15 Myr in 72% (DIVA-like; mean 13.8 Myr) or 97% (DIVA-like + J; mean 8.8 Myr) of the stochastic maps (Figure 4a,b). Using a MCC tree results in sharper, overconfident distributions of the inferred ages of dispersal to the islands. This is especially notable in the case of the model with founder-event speciation, where 88% of the replicates inferred that dispersal into the islands is a cladogenetic jump-dispersal with age equal to that of the *S. utriformis–S.peruensis* node in the MCC tree (Figure 4a). In analyses using 100 posterior trees as input, dispersal to the Galapagos has been inferred to occur more recently than 15 Myr in 57.8% (DIVA-like; mean 16 Myr) or 78.2% (DIVA-like + J; mean 12 Myr) of the replicates (Figure 4c,d). Thus, when using 100 posterior trees, inferred ages of the dispersal to Galapagos are older in average. In addition, the distribution of inferred ages is flatter, as it takes into account the uncertainty in node ages; this reduces the difference between models with and without founder-event speciation.

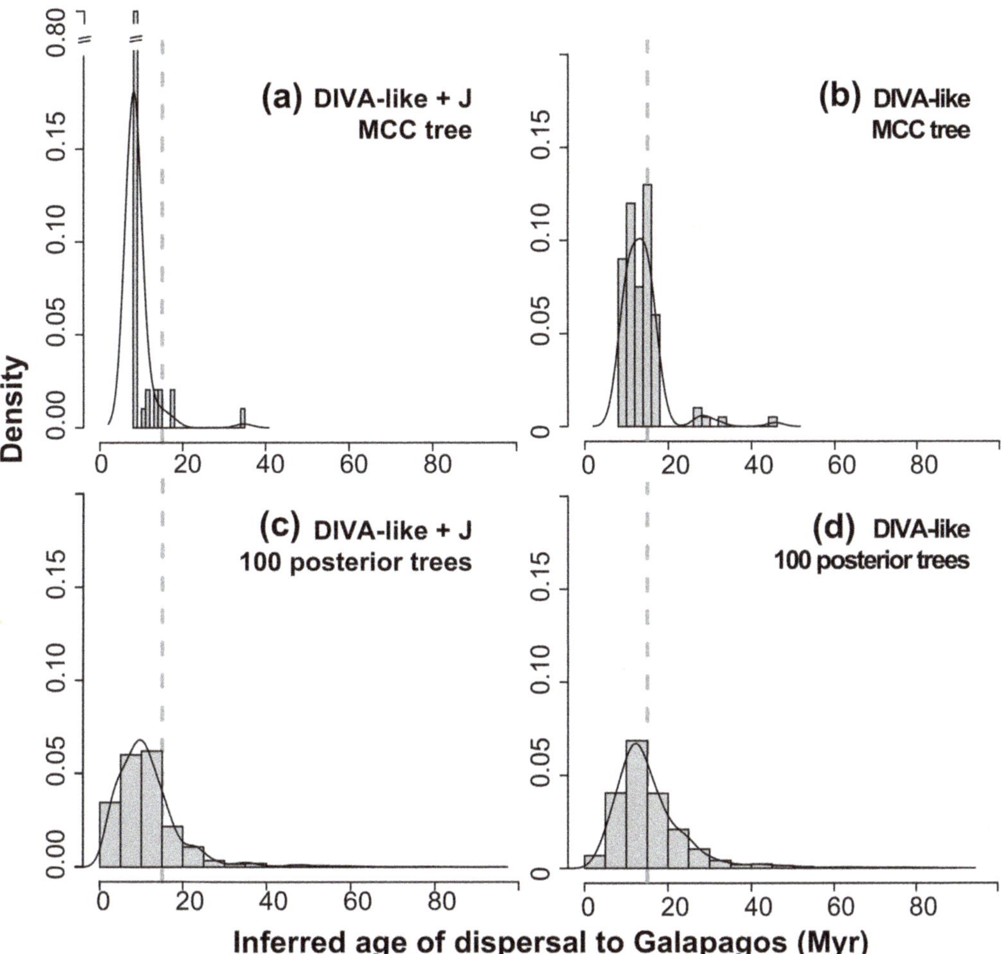

Figure 4. Histograms with the inferred age of dispersal of *Sicarius* to the Galapagos Islands. Analyses based on a maximum clade credibility tree (MCC tree) (**a,b**) produce sharper estimates that are closely tied to the age of split of *Sicarius utriformis* (from Galapagos) and its sister species. This effect is especially pronounced in the model including founder-event speciation (**a**), where 88% of the estimates have exactly the same age of that split. Accounting for uncertainty in topology and age estimates (**c,d**) reveals that the age of dispersal is much more uncertain. The dashed vertical line indicates the boundary between time slices in the time-stratified analysis and corresponds to the geological evidence of the oldest submerged paleo-islands of the Galapagos archipelago.

We then compared the fit of the data to unconstrained model vs. time-stratified model allowing dispersal only in the last 15 Myr. Using the MCC tree, the data fit better to a stratified model under DIVA-like, DIVA-like + J, and BAYAREA-like + J, while it fits better to an unconstrained model under DEC (stars in Figure 5). In all cases, however, the support for the preferred model is very weak, as the ratio between AICc weights of the preferred model ranges only between 1.29 to 2.58, and thus we cannot decisively reject any of the two models in favor of the other. When comparing models over 100 posterior trees, we observed an interesting pattern. Again, in most trees the data fit slightly better a time-stratified scenario under DIVA-like, DIVA-like + J, and BAYAREA-like + J (in 58, 68, and 58 of the 100 trees, respectively), and an unconstrained scenario under DEC (in 91 of the trees), but relative supports in these cases are once again ambiguous (1.67–2.91). However, in the fewer cases of trees where the data fit better an unconstrained model, the median relative support is higher and shows decisive support for this model (5.00–151.74) (Figure 5).

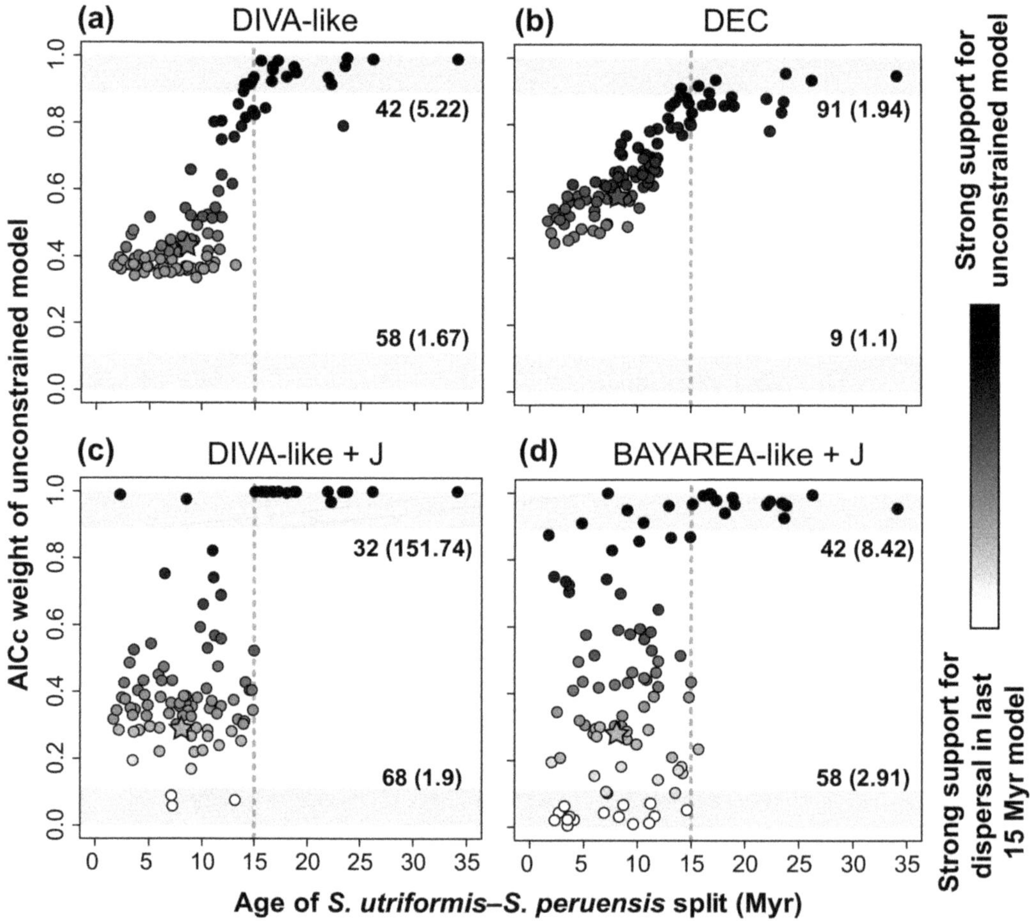

Figure 5. Relationship between AIC weight of the unconstrained biogeographic model allowing for dispersal of the Galapagos at any time (relative to the time-stratified model allowing dispersal only during the last 15 Myr) and the age of split of *Sicarius utriformis* (from Galapagos) and its sister species. Each dot represents an individual tree of a sample of 100 taken from the posterior distribution of a Bayesian analysis; the star represents the maximum clade credibility tree. Values of AIC weight close to 1 (darker shades) indicate strong support for the unconstrained model, while values close to 0 (lighter shades) indicate strong support for the time-stratified model. The light grey areas at the top and bottom of the graphs are the zones where the support for one of the alternative models is decisive. The dashed vertical line indicates the boundary between time slices in the time-stratified analysis. Numbers at the top and bottom of the graph are the number of trees supporting each model, and the median support relative to the alternative model. Trees in which the split between *S. utriformis* and its sister species are older than 15 Myr fit better to an unconstrained scenario allowing occupation of the Galapagos before that time.

We suspected that this pattern might be due to the age of split between *S. utriformis–S. peruensis*. The confidence interval of this age spans a wide range (1.2–22.2 Myr) and actually crosses the boundary between the two time slices of the time-stratified model (15 Myr). To investigate this, we plotted the AICc weight of the unconstrained model against the age of the split (Figure 5). The plot indicates that when the age of the split is younger than 15 Myr, there is weak support for the time-stratified model (DIVA-like and DIVA-like + J), to the unconstrained model (DEC), or the support to either of them is ambiguous (BAYAREA-like + J). On the other hand, the data fit better to the unconstrained model in all the 17 trees where the *S. utriformis–S. peruensis* split is older than the boundary of the time slice (15 Myr). While this is true for the four biogeographic models employed, the effect is much stronger in models including a founder-event speciation parameter, which tend to favor the unconstrained model more strongly: the relative support for the unconstrained model is higher in DIVA-like + J (776.2 ± 369) and BAYAREA-like + J (85.7 ± 110.86) than in DIVA-like (37.81 ± 37.97) and DEC (8.84 ± 4.73). Thus, in at least some of the trees from the posterior distribution, there is strong support for an unconstrained colonization of the Galapagos taking place before 15 Myr.

Finally, we compared the fit of the data to a model allowing dispersal to the Galapagos in the last 15 Myr (age of the oldest recorded paleo-islands) to a model allowing dispersal only in the last 3.5 Myr (age of the oldest emerged island). Under all biogeographic models, the model allowing dispersal in the last 15 Myr is strongly favored (Figure 6). In the few trees where this split is younger than 3.5 Myr, the 15 Myr model is still strongly favored by DIVA-like and DEC, while support to either scenario is ambiguous in DIVA-like + J and BAYAREA-like + J. Thus, our data indicate that dispersal of sand spiders to the Galapagos took place before the appearance of the oldest emerged island.

3.4. Ancestral Range Estimates of Particular Nodes in the Face of Topological Uncertainty

We estimated the most likely state in the root node (corresponding to the split between African *Hexophthalma* and Neotropical *Sicarius*). For simplicity, we only report the results under DIVA-like, which are similar to those of other biogeographic models explored here. Analyses using the MCC tree yield estimates for an ancestral range most likely including southern African scrublands and one of the temperate American deserts (Sechura, Atacama or Monte): most likely ranges are FOT (30.7%), FOS (30.3%), FTS (24.6%), FO (4.8%), FT (3.8%) and FS (3.8%), summing to a total of 98.3%. The most likely ranges across the 100 trees are FOT (27.8%), FOS (24.4%), FTS (22.4%), FO (5.5%), FT (4.2%) and FS (3.7%), total 88.1%. These latter ranges have their likelihoods slightly diminished because of an increase in the likelihood of ranges including Andean dry forests, namely FOD (2.3%), FSD (2%), and FTD (1.5%). Nonetheless, these changes are rather small, and, even in the face of phylogenetic uncertainty, we can be fairly confident that the ancestor of *Sicarius* + *Hexophthalma* lived in a range including deserts and xeric scrublands of southern Africa and southern or western South America.

We used stochastic maps to estimate the ancestral range that sourced species to the Brazilian Caatinga. In more than 99.9% of the maps, occupation of the Caatinga is the result of a single dispersal event, either jump-dispersal (DIVA-like + J) or anagenetic dispersal to a wide range including the Caatinga immediately followed by vicariance (DIVA-like). Using 100 stochastic maps resolved from the ancestral range estimates on the MCC tree, the most likely candidates for the area that originated this single dispersal are Mesoamerican dry forests, Caribbean dry forests, Argentinean Monte, Atacama desert or Andean dry forests (Table 1). When accounting for topological uncertainty using 100 stochastic maps for each of the 100 posterior trees, the results are similar but there is a substantial increase in the likelihood of Andean dry forests to be the source area, and this area actually becomes the most likely source (Table 1); using only the MCC tree underestimates this possibility. This increase in the likelihood is related to the presence of the rogue taxon *Sicarius andinus*, which inhabits Andean dry forests and is resolved as the sister taxon to the Caatinga clade in several trees of the posterior distribution.

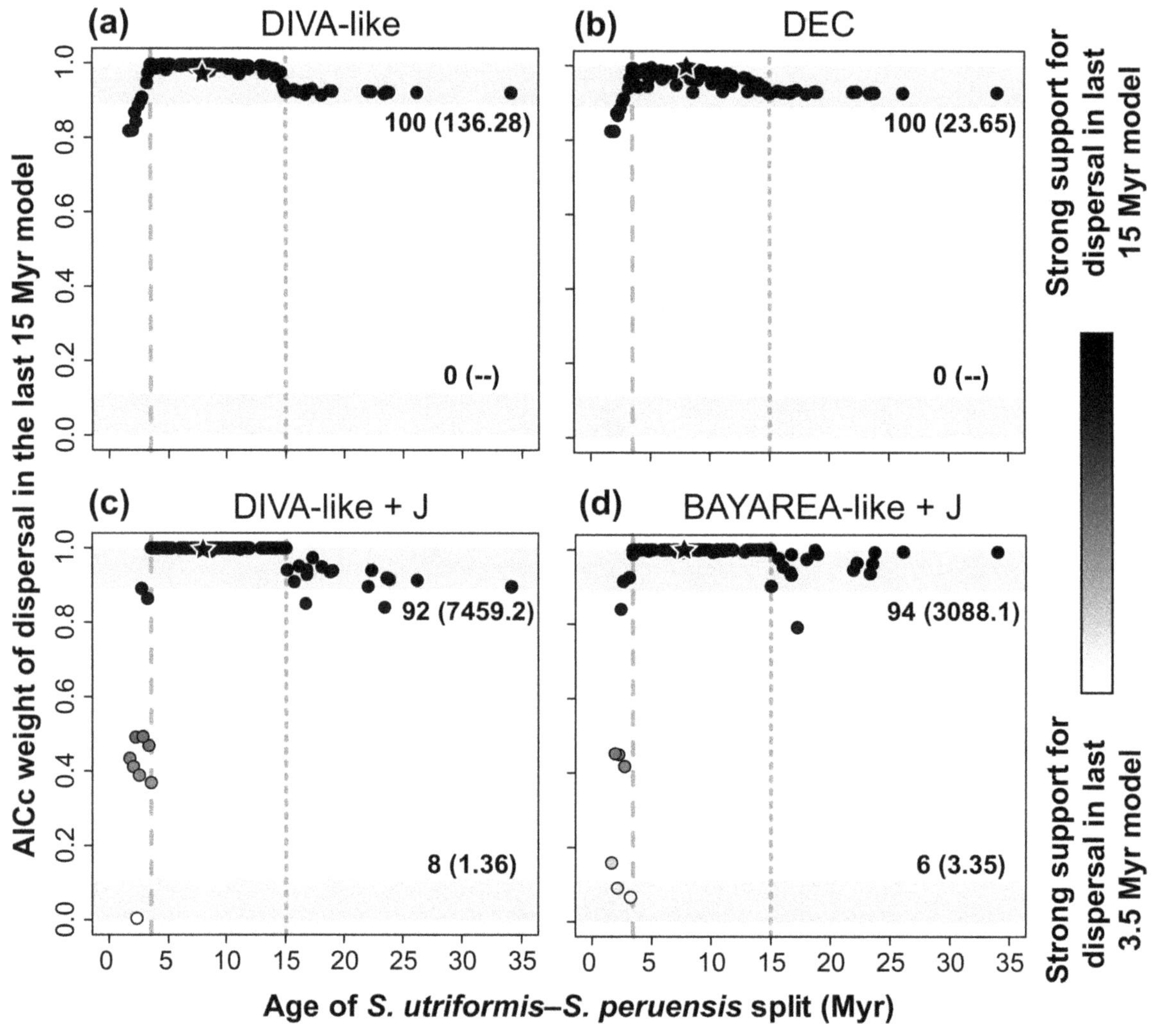

Figure 6. Relationship between AIC weight of the time-stratified biogeographic model allowing dispersal only during the last 15 Myr (relative to the time-stratified model allowing dispersal only during the last 3.5 Myr) and the age of split of *Sicarius utriformis* (from Galapagos) and its sister species. Each dot represents an individual tree of a sample of 100 taken from the posterior distribution of a Bayesian analysis; the star represents the maximum clade credibility tree. Values of AIC weight close to 1 (darker shades) indicate strong support for the 15 Myr model, while values close to 0 (lighter shades) indicate strong support for the 3.5 Myr model The light grey areas at the top and bottom of the graphs are the zones where the support for one of the alternative models is decisive. The dashed vertical lines indicate the boundary between time slices in each of the time-stratified models. Numbers at the top and bottom of the graph are the number of trees supporting each model, and the median support relative to the alternative model. With few exceptions, most trees fit decisively better to a model allowing for dispersal in the last 15 Myr, regardless of the age of the split.

Table 1. Geographic range of origin of clades dispersing into the Caatinga and their relative frequencies (in %) in different biogeographic stochastic maps. Taking topological uncertainty into account (by running stochastic maps in 100 posterior trees) increases the percentage of inferences of dispersal coming from Andean or Chiquitano dry forests (values marked in bold) while decreasing the probability of dispersal coming from Mesoamerican or Caribbean dry forests, or the Atacama desert (values marked in italics).

Source Trees (each with 100 BSM)	Biogeographic Model	Caribbean (B)	Chiquitano (C)	Andes (D)	Mesoamerica (M)	Monte (O)	Atacama (T)	Caribbean + Andes	Andes + Mesoamerica	Caribbean + Mesoamerica	Caribbean + Monte	Andes + Monte	Monte + Mesoamerica
MCCT	DVL + J	25	0	13	30	15	15	0	0	2	0	0	0
MCCT	DVL	22	0	9	19	19	19	0	1	2	2	4	3
100 trees	DVL + J	22	**2**	**24**	22	**16**	*11*	0	0	2	0	0	0
100 trees	DVL	*16.2*	**1.2**	**26.6**	*15.3*	*12.5*	*15.5*	**1.3**	1.3	2.7	1.9	*1.7*	2

We inferred the age of the dispersal event to the Caatinga and discovered a similar pattern to what we observed regarding the Galapagos. When using a MCC tree, the estimates are sharp and overconfident (Figure 7a,b), especially in the analysis including founder-event speciation, which shows four peaks closely tied to the ages of the cladogenetic events immediately leading to the Caatinga clade. Using 100 posterior trees to estimate the age of the dispersal event incorporates the uncertainty in the node ages, and reduces the differences between DIVA-like and DIVA-like + J (Figure 7c,d).

3.5. Summary of Biogeographic Events in the Face of Uncertainty

We were able to count the most frequently inferred biogeographic events (dispersals, extinctions, allopatry and within-area speciation) by taking into account uncertainty in topology and ancestral range estimates. The results are summarized in Figure 2 and Table 2. A substantial fraction of *Sicarius* diversity has been generated by within-area speciation, mainly in the Caatinga and the Atacama and, to a lesser extent, in the Monte scrubland, Sechura desert and Andean dry forests. Three of the dispersal events are robust to both types of uncertainty: one from the Sechura to Galapagos, one from the Atacama to Sechura, and one from the Andes to the Chiquitano dry forest. Other dispersal events are sensitive to uncertainty in topology and/or in ancestral range estimates, but generally involve geographically close areas, such as Mesoamerican and Caribbean dry forests, Atacama and Monte, and Atacama and Andes. The Caatinga has exchanged lineages with a single other area whose identity is uncertain, but candidates are Mesoamerican, Caribbean or Andean dry forests and, less likely, the Atacama desert and the Monte scrubland.

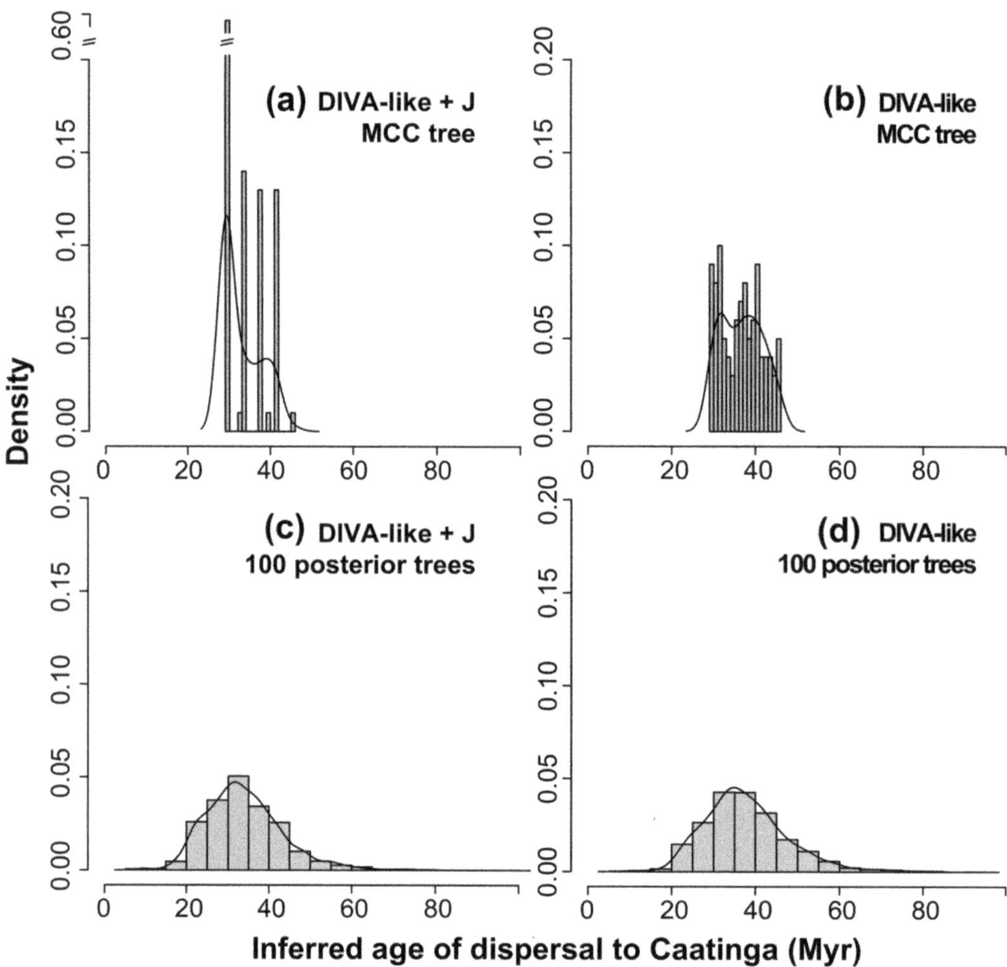

Figure 7. Histograms with the inferred age of dispersal of *Sicarius* to the Caatinga. Analyses based on a maximum clade credibility tree (MCC tree) (**a,b**) produce sharper estimates closely tied to the ages of cladogenetic events in this tree. This is especially notable in the model including founder-event speciation (**a**), which displays four peaks, each associated with the age of the four successive nodes in the tree that could have originated the founder-event dispersal to the Caatinga. Accounting for uncertainty in topology and age estimates (**c,d**) reveals that the age of dispersal is much more uncertain.

Table 2. List of the most common biogeographic events inferred under DIVA-like, averaged over 100 trees from the posterior distribution.

Starting Range	Ending Range	Average Events	Event Type
I	I	10.0948	In-situ speciation
T	T	7.5704	In-situ speciation
F	F	3.9997	In-situ speciation
S	S	2.5332	In-situ speciation
O	O	2.168	In-situ speciation
D	D	1.6566	In-situ speciation
T	TS	1.0111	Dispersal
MB	M	0.9475	Vicariance
MB	B	0.9475	Vicariance
D	DC	0.8726	Dispersal
SG	S	0.8004	Vicariance
SG	G	0.8004	Vicariance
S	SG	0.7997	Dispersal
OT	O	0.6377	Vicariance
OT	T	0.636	Vicariance
TS	S	0.5241	Vicariance
TS	T	0.5228	Vicariance

4. Discussion

4.1. Inference of Biogeographic Events in the Face of Uncertainty

In recent times, we have learned that tree topology may reach stability with more sequence data, but some clades remain elusive even in massive phylogenomic datasets [14,15]. In addition, estimates of ages of divergence are based on limited data and many assumptions (e.g., the placement of fossil calibrations) and will always be uncertain; as Bromham [18] neatly pointed out, "paleontological evidence and molecular dates paint history with a broad brush, not fine pen work". In this scenario, it seems unwise to put too much faith in a single tree, which represents only one of many similarly probable topologies and age estimates. Several studies have successfully incorporated such uncertainty in biogeographic inference [27,29]. Our results corroborate this: when combining biogeographic stochastic maps with a sample of trees, there are valuable insights to be gained.

We illustrate this concept by studying particular biogeographic events. Biogeographic maps can be used to estimate both transitions among areas and ages of biogeographic events [2]. However, they are tied to the particular tree that is used as a base, which may bias the biogeographic inferences. When we estimated biogeographic transitions between the Caatinga and other dry Neotropical areas, analyses using only the MCC tree underestimated the possibility of dispersal coming from the Andes when compared to the analyses using 100 posterior trees (Table 1). Perhaps more importantly, the inferred ages of biogeographic events are estimated with overconfidence when stochastic maps are run on a single tree (Figures 3a,b, 4a,b and 7a,b). This is because stochastic maps can only infer the ages of events along the branches and nodes of that particular tree. This overconfidence is potentially problematic, because many biogeographic studies are aimed at linking phylogenetic and geological history; the correlation (or lack thereof) of node ages (or biogeographic events) with geological events is often used to reach biological conclusions (e.g., [50]). If such uncertainty is not taken into account, researchers may achieve inaccurate results. We show that running stochastic maps on a single tree does not fully capture all the possible biogeographic histories of a clade, and thus strongly encourage researchers to use this method in combination with a sample of trees to take advantage of its full potential.

Additionally, it seems that using several trees reduces the differences between different biogeographic models (e.g., DIVA-like and its +J variant). We find that lineages through time by area and inferred ages of dispersal are more similar among models when phylogenetic uncertainty is taken into account (Figures 3c,d, 4c,d and 7c,d). Thus, at least in some cases, phylogenetic uncertainty might be more influential than uncertainty in the choice of a particular biogeographic model.

On a lighter note, our results suggest that phylogenetic uncertainty is not crucial for model selection. We show that the outcome of model selection (DEC, DIVA-like, BAYAREA-like, and their +J variants) is the same regardless if the comparison is performed using the MCC tree or 100 posterior trees (Online Supplementary Figure S12). Thus, it seems that model selection can be performed using the MCC tree, and biogeographic stochastic maps can be run on a sample of posterior trees using only the preferred model.

4.2. Cladogenetic Events, Founder-Event Speciation, and Uncertainty

We find that overconfidence in age estimates is stronger for cladogenetic events (relative to anagenetic events) when a single tree is used. Models including founder-event speciation disproportionately favor cladogenetic events [5], and thus are more prone to overconfidence, especially when inferring ages of biogeographic events. Such models have been introduced by Matzke [10,11] as a way to model the possibility of dispersal to a new area being followed by speciation over the course of very few generations—almost instantaneously in an evolutionary timescale. In practice, this parameter allows dispersal to happen simultaneously with cladogenetic events, i.e., at tree nodes. This is opposed to anagenetic dispersal, which happens along the branches. Stochastic maps resolve the age of an anagenetic dispersal at any point of such a branch, but founder-event dispersal is always tied to the age of a particular node. Because of this, ages of events inferred from models including founder-event speciation are estimated with more overconfidence and closely tied to the ages of nodes when a single tree is used to run the stochastic maps (Figures 3a, 4a and 7a). This overconfidence is partly smoothed by the variability in ages that can be incorporated by using a sample of posterior trees (Figures 3c,d, 4c,d and 7c,d). Thus, we recommend incorporating phylogenetic uncertainty when running stochastic maps especially when using models including founder event-speciation. It is especially important to have this in mind since model selection often favors such models [5,11,49]. It should also be noted, however, that even models favoring anagenetic dispersal also suffer from overconfidence when using a single tree.

4.3. Lineages through Time by Area

We here explore the concept of "lineages through space and time" plots. Former implementations of such plots have been independently conceived by Spriggs et al. [51] and Ceccarelli et al. [52], who were interested in tracking changes in diversity of two different areas through time. These earlier approaches equated the most likely area estimate at nodes as the "true" ancestral range. A later development by Skeels [53] takes advantage of biogeographic stochastic maps to take the uncertainty in ancestral range estimations into account, but this script was designed to work on the results based on a single tree. We here use a similar approach that also takes phylogenetic uncertainty into account. By dividing the tree into user-defined time slices, it is possible to count the number of lineages occupying each area in each time period. Diversity increases through time as a combination of in-situ speciation and new dispersals into the area. Additionally, diversity can also decrease if the model infers extinctions in a particular area (Supplementary Figure S13), or if the taxon sampling includes fossil tips (ILFM, unpublished data).

These plots may be used as rough approximations to detect areas which might be under special processes. For instance, our data on sand spiders indicate that the Caatinga rapidly accumulated diversity after the initial dispersal to this area (Figure 3), which is consistent with observations that this is one of the most diverse dry forests in the Neotropical region [37]. Nevertheless, it should be noted that these plots present limitations. It is known that DEC and its variants vastly and consistently underestimate extinctions [6]; this has been observed in our data, as extinction was estimated to be close to 0 under some models (Online Supplementary Table S9). Known but unsampled species are effectively "extinct" for the purposes of the model and will not be considered, which may affect the results of the plot, particularly if the sampling is uneven across areas. This is clear when we take a look at the diversification of lineages from Africa: only four in-situ speciation events are inferred (Figure 3, Table 2), even though the genus *Hexophthalma* currently includes 8 species [54]. It is clear that diversification in this area is underestimated by our sampling, which only includes three species. Thus, we recommend that such plots are only considered when the sampling of the group is complete, or at least when the unsampled species are not biased to a particular area. Second, even if the sampling is complete, these plots will not account for truly extinct species. Researchers interested in comparing diversification rates among areas, while accounting for extinct lineages, should refer to methods specifically designed for modelling this (e.g., BAMM; [55]). On a lighter note, because the approach we take can accommodate phylogenetic uncertainty, one can include taxa with unknown phylogenetic placement by incorporating them to a sample of trees based only on taxonomic information, as undertaken by [29], thus possibly alleviating the effect of missing taxa.

4.4. Sand Spiders Dispersed to Galapagos Paleo-Islands

The Galapagos Islands are volcanic in origin, and while the oldest emerged island is 3–4 million years old [34], there is evidence of submerged paleo-islands to the east [35] and north [36] of the current archipelago. The particular biota of the islands has affinities with those of South America, the Greater Antilles and other Pacific islands [56,57]. It is clear that the only sand spider of the islands, *S. utriformis*, is of South American origins, as its closest relative lives on the coast of Peru (Figure 2). The estimated age of split of these two species (1.2–22 Myr, 95% highest posterior density interval, median 9.7) exceeds the age of the oldest emerged islands. Accordingly, a biogeographic model allowing for dispersal to the islands in the last 15 Myr is strongly favored in a relation to a model allowing dispersal only in the last 3.5 Myr (Figure 6). In most trees the *S. utriformis–S. peruensis* split is older than 3.5 Myr; thus the second scenario requires that *S. utriformis* had originated in coastal Peru through in-situ speciation before 3.5 Myr, dispersed to the islands after their formation, and then went extinct in the continent (Online Supplementary Figure S13). In contrast, the first scenario only requires one dispersal event from the ancestor of the pair of species to the islands, followed by a (costless) allopatric event. Thus, our data fit better a model in which *Sicarius* reached the Galapagos when the archipelago was formed by currently submerged paleo-islands.

Interestingly, the oldest paleo-island reported by Christie et al. [35] is only ~530 km distant from the continent, while the current islands lie at ~940 km away from the continent. Thus, the paleo-islands were closer to the continent, which could help explaining how a group with poor dispersal capabilities reached a volcanic archipelago. This scenario is similar to the metapopulation vicariance model that has been proposed to explain the presence of ancient, poorly dispersing groups in recent volcanic archipelagos, particularly in the Galapagos [57,58]. It could also explain other instances of clades whose estimated ages are older than the islands they occupy, such as sheet-web weavers in the Juan Fernández Islands [59].

Is it possible that there were even older paleo-islands in the Galapagos? Christie et al. [35] suggested that it is very possible that the history of the archipelago could be as old as that of the volcanic hotspot, spanning 80–90 million years, and thus other uncharted paleo-islands might exist. Further evidence for this hypothesis has been recently reviewed by Heads & Grehan [57]. To account for this possibility, we compared a model where dispersal to the Galapagos was only possible in last 15 Myr to an unconstrained model where dispersal could happen at any moment—thus, considering the possibility of even older paleo-islands. The data never provide definite support to the 15 Myr model, and in those trees where the split between *S. utriformis* and *S. peruensis* is older than 15 Myr, it strongly supports the unconstrained model (Figure 5). In addition, the inferred age of dispersal to Galapagos in the unconstrained model has some probability of being older than 15 Myr (Figure 4). Thus, our data are unable to reject dispersal to the Galapagos before the age of the oldest recorded paleo-islands. This is in line with the opinion that the archipelago must be very old and uncharted paleo-islands exist [35].

Interestingly, we found a correlation between the support of alternative models and the ages of split (Figure 5). There is a positive correlation between age of split and AIC weight of the unconstrained model, stronger in the models without founder-event speciation (DEC, r = 0.86; DIVA-like, r = 0.83; DIVA-like + J, r = 0.63, BAYAREA-like + J, r = 0.50). This is probably because these models favor anagenetic events, and thus the amount of time passed is correlated with opportunity for dispersal. This means that even if the split is younger than 15 Myr, but very close to the limit between time slices, the window of opportunity for a dispersal is very narrow, and thus even in these cases the unconstrained model is favored.

4.5. Ancient Connections among Neotropical Dry Forests

Our data indicate that the Caatinga clade reached this area as a result of a single dispersal. Such dispersal most likely originated from northern dry forests in the Caribbean and Mesoamerica, or from Andean dry forests. Interestingly, this second possibility only appears as a likely candidate when topological uncertainty is taken into account (Table 1). Such connections seem to support the hypothesis by Pennington et al. [38] that some areas of present-day Amazonia have been replaced by xeric vegetation in the past. They compiled detailed evidence that compellingly indicates dry conditions in the region during the Quaternary. Phylogenetic patterns of other organisms, such as birds, support the hypothesis of recent connections among Neotropical dry forests (e.g., [60]). Our data, on the other hand, indicate that dispersal of sand spiders to the Caatinga happened as early as in the Oligocene (Figure 7), with no recent dispersals to other areas. Other groups of organisms display similarly ancient histories in this biome (e.g., geckos; [61]). Thus, it may be plausible that currently disjunct Neotropical dry areas might have gone through many periods of connections over the last 30 million years, and that groups with different dispersing capabilities could have responded idiosyncratically to such connections.

4.6. Script Availability

The script for replicating the analyses is available as Supplementary Materials S1, or from GitHub (https://github.com/ivanlfm/BGB_BSM_multiple_trees, accessed on 20 June 2021). The code is thoroughly commented and documented, with explanations for each of the options. It has been successfully tested with two additional datasets, one of them including fossil tips, and thus we expect it to be easily adaptable to other researchers' needs. These datasets varied between ~30 and ~100 taxa, ~10 areas (with maximum range size set to ~3) and runs included 2–3 free parameters. In each case, analyses run over 100 posterior trees were usually completed between 8 and 12 h on a standard personal computer (Intel ® i5-5200U 2.20 GHz with 4 GB of RAM). Thus, we expect that the analyses outlined above are not too demanding computationally. As it relies on BioGeoBEARS, the script can be used with dated trees produced by any phylogenetic software.

4.7. Concluding Remarks

(1) Biogeographic hypothesis-testing, inferences of transitions among areas and age of biogeographic events using biogeographic stochastic maps benefit from running the analysis over a sample of trees, instead of a single, target tree, since they incorporate uncertainty in topology and age estimates that might be relevant to the questions at hand. We provide a broadly customizable R script to run such analyses.

(2) Including phylogenetic uncertainty is especially important in models including a founder-event speciation parameter; ages of biogeographic events estimated under these models are tightly tied to cladogenetic events, such that using a single tree results in overconfident estimates that disregard the uncertainty in age estimates present in trees from the posterior distribution.

(3) Our data strongly suggest *Sicarius* most likely dispersed to the Galapagos Islands before the formation of the oldest emerged island. This is congruent with geological findings that indicate that seamounts along the volcanic hotspot are former paleo-islands of the archipelago that are now submerged.

(4) Our data indicate *Sicarius* dispersed into the Caatinga around 30 Mya, suggesting an ancient colonization of this area. The route of dispersal is unclear due to topological uncertainty, but most likely consisted of a northern route connecting the Caatinga to the Caribbean and Mesoamerican dry forests, or of a southern route connecting the Caatinga to the Andean dry forests.

Supplementary Materials: The following are available online at https://www.mdpi.com/article/10.3390/d13090418/s1. File S1. R script for running and summarizing ancestral range estimates and biogeographic stochastic maps in a sample of trees. Updated versions can be found in https://github.com/ivanlfm/BGB_BSM_multiple_trees, (accessed on 15 June 2021). File S2. Maximum clade credibility phylogenetic tree representing relationships among *Sicarius* and *Hexophthalma* from the analysis of Magalhaes et al. (2019). File S3. 27,000 trees from the posterior distribution from the analysis of Magalhaes et al. (2019). File S4. Phylip-formatted file with distribution ranges of *Sicarius* and *Hexophthalma*. File S5. R script for performing biogeographic model selection in our dataset. Files S6–S8. Inputs for performing the time-stratified analysis in BioGeoBEARS allowing occupation of Galapagos only in the last 15 Myr or 3.5 Myr. Files S9–S12. Supplementary Table S9 and Figures S10–S13.

Author Contributions: Conceptualization, I.L.F.M. and M.J.R.; Data collection, I.L.F.M. and A.J.S.; Scripts, I.L.F.M. and M.J.R.; Formal Analysis, I.L.F.M.; Writing—Original Draft Preparation, I.L.F.M.; Writing—Review & Editing, all authors; Funding Acquisition, A.J.S. and M.J.R. All authors have read and agreed to the published version of the manuscript.

Funding: This work was supported by a CONICET post-doctoral fellowship to I.L.F.M., a PICT-2015-0283 grant from ANPCyT to M.J.R., and FAPEMIG (PPM-00605-17), CNPq (405795/2016-5; 307731/2018-9), and Instituto Nacional de Ciência e Tecnologia dos Hymenoptera Parasitóides da Região Sudeste Brasileira (http://www.hympar.ufscar.br, accessed on 23 May 2021, CNPq 465562/2014-0, FAPESP 2014/50940-2) to A.J.S.

Data Availability Statement: All the data and scripts used in our analyses is available as Online Supplementary Materials.

Acknowledgments: We thank N. Matzke for his diligence and support to the BioGeoBEARS community through the online forums. Earlier versions of the text benefited from a critical revision by U. Oliveira, three anonymous reviewers, and the editor, Matjaž Kuntner.

Conflicts of Interest: The authors declare no conflict of interest.

References

1. Huelsenbeck, J.P.; Imennov, N.S. Geographic origin of human mitochondrial DNA: Accommodating phylogenetic uncertainty and model comparison. *Syst. Biol.* **2002**, *51*, 155–165. [CrossRef]
2. Dupin, J.; Matzke, N.J.; Särkinen, T.; Knapp, S.; Olmstead, R.G.; Bohs, L.; Smith, S.D. Bayesian estimation of the global biogeographical history of the Solanaceae. *J. Biogeogr.* **2017**, *44*, 887–899. [CrossRef]
3. Ronquist, F. Dispersal-vicariance analysis: A new approach to the quantification of historical biogeography. *Syst. Biol.* **1997**, *46*, 195–203. [CrossRef]
4. Bremer, K. Ancestral areas: A cladistic reinterpretation of the center of origin concept. *Syst. Biol.* **1992**, *41*, 436–445. [CrossRef]
5. Ree, R.H.; Sanmartín, I. Conceptual and statistical problems with the DEC+J model of founder-event speciation and its comparison with DEC via model selection. *J. Biogeogr.* **2018**, *45*, 741–749. [CrossRef]
6. Ree, R.H.; Smith, S.A. Maximum likelihood inference of geographic range evolution by dispersal, local extinction, and cladogenesis. *Syst. Biol.* **2008**, *57*, 4–14. [CrossRef]
7. Ree, R.H.; Moore, B.R.; Webb, C.O.; Donoghue, M.J. A likelihood framework for inferring the evolution of geographic range on phylogenetic trees. *Evolution* **2005**, *59*, 2299–2311. [CrossRef] [PubMed]
8. Yu, Y.; Harris, A.J.; Blair, C.; He, X. RASP (Reconstruct Ancestral State in Phylogenies): A tool for historical biogeography. *Mol. Phylogenet. Evol.* **2015**, *87*, 46–49. [CrossRef] [PubMed]
9. Matzke, N.J. BioGeoBEARS: BioGeography with Bayesian (and Likelihood) Evolutionary Analysis in R Scripts. 2013. Available online: https://github.com/nmatzke/BioGeoBEARS (accessed on 13 October 2020).

10. Matzke, N.J. Probabilistic historical biogeography: New models for founder-event speciation, imperfect detection, and fossils allow improved accuracy and model-testing. *Front. Biogeogr.* **2013**, *5*. [CrossRef]
11. Matzke, N.J. Model selection in historical biogeography reveals that founder-event speciation is a crucial process in island clades. *Syst. Biol.* **2014**, *63*, 951–970. [CrossRef]
12. Turk, E.; Bond, J.; Cheng, R.-C.; Čandek, K.; Hamilton, C.A.; Kralj-Fišer, S.; Kuntner, M. A natural colonization of Asia: Phylogenomic and biogeographic history of coin spiders (Araneae: Nephilidae: *Herennia*). *Diversity* **2021**, in press.
13. Hortal, J.; De Bello, F.; Diniz-Filho, J.A.F.; Lewinsohn, T.M.; Lobo, J.M.; Ladle, R.J. Seven Shortfalls that Beset Large-Scale Knowledge of Biodiversity. *Annu. Rev. Ecol. Evol. Syst.* **2015**, *46*, 523–549. [CrossRef]
14. Ballesteros, J.A.; Sharma, P.P. A critical appraisal of the placement of Xiphosura (Chelicerata) with account of known sources of phylogenetic error. *Syst. Biol.* **2019**, *68*, 896–917. [CrossRef] [PubMed]
15. Suh, A. The phylogenomic forest of bird trees contains a hard polytomy at the root of Neoaves. *Zool. Scr.* **2016**, *45*, 50–62. [CrossRef]
16. Ho, S.Y.W.; Duchêne, S. Molecular-clock methods for estimating evolutionary rates and timescales. *Mol. Ecol.* **2014**, *23*, 5947–5965. [CrossRef] [PubMed]
17. Drummond, A.J.; Ho, S.Y.W.; Phillips, M.J.; Rambaut, A. Relaxed phylogenetics and dating with confidence. *PLoS Biol.* **2006**, *4*, e88. [CrossRef]
18. Bromham, L. Six impossible things before breakfast: Assumptions, models, and belief in molecular dating. *Trends Ecol. Evol.* **2019**, *34*, 474–486. [CrossRef]
19. Parham, J.F.; Donoghue, P.C.J.; Bell, C.J.; Calway, T.D.; Head, J.J.; Holroyd, P.A.; Inoue, J.G.; Irmis, R.B.; Joyce, W.G.; Ksepka, D.T.; et al. Best practices for justifying fossil calibrations. *Syst. Biol.* **2012**, *61*, 346–359. [CrossRef]
20. Warnock, R.C.M.; Parham, J.F.; Donoghue, P.C.J.; Joyce, W.G.; Lyson, T.R. Calibration uncertainty in molecular dating analyses: There is no substitute for the prior evaluation of time priors. *Proc. R. Soc. B Biol. Sci.* **2015**, *282*, 20141013. [CrossRef]
21. Huelsenbeck, J.P.; Rannala, B.; Masly, J.P. Accommodating phylogenetic uncertainty in evolutionary studies. *Science* **2000**, *288*, 2349–2350. [CrossRef]
22. Nylander, J.A.A.; Olsson, U.; Alström, P.; Sanmartín, I. Accounting for phylogenetic uncertainty in biogeography: A Bayesian approach to dispersal-vicariance analysis of the thrushes (Aves: *Turdus*). *Syst. Biol.* **2008**, *57*, 257–268. [CrossRef] [PubMed]
23. Yu, Y.; Harris, A.J.; He, X. S-DIVA (Statistical Dispersal-Vicariance Analysis): A tool for inferring biogeographic histories. *Mol. Phylogenet. Evol.* **2010**, *56*, 848–850. [CrossRef] [PubMed]
24. Baker, C.M.; Boyer, S.L.; Giribet, G. A well-resolved transcriptomic phylogeny of the mite harvestman family Pettalidae (Arachnida, Opiliones, Cyphophthalmi) reveals signatures of Gondwanan vicariance. *J. Biogeogr.* **2020**, *47*, 1345–1361. [CrossRef]
25. Santaquiteria, A.; Siqueira, A.C.; Duarte-Ribeiro, E.; Carnevale, G.; White, W.; Pogonoski, J.; Baldwin, C.C.; Ortí, G.; Arcila, D.; Betancur, R. Phylogenomics and historical biogeography of seahorses, dragonets, goatfishes, and allies (Teleostei: Syngnatharia): Assessing the factors driving uncertainty in biogeographic inferences. *Syst. Biol.* **2021**, in press. [CrossRef]
26. Landis, M.J.; Freyman, W.A.; Baldwin, B.G. Retracing the Hawaiian silversword radiation despite phylogenetic, biogeographic, and paleogeographic uncertainty. *Evolution* **2018**, *72*, 2343–2359. [CrossRef]
27. Landis, M.J.; Eaton, D.A.R.; Clement, W.L.; Park, B.; Spriggs, E.L.; Sweeney, P.W.; Edwards, E.J.; Donoghue, M.J. Joint phylogenetic estimation of geographic movements and biome shifts during the global diversification of *Viburnum*. *Syst. Biol.* **2021**, *70*, 67–85. [CrossRef]
28. Matos-Maraví, P.; Wahlberg, N.; Freitas, A.V.L.; Devries, P.; Antonelli, A.; Penz, C.M. Mesoamerica is a cradle and the Atlantic Forest is a museum of Neotropical butterfly diversity: Insights from the evolution and biogeography of Brassolini (Lepidoptera: Nymphalidae). *Biol. J. Linn. Soc.* **2021**, *133*, 704–724. [CrossRef]
29. Yan, Y.; Davis, C.C.; Dimitrov, D.; Wang, Z.; Rahbek, C.; Borregaard, M.K. Phytogeographic history of the tea family inferred through high-resolution phylogeny and fossils. *Syst. Biol.* **2021**, in press. [CrossRef]
30. Magalhaes, I.L.F.; Neves, D.M.; Santos, F.R.; Vidigal, T.H.D.A.; Brescovit, A.D.; Santos, A.J. Phylogeny of Neotropical *Sicarius* sand spiders suggests frequent transitions from deserts to dry forests despite antique, broad-scale niche conservatism. *Mol. Phylogenet. Evol.* **2019**, *140*, 106569. [CrossRef]
31. Binford, G.J.; Callahan, M.S.; Bodner, M.R.; Rynerson, M.R.; Núñez, P.B.; Ellison, C.E.; Duncan, R.P. Phylogenetic relationships of *Loxosceles* and *Sicarius* spiders are consistent with Western Gondwanan vicariance. *Mol. Phylogenet. Evol.* **2008**, *49*, 538–553. [CrossRef] [PubMed]
32. Magalhaes, I.L.F.; Oliveira, U.; Santos, F.R.; Vidigal, T.H.D.A.; Brescovit, A.D.; Santos, A.J. Strong spatial structure, Pliocene diversification and cryptic diversity in the Neotropical dry forest spider *Sicarius cariri*. *Mol. Ecol.* **2014**, *23*, 5323–5336. [CrossRef]
33. Magalhaes, I.L.F.; Brescovit, A.D.; Santos, A.J. Phylogeny of Sicariidae spiders (Araneae: Haplogynae), with a monograph on Neotropical Sicarius. *Zool. J. Linn. Soc.* **2017**, *179*, 767–864.
34. White, W.M.; McBirney, A.R.; Duncan, R.A. Petrology and geochemistry of the Galápagos Islands: Portrait of a pathological mantle plume. *J. Geophys. Res.* **1993**, *98*, 533–563. [CrossRef]
35. Christie, D.M.; Duncan, R.A.; McBirney, A.R.; Richards, M.A.; White, W.M.; Harpp, K.S.; Fox, C.G. Drowned islands downstream from the Galapagos hotspot imply extended speciation times. *Nature* **1992**, *355*, 246–248. [CrossRef]
36. Werner, R.; Hoernle, K.; Van Den Bogaard, P.; Ranero, C.; Von Huene, R.; Korich, D. Drowned 14-m.y.-old Galapagos archipelago off the coast of Costa Rica: Implications for tectonic and evolutionary models. *Geology* **1999**, *27*, 499–502. [CrossRef]

37. DRYFLOR. Plant diversity patterns in neotropical dry forests and their conservation implications. *Science* **2016**, *353*, 1383–1388. [CrossRef] [PubMed]

38. Pennington, R.T.; Prado, D.E.; Pendry, C.A. Neotropical seasonally dry forests and Quaternary vegetation changes. *J. Biogeogr.* **2000**, *27*, 261–273. [CrossRef]

39. Prado, D.E.; Gibbs, P.E. Patterns of species distributions in the dry seasonal forests of South America. *Ann. Missouri Bot. Gard.* **1993**, *80*, 902–927. [CrossRef]

40. R Core Team. *R: A Language and Environment for Statistical Computing. R Foundation for Statistical Computing*; R Core Team: Vienna, Austria, 2020.

41. Revell, L.J. phytools: An R package for phylogenetic comparative biology (and other things). *Methods Ecol. Evol.* **2012**, *3*, 217–223. [CrossRef]

42. Paradis, E.; Schliep, K. ape 5.0: An environment for modern phylogenetics and evolutionary analyses in R. *Bioinformatics* **2019**, *35*, 526–528. [CrossRef] [PubMed]

43. Lüdecke, D. sjmisc: Data and variable transformation functions. *J. Open Source Softw.* **2018**, *3*, 754. [CrossRef]

44. Akaike, H. Information theory and an extension of the maximum likelihood principle. In Proceedings of the Second International Symposium on Information Theory, Tsahkadsor, Armenia, 2–8 September 1971; Petrov, B.N., Caski, F., Eds.; Akademiai Kiado: Budapest, Hungary, 1973; pp. 267–281.

45. Wagenmakers, E.J.; Farrell, S. AIC model selection using Akaike weights. *Psychon. Bull. Rev.* **2004**, *11*, 192–196. [CrossRef] [PubMed]

46. Cala-Riquelme, F.; Gutiérrez-Estrada, M.; Florez-Daza, A.E.; Agnarsson, I. A new six-eyed sand spider *Sicarius* Walckenaer, 1847 (Araneae: Haplogynae: Sicariidae) from Colombia, with information on its natural history. *Arachnology* **2017**, *17*, 176–182. [CrossRef]

47. Echeverría-Londoño, S.; Enquist, B.J.; Neves, D.M.; Violle, C.; Boyle, B.; Kraft, N.J.B.; Maitner, B.S.; McGill, B.; Peet, R.K.; Sandel, B.; et al. Plant functional diversity and the biogeography of biomes in North and South America. *Front. Ecol. Evol.* **2018**, *6*, 1–12. [CrossRef]

48. Nash, J.C. On best practice optimization methods in R. *J. Stat. Softw.* **2014**, *60*, 1–14. [CrossRef]

49. Matzke, N.J. Statistical Comparison of DEC and DEC+J Is Identical to Comparison of Two ClaSSE Submodels, and Is Therefore Valid. *OSF Prepr.* **2021**, 1–40. [CrossRef]

50. Renner, S.S. Available data point to a 4-km-high Tibetan Plateau by 40 Ma, but 100 molecular-clock papers have linked supposed recent uplift to young node ages. *J. Biogeogr.* **2016**, *43*, 1479–1487. [CrossRef]

51. Spriggs, E.L.; Clement, W.L.; Sweeney, P.W.; Madriñán, S.; Edwards, E.J.; Donoghue, M.J. Temperate radiations and dying embers of a tropical past: The diversification of *Viburnum*. *New Phytol.* **2015**, *207*, 340–354. [CrossRef]

52. Ceccarelli, F.S.; Koch, N.M.; Soto, E.M.; Barone, M.L.; Arnedo, M.A.; Ramírez, M.J. The grass was greener: Repeated Evolution of specialized morphologies and habitat shifts in ghost spiders following grassland expansion in South America. *Syst. Biol.* **2019**, *68*, 63–77. [CrossRef]

53. Skeels, A. Lineages through space and time plots: Visualising spatial and temporal changes in diversity. *Front. Biogeogr.* **2019**, *11*, 1–6. [CrossRef]

54. Lotz, L.N. An update on the spider genus *Hexophthalma* (Araneae: Sicariidae) in the Afrotropical region, with descriptions of new species. *Eur. J. Taxon.* **2018**, *2018*, 475–494. [CrossRef]

55. Rabosky, D.L. Automatic detection of key innovations, rate shifts, and diversity-dependence on phylogenetic trees. *PLoS ONE* **2014**, *9*, e89543. [CrossRef]

56. Grehan, J. Biogeography and evolution of the Galapagos: Integration of the biological and geological evidence. *Biol. J. Linn. Soc.* **2001**, *74*, 267–287. [CrossRef]

57. Heads, M.; Grehan, J.R. The Galápagos Islands: Biogeographic patterns and geology. *Biol. Rev.* **2021**, *96*, 1160–1185. [CrossRef] [PubMed]

58. Heads, M. Metapopulation vicariance explains old endemics on young volcanic islands. *Cladistics* **2018**, *34*, 292–311. [CrossRef]

59. Arnedo, M.A.; Hormiga, G. Repeated colonization, adaptive radiation and convergent evolution in the sheet-weaving spiders (Linyphiidae) of the south Pacific Archipelago of Juan Fernandez. *Cladistics* **2021**, *37*, 317–342. [CrossRef]

60. Corbett, E.C.; Bravo, G.A.; Schunck, F.; Naka, L.N.; Silveira, L.F.; Edwards, S.V. Evidence for the Pleistocene Arc Hypothesis from genome-wide SNPs in a Neotropical dry forest specialist, the Rufous-fronted Thornbird (Furnariidae: *Phacellodomus rufifrons*). *Mol. Ecol.* **2020**, *29*, 4457–4472. [CrossRef]

61. Werneck, F.P.; Gamble, T.; Colli, G.R.; Rodrigues, M.T.; Sites, J.W. Deep diversification and long-term persistence in the south american "dry diagonal": Integrating continent-wide phylogeography and distribution modeling of geckos. *Evolution* **2012**, *66*, 3014–3034. [CrossRef]

 *diversity*

Article

A Natural Colonisation of Asia: Phylogenomic and Biogeographic History of Coin Spiders (Araneae: Nephilidae: *Herennia*)

Eva Turk [1,2,*], Jason E. Bond [3], Ren-Chung Cheng [4], Klemen Čandek [5], Chris A. Hamilton [6], Matjaž Gregorič [1], Simona Kralj-Fišer [1] and Matjaž Kuntner [1,5,7,8,*]

[1] Jovan Hadži Institute of Biology, Research Centre of the Slovenian Academy of Sciences and Arts, 1000 Ljubljana, Slovenia; matjaz.gregoric@zrc-sazu.si (M.G.); simona.kralj-fiser@zrc-sazu.si (S.K.-F.)
[2] Department of Biology, Biotechnical Faculty, University of Ljubljana, 1000 Ljubljana, Slovenia
[3] Department of Entomology and Nematology, University of California, Davis, CA 95616, USA; jbond@ucdavis.edu
[4] Department of Life Sciences, National Chung Hsing University, Taichung 40227, Taiwan; bolasargiope@email.nchu.edu.tw
[5] Department of Organisms and Ecosystems Research, National Institute of Biology, 1000 Ljubljana, Slovenia; klemen.candek@nib.si
[6] Department of Entomology, Plant Pathology and Nematology, University of Idaho, Moscow, ID 83844, USA; hamiltonlab@uidaho.edu
[7] Department of Entomology, National Museum of Natural History, Smithsonian Institution, Washington, DC 20560, USA
[8] Centre for Behavioural Ecology and Evolution, College of Life Sciences, Hubei University, Wuhan 430011, China
*   Correspondence: eva.turk@zrc-sazu.si (E.T.); matjaz.kuntner@nib.si (M.K.)

Citation: Turk, E.; Bond, J.E.; Cheng, R.-C.; Čandek, K.; Hamilton, C.A.; Gregorič, M.; Kralj-Fišer, S.; Kuntner, M. A Natural Colonisation of Asia: Phylogenomic and Biogeographic History of Coin Spiders (Araneae: Nephilidae: *Herennia*). *Diversity* **2021**, *13*, 515. https://doi.org/10.3390/d13110515

Academic Editors: Mark Harvey and Michael Wink

Received: 10 September 2021
Accepted: 19 October 2021
Published: 22 October 2021

Publisher's Note: MDPI stays neutral with regard to jurisdictional claims in published maps and institutional affiliations.

Abstract: Reconstructing biogeographic history is challenging when dispersal biology of studied species is poorly understood, and they have undergone a complex geological past. Here, we reconstruct the origin and subsequent dispersal of coin spiders (Nephilidae: *Herennia* Thorell), a clade of 14 species inhabiting tropical Asia and Australasia. Specifically, we test whether the all-Asian range of *Herennia multipuncta* is natural vs. anthropogenic. We combine Anchored Hybrid Enrichment phylogenomic and classical marker phylogenetic data to infer species and population phylogenies. Our biogeographical analyses follow two alternative dispersal models: ballooning vs. walking. Following these assumptions and considering measured distances between geographical areas through temporal intervals, these models infer ancestral areas based on varying dispersal probabilities through geological time. We recover a wide ancestral range of *Herennia* including Australia, mainland SE Asia and the Philippines. Both models agree that *H. multipuncta* internal splits are generally too old to be influenced by humans, thereby implying its natural colonisation of Asia, but suggest quite different colonisation routes of *H. multipuncta* populations. The results of the ballooning model are more parsimonious as they invoke fewer chance dispersals over large distances. We speculate that coin spiders' ancestor may have lost the ability to balloon, but that *H. multipuncta* regained it, thereby colonising and maintaining larger areas.

Keywords: coin spider; Nephilidae; phylogenomics; biogeography; dispersal probability

1. Introduction

Biogeography is a scientific field that integrates evolutionary hypotheses, contemporary and fossil taxonomic distributions and time calibrated phylogenies. Nevertheless, modern biogeography has struggled to become an exact science for several reasons. First, time calibrated phylogenies often yield unreliable topologies and/or divergence times [1] or produce very wide margins of error [2]. Consequently, time estimates of divergence and speciation events remain vague, and hypothesis testing imprecise (but, see

Magalhaes et al. [3]). Second, organism-specific biology is usually not accounted for within historical biogeographic reconstructions. When it is, the organism-specific resolution rarely goes beyond basic trait-state binning [4], e.g., winged versus pedestrian versus aquatic animals, etc. Third, although the Earth's tectonic and climatic histories represent essential variables for the distribution of organisms, their precise reconstructions through insets of time have not been integrated into biogeographic algorithms.

We have recently discussed this gap in methodology in a biogeographic study of a globally distributed spider family [5]. We suggested and demonstrated a novel method of fine-tuning biogeographical analyses by combining a robust phylogeny and specific organismal biology with dispersal probability estimates, based on concrete measurements between geographical regions in the geological past. This approach proved suitable for analyses across large geographical areas, where geological reconstructions are sufficiently accurate and for organisms whose dispersal biology is well understood. When the geological past of the area of interest is not as clear and species biology is unknown, however, this methodology has to be modified. Here, we produce a comprehensive phylogeny of coin spiders (Nephilidae: *Herennia*) using phylogenomic data and test and discuss an alternative approach to biogeographical inference to the one proposed previously.

Among the golden orbweaver spiders of the family Nephilidae (catalogued as Nephilinae-Araneidae; we here follow the family classification proposed by [6]), coin spiders (genus *Herennia* Thorell, 1877 [7]) are the most species-rich genus with over a dozen species distributed in tropical Asia and Australasia. As all nephilids, they exhibit extreme sexual size dimorphism with males but a fraction of female size [6]. Unlike the remainder of genera though, all *Herennia* build arboricolous ("tree-hugging") ladder webs on tree trunks [6,8]. A newly updated *Herennia* taxonomy (in preparation) recognises 14 species of coin spiders (three of which have not yet been formally described and thus feature manuscript names in quotation marks). With the exception of *H. multipuncta* Doleschall 1859 [9], they are distributed narrowly, mostly as island endemics. *Herennia gagamba* Kuntner 2005 [8] and *H. tone* Kuntner 2005 are found in the Philippines, *H.* "tsoi" Kuntner et al., (in preparation) in Taiwan, *H.* "maj" Kuntner (in preparation) in Vietnam, *H. etruscilla* Kuntner 2005 in Java, *H.* "eva" Kuntner (in preparation) in Sulawesi, *H. deelemanae* Kuntner 2005 in Borneo, *H. jernej* Kuntner 2005 in Sumatra, *H. sonja* Kuntner 2005 in Borneo and Sulawesi, *H. papuana* Thorell 1881 [10] in New Guinea and Australia, *H. agnarssoni* Kuntner 2005 in the Solomon Islands, *H. milleri* Kuntner 2005 in New Guinea and New Britain and *H. oz* Kuntner 2005 in northern Australia (Figure 1). In contrast, *H. multipuncta* is distributed throughout southern India, Indochina and the Philippine and Indonesian archipelagos (Figure 1). Unlike other species, which are obligatory arboricoles in pristine forests, *H. multipuncta* is synanthropic, frequently found in managed habitats, and lives in sympatry with other, narrower endemic species of the genus [8]. This fact has sparked speculation on the invasive origin of *H. multipuncta* super-range [8], but this hypothesis has remained untested.

Although coin spiders exhibit several intriguing biological features, many aspects of their biology remain unexplored. A prior revision of the genus discussed the taxonomy, biology and biogeography of the 11 then-known species [8]. It suggested purely Australasian speciation of coin spiders and proposed the "*Herennia* line", west of Wallace's and Huxley's lines, which was only crossed by one then-known species, *H. multipuncta*. Since then, several new species have been recognised, some of them inhabiting areas west of the proposed line (Kuntner et al., in preparation). A recent nephilid biogeographic study [5] inferred historical biogeography of 10 species with available genetic data; however, that study was global and, thus, its biogeographical resolution was necessarily insufficient to resolve the *Herennia* biogeographic history. Nonetheless, Turk et al. [5] suggested an Indomalayan origin of the genus with recent colonisation of Australasia by the ancestor of *H. milleri* and *H. multipuncta*.

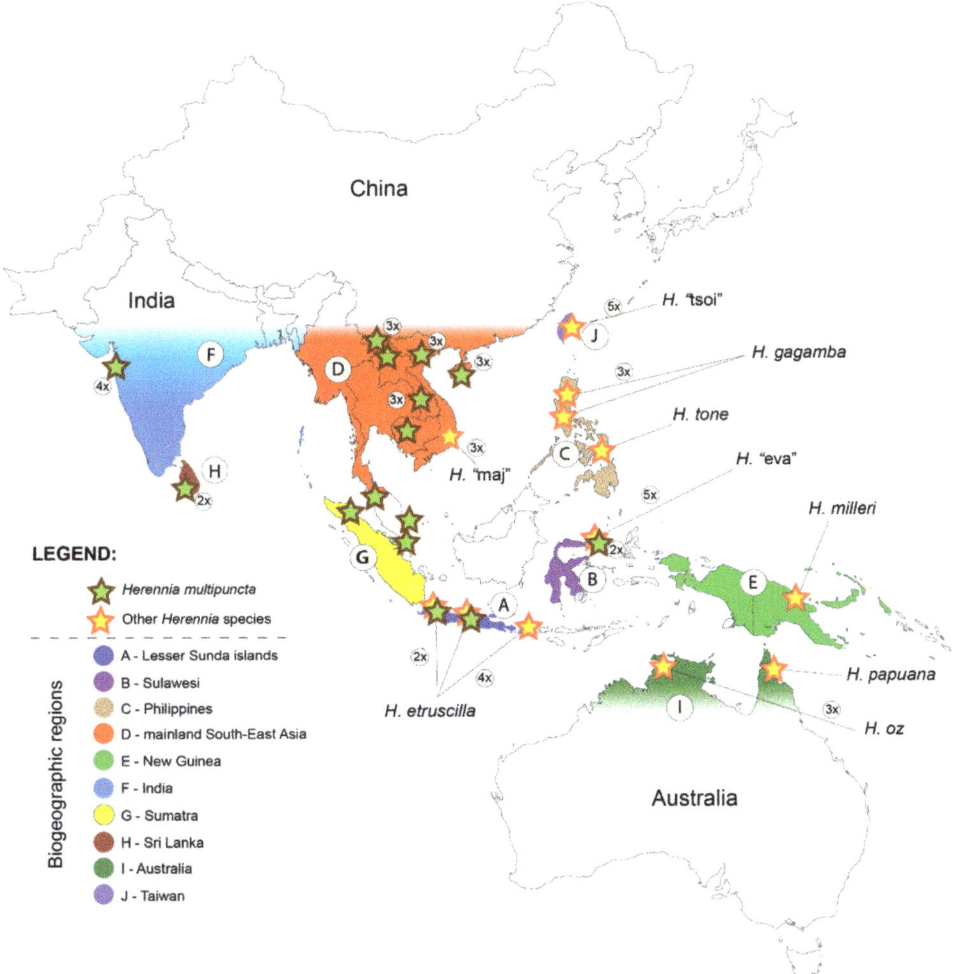

Figure 1. Sampling locations of *Herennia* individuals used in this study. Encircled numbers denote the number of individuals from the same sampling location. Geographic areas, used for historical biogeography inference (**A–J**), are colour-coded.

Here, we aimed to answer the following three main questions: (i) what is the sequence and chronology of coin spider dispersal from their origin to the present distribution, (ii) how would alternative dispersal biologies influence this pattern and (iii) is the unusually large range of *H. multipuncta* a result of human activity, and is the species invasive? We used phylogenomic data to construct a species-level phylogenetic scaffold, and then used classical phylogenetic markers to infer the most comprehensive population-level phylogeny and chronogram of coin spiders to date. We used this reference phylogeny to infer coin spiders' historical biogeography by adapting the methodology proposed by Turk et al. [5]. We tested two alternative models, each assuming a different type of dispersal, while accounting for the complex geological past of Australasia (Figure 2). The first model (A) assumes active dispersal via ballooning [11]. Ballooning behaviour has been observed in other nephilid species [12], but not yet in coin spiders. This type of dispersal promotes island colonisation, but also facilitates gene flow maintenance across large distances, thus inhibiting island endemism. The second model (B) assumes short distance random walking dispersal during the search for vacant habitat as the main method of dispersal. It allows for passive dispersal over long distances of connected lands given enough time. Neither model completely excludes rare chance occurrences of long-distance dispersal with wind currents.

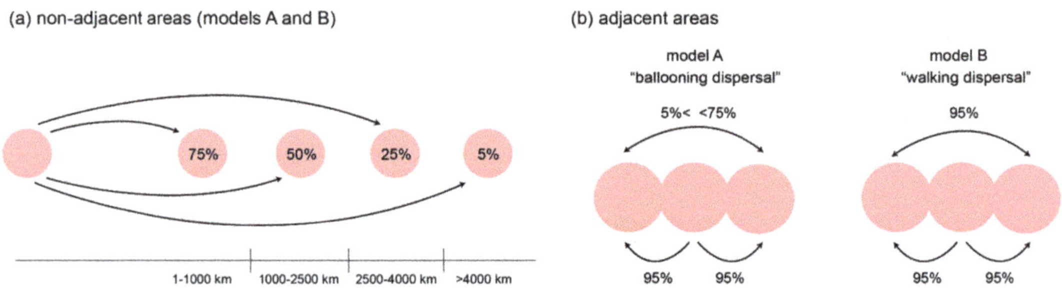

Figure 2. Methods of dispersal probability attribution. Each pink circle represents a geographical area. (**a**) For non-adjacent pairs of areas, both models attributed dispersal probabilities based on physical distances between the areas. Probabilities were inversely proportionate to the distance and binned into four categories. (**b**) When areas were in physical contact, model A attributed the maximal, 95% dispersal probability only between directly neighbouring areas, while indirectly connected areas were scored as non-adjacent pairs of areas. In contrast, model B treated all physically connected areas as a single area, attributing a 95% dispersal probability to all pairs of areas.

To address our main questions highlighted above, we inferred the most comprehensive species- and population-level phylogenies of coin spiders to date and calculated the most plausible biogeographic histories and dispersal trajectories using the two alternative models. If our data revealed *H. multipuncta* intraspecific splits are too recent to be resolved via phylogenetic time calibration, this would speak for a human-induced dispersal of *H. multipuncta* into other *Herennia* species ranges. In contrast, if intraspecific splits were resolved, and reconstructed as more ancient than human presence in this area (over 50,000 years [13]), then a natural colonisation would be implied. Generally, the estimated ages of nodes within *H. multipuncta* did not support the human-driven dispersal hypothesis. We speculate that ballooning ability and/or propensity was lost in coin spiders (except for *H. multipuncta*), resulting in a high occurrence of island endemism in the genus.

2. Materials and Methods

2.1. Species-Level Phylogeny

Currently, tissue material and genetic data are available for ten of the 14 known *Herennia* species, while the remaining four (*H. agnarssoni*, *H. deelemanae*, *H. jernej* and *H. sonja*) are only known from holotypes and could not be included here. Prior to our study, the best supported nephilid phylogeny used Anchored Hybrid Enrichment (AHE) data to resolve phylogenetic relationships among 22 species including five species of *Herennia* [6]. We here expand this taxon coverage to include eight *Herennia* species, of which *H. multipuncta* was represented by a specimen from Sri Lanka and one from Laos in order to test this widespread species monophyly (Supplementary Spreadsheet S1).

We employed the AHE targeted-sequencing approach for spiders (outlined in Hamilton et al. [14]) to target 585 single copy orthologous loci from across the genome. These loci have been shown to possess sufficient variation for resolving both shallow and deep-scale evolutionary relationships throughout the Araneae, e.g., [14,15]. These data have also been used to recover inter- and intrageneric relationships, as well as inter- and intraspecies relationships within a range of spider families [6,16–21]. Library preparation, enrichment, sequencing, assembly, alignment and phylogeny construction from AHE data followed the procedures described in Kuntner et al. [6]. The full AHE matrix is available as Supplementary File S1.

For two species, *H. papuana* and *H. tone*, we were unable to recover AHE data. To identify their phylogenetic placement, we ran maximum likelihood (RAxML) and Bayesian (MrBayes) phylogenetic inference analyses only on COI sequences, obtained for the ten species. The primers and PCR amplification protocols to obtain partial COI sequences followed Kuntner et al. [22] (matrix available as Supplementary File S2; COI GenBank accession codes OK017092 to OK017142). The placements of *Herennia* "maj" and *H. milleri*, the latter likely due to low COI sequence quality, were poorly supported and inconsistent with the AHE topology. To eliminate the influence of these "rogue taxa" on *H. papuana* (topological placement of *H. tone* was consistent regardless of their presence), these two terminals were removed from the COI only dataset and the analyses were repeated. Results produced by MrBayes were inspected with Tracer to ensure effective sample sizes were >200.

Divergence dating calibration was performed in BEAST2 [23], again using only COI sequences. We used a relaxed log normal clock and set the bModelTest [24] as the nucleotide substitution model and a birth-death tree prior. The ucldMean prior was set as normally distributed with a mean of 0.0199 and a standard deviation of 0.001 following Bidegaray-Batista and Arnedo [25]. The topology was constrained as described above, whereas clade ages and their confidence intervals were constrained in three nodes that correspond directly to those acquired by Kuntner et al. [6]. BEAST ran on four MCMC chains for 10 million generations. Results were checked with Tracer and summarised with TreeAnnotator with a 10% burnin.

2.2. Population-Level Phylogeny

In the population-level phylogeny, species were represented by a varying number of samples, ranging from one in *H. tone*, *H. milleri* and *H. papuana* to 26 in *H. multipuncta*. Again, we inferred the phylogeny in BEAST2, using COI and 28S sequences. For 28S sequences, primers and PCR amplification protocols again followed Kuntner et al. [22] (concatenated matrix available as Supplementary File S3; 28S GenBank accession codes OK017174 to OK017212). Where an individual lacked data for both genes, only one was used. We employed a relaxed log normal clock and set the bModelTest. The Coalescent Bayesian Skyline prior, allowing for stochastic changes in population sizes through time, was chosen as the tree prior following Ritchie et al. [26]. COI priors were set as before, while the ucldMean prior for 28S was set as normally distributed with a mean of 0.0011 and standard deviation of 0.0003 after Bidegaray-Batista and Arnedo [25]. The ages of all species-level splits were constrained to those recovered in the previous phylogeny. BEAST ran on four MCMC chains for 70 million generations to ensure large enough effective sample sizes. Again, a consensus tree was obtained with TreeAnnotator with a 10% burnin.

2.3. Inference of Biogeographic History

For biogeographic analyses, we pruned the reference population-level phylogeny so that each population (locality) was represented by a single specimen (hereafter referred to as "pruned phylogeny"). The individual with the most complete sequence was chosen as the representative of each population. This narrowed the population tree to 29 tips, with the number of representatives per species ranging from one in *H. oz*, *H. milleri*, *H. papuana* and *H. tone* to 13 in *H. multipuncta*. We treated the known areas of species occupancies within 10 biogeographic regions. These consisted of three continental landmasses—mainland South-East Asia, Australia and India—and seven islands or archipelagos—Sulawesi, Sumatra, New Guinea, Sri Lanka, Taiwan, the Philippines and the island pair Java and Lombok, for simplicity referred to as Lesser Sunda islands. As the tips in the phylogenetic tree are individual samples, necessarily only inhabiting a single biogeographic region, each tip was attributed one region.

Following the rationale developed in Turk et al. [5], dispersal probabilities were fine-tuned to reflect the varying geographical configuration of biogeographic regions during the area's lively geological past. Physical distance among landmasses was used as a proxy for dispersal probabilities, scored separately in six time slices, each spanning 5 million years. As the geological history of the area, especially Indonesian islands, is extremely complex and thus difficult to reconstruct with precision, we binned dispersal probabilities into five categories. Following the argumentation in Turk et al. [5], we attributed a 95% dispersal probability to pairs of geographic regions in physical contact, where dispersal is likely but not necessary, and a 5% dispersal probability where distances exceed 4000 kilometres. For distances between regions of 1000 kilometres or less, a 75% dispersal probability was assigned, for distances between 1000 and 2500 kilometres, a 50% dispersal probability, and for distances between 2500 and 4000 kilometres, a 25% dispersal probability (Figure 2a). If a region had not yet emerged or was sunk during a time slice, it was disallowed in the "areas allowed" option in RASP (see below).

In contrast to the previous paper, however, we attributed these probabilities in two ways, differing in the definition of physical contact among regions (Figure 2b). In time slices where, for example, three regions were consecutively connected by land, model A attributed a 95% dispersal probability only between the middle and marginal areas, but not between the marginal areas themselves, thus accounting for the larger physical distance between them, despite physical connection via the middle area. This model, which we term "ballooning dispersal", puts emphasis on long-distance dispersal via ballooning as the main method of dispersal in coin spiders. Model B, in contrast, attributed a 95% dispersal probability for all three pairs of regions in the previous example. Biologically, model B assumes coin spiders largely disperse over land with small-step, gradual expansion; we term this "walking dispersal". Given enough time, this model assumes that spiders can reach all physically connected areas equally likely.

Geological data (Supplementary Spreadsheet S1) were compiled from a tectonic reconstruction model [27] via GPlates plate tectonics visualisation software [28] and geological literature [29–33]. We reconstructed the genus' historical biogeography with RASP 4.0 beta [34], comparing all six included biogeographical models. The maximum number of ancestral areas occupied was set to three. We evaluated model fit through weighted AIC_c values (AIC_c_wt), expressing the model's relative probability, corrected for small sample sizes.

3. Results

3.1. Phylogenies

The AHE phylogeny placed eight *Herennia* species unequivocally and with overwhelming support in all nodes (Figure 3a). The oldest split from the MRCA of all coin spiders was recovered in the Vietnamese *H.* "maj". *Herennia* "eva" and *H. gagamba* then branched off as sister species, followed by *H.* "tsoi" and *H. multipuncta*, *H. milleri* and finally *H. etruscilla* and *H. oz* as sister species. As for the COI sequences, both RAxML and MrBayes consistently placed *H. tone* as sister to *H.* "eva" with a 99% support in MrBayes (Figure 3b). After the elimination of "rogue" taxa that interfered with topological stability, *H. papuana* was placed as sister to ((*H.* "tsoi", *H. multipuncta*), (*H. milleri*, *H. etruscilla*, *H. oz*)) consistently using both methods, again with a high node support of 92% in MrBayes (Figure 3b).

In the population-level phylogeny (Supplementary Figure S1), samples of the same species always grouped together; however, samples from the same locality often did not (e.g., *H. etruscilla* populations from Java, and *H. multipuncta* populations from Laos, Vietnam, Malaysia, Yunnan and Hainan). In the pruned phylogeny containing one tip per population, used in biogeographical reconstruction (see Figure 4), the divergence dating revealed frequent within-species cladogenesis during the last few million years.

3.2. Biogeographical Reconstruction: Model A

RASP identified DIVALIKE+j as the best model for the data (Table 1, Figure 4). The node uniting all *Herennia* taxa received strong support for a wide ancestral distribution in Australia, mainland SE Asia and the Philippines (61%). Although *H.* "maj" persisted only in mainland SE Asia, the most recent common ancestor (MRCA) of all other species persisted in the Philippines and Australia (65%). The clade containing *H. gagamba*, *H. tone* and *H. "eva"* remained in the Philippines, with the latter species colonising Sulawesi sometime during the last four million years.

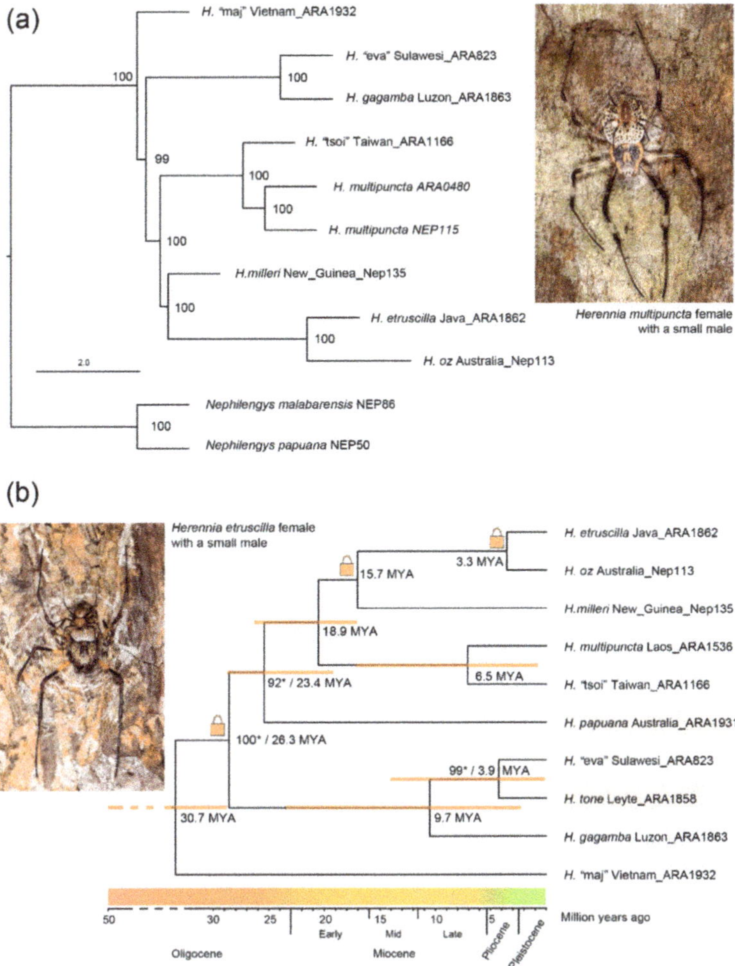

Figure 3. Species-level phylogenies of coin spiders. (**a**) AHE-only phylogeny produced with AS-TRAL [35], resolving the relationships between eight species of *Herennia* with *Nephilengys* as the outgroup as per Kuntner et al. [6]; (**b**) Species-level phylogeny of the available ten species of *Herennia*, calculated from COI data. Highlighted are two species lacking AHE data, *H. papuana* and *H. tone*. The lock symbols denote age-constrained nodes. Supports for nodes not present in the AHE-only phylogeny (marked with an asterisk) were recovered by MrBayes.

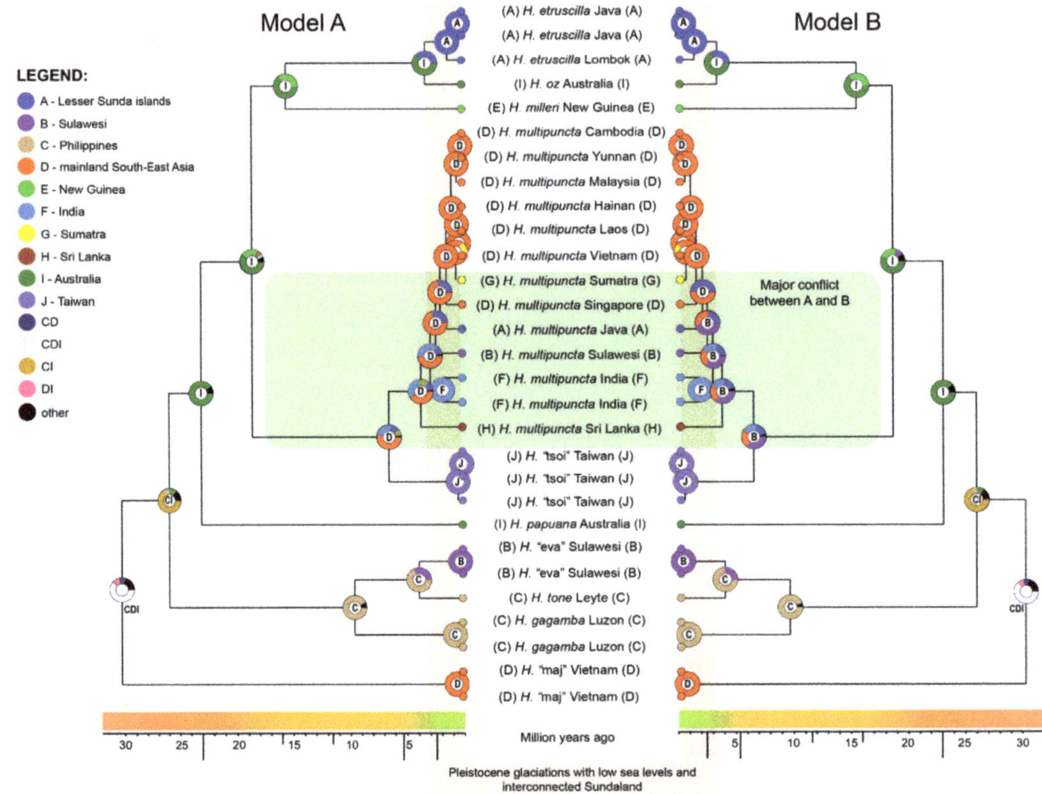

Figure 4. RASP ancestral area reconstruction of two alternative biogeographic models on a COI + 28S population-level phylogeny of *Herennia*. Model A assumes long-distance dispersal via ballooning as the main method of dispersal in coin spiders, while model B predicts that coin spiders mainly disperse over land with small-step, gradual expansion. Encircled letters signify the likeliest ancestral area in that node, with a combination of several letters indicating an inferred distribution in all those areas. Green area marks the major conflicts in ancestral area reconstruction between the two models.

Table 1. RASP model scores for models A and B. Best supported RASP model is shown in bold. An asterisk (*) in the last column indicates that the model variant allowing jump dispersal (+j) was a significantly better fit for the data than the variant without it.

Model A	LnL	AICc	AICc_wt	Δ −j/+j
DEC	−62.28	129	2.5×10^{-7}	*
DEC+j	−47.08	101.1	0.29	
DIVALIKE	−55.86	116.2	0.0002	*
DIVALIKE+j	**−46.19**	**99.33**	**0.7**	
BAYAREALIKE	−72.01	148.5	1.5×10^{-11}	*
BAYAREALIKE+j	−50.27	107.5	0.012	
Model B				
DEC	−62.86	130.2	3.1×10^{-7}	*
DEC+j	−48	103	0.25	
DIVALIKE	−56.68	117.8	0.0001	*
DIVALIKE+j	**−46.92**	**100.8**	**0.74**	
BAYAREALIKE	−73.4	151.3	8.1×10^{-12}	*
BAYAREALIKE+j	−50.89	108.7	0.014	

The MRCA of all other species remained in Australia only, with *H. papuana* persisting in the same area until the present. The MRCA of *H.* "tsoi" and *H. multipuncta* shifted from Australia to mainland SE Asia, from where *H.* "tsoi" colonised Taiwan, while *H. multipuncta* remained in SE Asia with consecutive colonisations of Sri Lanka, India, Sulawesi, Java and Sumatra during the last four million years. From the remaining Australian MRCA, *H. milleri* colonised New Guinea, while its sister Australian clade diversified into the Australian *H. oz* and *H. etruscilla*, which shifted to the Lesser Sunda islands in the last few million years.

3.3. Biogeographical Reconstruction: Model B

Again, DIVALIKE+j received the highest statistical support in RASP (Table 1, Figure 4). All nodes except the MRCA of *H.* "tsoi" and *H. multipuncta* and their intra-species nodes received nearly identical support as in model A. Here, the reconstruction for the MRCA of *H.* "tsoi" and *H. multipuncta* was ambiguous, with similar support for Sulawesi (35%), mainland SE Asia (23%) and India (21%). The basal *H. multipuncta* was attributed the same three regions in almost identical shares (36, 23 and 20%, respectively). The species was reconstructed to have colonised Sri Lanka and India from Sulawesi. Next, it colonised (and remained in) the Lesser Sunda islands (49%) or mainland SE Asia (51%), and then dispersed across the SE Asian mainland and to Sumatra.

Comparing the fit of DIVALIKE+j between the two models through the calculation of the Bayes factor from the log of the likelihood scores ($B = 0.98$; after Kass and Raftery [36]) revealed a non-significant difference.

3.4. Human-Induced Dispersal of H. multipuncta?

Neither of the two models supported the hypothesis of a recent, human-induced dispersal of *H. multipuncta*. Node ages, recovered for phylogenetic splits among the included populations in the pruned phylogeny, ranged from 3.72 million years ago (mya) in the MRCA of all included populations to 0.13 mya (132,200 years) in the split between the Yunnanese and Cambodian populations. In the latter, the confidence interval ranged from 516,800 to 800 years ago, a time frame potentially compatible with human-induced dispersal, albeit between two areas in relatively close proximity. The only other node whose estimated age fit within the timeframe of human presence in the area was the split between the Vietnamese and Lao populations. It was dated to 277,300 years ago, with a confidence interval ranging from 757,500 to 26,900 years ago. The two areas are adjacent with no obvious barriers between them, implying that natural dispersal is very likely.

4. Discussion

In this study, we provide a test case of what next generation biogeographic inference should optimally encompass: a robust phylogenetic/phylogenomic framework for a comprehensive population-level ancestral area reconstruction that at the same time accounts for geological dynamics and species biology. To this end, we used a modified version of our previously proposed methodology [5], comparing two methods of dispersal probability quantification. Both models suggested a wide ancestral range and relatively old splits (from 3.72 to 0.13 mya) between terminals of *H. multipuncta*, strongly implicating a natural, not anthropogenic colonisation of the areas that constitute an extremely wide contemporary range of this species.

4.1. Phylogenetic Placements

The species-level phylogeny recovered unexpected relationships among species with overwhelming support (Figure 3a,b). Within the available taxon sample, *Herennia* "maj" was the first species to split from the coin spider MRCA, a placement that was not recovered in a prior COI-based phylogeny in Turk et al. [5]. As surprising as this may be, we believe the relationship is not artefactual, given our understanding of the robustness of AHE phylogenomic topologies in nephilids (*H.* "maj" was not included in Kuntner et al. [6]).

In the population-level phylogeny, species were represented by a varying number of specimens. In *H. multipuncta*, the largest sample, specimens from the same population failed to group together (Supplementary Figure S1). Perhaps genetic differences between them are not (yet) large enough, as they maintain modest gene flow among populations. This matches our emerging hypothesis of heightened dispersal propensity in this species, relative to others, as proposed below.

4.2. Biogeographic Inference with Two Models

Unlike our previous study [5], we only have limited knowledge about the dispersal biology of coin spiders. As in all motile animals, they can be assumed to, at least, undergo gradual, slow dispersal through suitable habitat during stochastic search for vacant space. On the other hand, we might expect coin spiders to practice active ballooning, as observed in the related *Nephila pilipes* [12], but no field or experimental evidence that would support this assertion currently exists. Even in the absence of ballooning, however, chance dispersal across geographic barriers, such as the sea, have to occur, otherwise they would not be able to simultaneously inhabit landmasses that have not been connected during the relevant timeframe, such as mainland SE Asia and Australia. The two proposed types of dispersal have different consequences: ballooning would make it easier to colonise new areas and maintain gene flow across the entire range, but the evolution of island endemism is less likely. Conversely, relying on "walking" dispersal with occasional, chance long-distance dispersal would facilitate genetic isolation and thus speciation and island endemism, but would restrict gene flow maintenance across large ranges. Regardless, we do not assume that all coin spider species necessarily exhibit identical dispersal behaviours and propensities.

Thus, in the absence of a better understood, and experimentally tested, organism-specific dispersal biology, we resorted to engaging in two biogeographical approaches (Figure 2). Model A assumed less likely dispersal between areas further apart, even if connected by land, because ballooning was taken as the default method of dispersal. In contrast, model B assigned equal dispersal probabilities to all physically connected areas, as it assumed spiders spread "on foot", gradually, over millions of years. Although such passive, "walking" dispersal is much slower, range expansion of, for example, only 10 m per year (which we deem as a distance easily overcome by orbweaving spiders) would theoretically allow for a spread across 10,000 kilometres over the course of a single million years, which is approximately the extent of the entire extant genus range. While the comparison of the two best-fitting models using the Bayes factor did not show significant differences between them, they nonetheless provided somewhat discrepant results.

Deep nodes near the root of the phylogeny were reconstructed with nearly identical probability proportions for ancestral areas by both models (Figure 4). A wide ancestral distribution over mainland SE Asia, Australia and the Philippines fragmented, with the ancestral mainland population leading to the Vietnamese *H.* "maj", and the Australian and Philippine populations giving rise to all other species. Interestingly, mainland SE Asia was not re-colonised until 24–29 million years later, depending on the model. In both models, *H. tone* and *H. gagamba* maintain the ancestral range in the Philippines, while *H.* "eva" disperses to Sulawesi sometime during the last four million years. This inferred colonisation is plausible assuming either model of dispersal, requiring an active or chance dispersal across approximately 600 km of sea [32]. Either type of dispersal on this route may have been further facilitated by a possible island chain connecting Sulawesi to the Philippines in the Neogene [33].

The most likely ancestral areas of the (*H.* "tsoi" + *H. multipuncta* + *H. milleri* + *H. oz* + *H. etruscilla*) and (*H. milleri* + *H. oz* + *H. etruscilla*) MRCAs are reconstructed to Australia by both models (Figure 4). The latter clade features another recent over-water colonisation of *H. etruscilla* from Australia to Lesser Sunda islands. A salient difference between the two models is the inferred dispersal of *H. multipuncta* (Figure 4). In model A, the MRCA of *H. multipuncta* and *H.* "tsoi" is already distributed in mainland SE Asia, from where *H.* "tsoi" colonises Taiwan during the last 6 million years. This does not exceed the understood age of the island (approximately 9 million years [29]), which has remained close to, and even connected with, the Asian mainland by a land bridge during Pleistocene glaciations [30]. In *H. multipuncta*, model A suggests gradual expansion from mainland SE Asia to Sri Lanka, India, across Indonesian islands and finally within mainland SE Asia itself. On the other hand, model B inconclusively places the MRCA of *H. multipuncta* and *H.* "tsoi" to Sulawesi, a challenging proposition. At 6 mya, the distance between Sulawesi and Taiwan, which *H.* "tsoi" supposedly crossed, was not considerably shorter than today (approximately 2300 kilometres), having been separated by the Philippine archipelago.

Deeper nodes within *H. multipuncta* are also ambiguous in model B, with similar probabilities for several ancestral areas, but it generally suggests gradual dispersal of the species from Sulawesi to Sri Lanka and India, colonisation of Java directly or via mainland SE Asia, and finally a spread within SE Asia and colonisation of Sumatra. Dispersal from Sulawesi to Sri Lanka during this time would require crossing a >4000-kilometre distance either by chance dispersal, which is rarely observed on such a scale in spiders in general (as discussed in Turk et al. [5]), or over land. As opposed to islands such as Sri Lanka, New Guinea and Taiwan, however, it is unclear whether Sulawesi formed land connections with the mainland or mainland-connected islands, such as Borneo, during the time of Plio-Pleistocene sea level changes [33]. Furthermore, a subsequent colonisation of India, requiring another >4000-kilometre dispersal event shortly after the first, seems highly unlikely. On the other hand, all sampled *H. multipuncta* populations except those from Sulawesi, India and Sri Lanka were connected by land during the Pleistocene on the landmass Sundaland [32]. Dispersing around this landmass could indeed be performed by "walking" dispersal, but, as seen above, even model B infers several oversea dispersals, either by chance or actively.

Curiously, species such as *H.* "maj" seem to maintain a narrow distribution despite living on the Asian mainland, where they could, across millions of years, passively disperse over a much wider area (assuming our model B with no other limiting factors). We speculate such species are either restricted to specific habitats that are not continuous, decreasing the chance of successful active or passive dispersal, or are confined to their range through ecological competition with other species and genera in adjacent areas.

4.3. A Natural "Coinquest" of H. multipuncta

The pruned population-level phylogeny showed relatively recent splitting between the sampled populations of *H. multipuncta* (median node ages ranging from 3.72 to 0.13 mya), however, not recent enough to infer human-related dispersal. If that were the case, nucleotide sequence divergencies would have been predictively low, having only accumulated changes over the past few thousand years (contra to what we see in our phylogeny). In the two cases where confidence intervals of node ages do overlap with human presence in the species' range, the pairs of geographical areas are either adjacent or in close proximity, making the exclusion of natural colonisation difficult.

Why, then, is a super-wide present-day range of *H. multipuncta* unique among all *Herennia*? We hypothesise that although nephilid ancestors perform active ballooning, retained in *Nephila* and *Trichonephila* [9,34,35], coin spiders secondarily lost the ability to balloon. Such a loss of dispersal ability is a common phenomenon in island spider biology, severely limiting gene flow maintenance and often leading to single-island endemism [37,38]. *Herennia multipuncta* might have regained this ability, allowing it to disperse across suitable habitat and inhabit most of the genus range, sometimes overlapping with more ancient *Herennia* spp ranges. This would also allow it to sustain some degree of gene flow among populations, which would, in turn, explain the recovered phylogenetic picture of non-monophyly of sympatric specimens of certain populations (from Vietnam, Laos, Malaysia, Hainan and Yunnan). In many parts of its range, *H. multipuncta* is sympatric with other species (*H. etruscilla* in Java, *H.* "eva" in Sulawesi, *H.* "maj" in Vietnam, *H. jernej* in Sumatra, etc.) and, according to our dated population phylogeny, this has been the case for millions of years. This pattern suggests that *H. multipuncta* does not outcompete other sympatric species, perhaps due to subtle differences in ecological niches. In fact, adaptiveness to different habitats might be another trait, specific to *H. multipuncta*, that enabled its easier and more successful dispersal. These interpretations could be tested in an ecological framework in the near future, as could the ability of *H. multipuncta*, but not other *Herennia*, to disperse via ballooning.

4.4. Limitations in Methodology and Future Work

One source of statistical bias in the present study is the incomplete representation of coin spider species in the phylogeny and with it the lack of representation of the missing species' geographical distributions. The precise phylogenetic placement of the Bornean *H. deelemanae*, Sumatran *H. jernej*, Bornean and Sulawesian *H. sonja*, as well as *H. agnarssoni* from the Solomon Islands has never been tested outside of a morphological, cladistic framework [8]; therefore, the influence of their distributions on biogeographical reconstruction remains unclear. Furthermore, at the population level, species are not equally represented in terms of specimen number and range cover, also potentially biasing evolutionary relationships and divergence times between populations. That said, specimen collections of coin spiders are scarce. The present study was performed with all genetic material currently available to us.

One of the topics addressed in the study was the dispersal behaviour of coin spiders. Ideally, the presence or absence of active ballooning ought to be tested experimentally. Considering that ballooning is difficult to observe in nature, future research could include subjecting juvenile coin spiders to wind tunnel experiments [12] in a laboratory environment. If performed on multiple species, such an experiment might serve as a test of our hypothesised regained ballooning behaviour in *H. multipuncta*, but not in other species.

5. Conclusions

We have demonstrated the importance of the understanding of organismal biology in biogeographic reconstructions. In organisms where dispersal is not well understood, testing alternative modes of dispersal through parallel statistical models might prove helpful in uncovering the most likely dispersal biology without direct field observation. By modifying our previously proposed pipeline to account not only for the specifics of the geological history of the area, but also dispersal specifics of the studied organisms, we are further contributing to the development of biogeographic methodology.

Supplementary Materials: The following are available online at https://www.mdpi.com/article/10.3390/d13110515/s1, File S1: AHE matrix (Fasta) with partition file, File S2: COI species-level matrix (Nexus), File S3: concatenated COI and 28S population-level matrix (Nexus), Figure S1: full population-level phylogeny, Spreadsheet S1: geology and geography data.

Author Contributions: Conceptualisation, E.T. and M.K.; methodology, E.T., K.Č., C.A.H. and M.K.; formal analysis, E.T., K.Č., C.A.H., J.E.B., R.-C.C. and M.K.; investigation, all authors; writing—original draft preparation, E.T. and M.K.; writing—review and editing, all authors; visualisation, K.Č.; funding acquisition, M.K. and S.K.-F. All authors have read and agreed to the published version of the manuscript.

Funding: This research was funded by the Slovenian Research Agency, grants P1-0236, P1-0255, J1-9163 and 1000-17-0618.

Institutional Review Board Statement: Not applicable.

Informed Consent Statement: Not applicable.

Data Availability Statement: The new genetic sequences are available on GenBank (accession codes: OK017092 to OK017142 (COI), OK017174 to OK017212(28S)) and also as a supplement to this paper. AHE and geological data are available as a supplement to this paper.

Acknowledgments: We thank A. V. Abramov, I. Agnarsson, G. J. Anderson, M. Bedjanič, D. Court, A. Harmer, M. S. Harvey, S. Huber, P. Jaeger, D. Li, W. Maddison, W. Schawaller and I.-M. Tso for their kind assistance in field work and/or specimen acquisition. We thank I. Magalhaes and two anonymous referees for constructive criticism.

Conflicts of Interest: The authors declare no conflict of interest.

References

1. Jin, Y.; Brown, R.P. Partition number, rate priors and unreliable divergence times in Bayesian phylogenetic dating. *Cladistics* **2018**, *34*, 568–573. [CrossRef]
2. Brown, J.W.; Smith, S.A. The Past Sure is Tense: On Interpreting Phylogenetic Divergence Time Estimates. *Syst. Biol.* **2018**, *67*, 340–353. [CrossRef]
3. Magalhaes, I.L.F.; Santos, A.J.; Ramírez, M.J. Incorporating topological and age uncertainty into event-based biogeography supports paleo-islands in Galapagos and ancient connections among Neotropical dry forests. *Diversity* **2021**, *13*, 418, In press. [CrossRef]
4. Crews, S.C.; Esposito, L.A. Towards a synthesis of the Caribbean biogeography of terrestrial arthropods. *BMC Evol. Biol.* **2020**, *20*, 1–27. [CrossRef]
5. Turk, E.; Čandek, K.; Kralj-Fišer, S.; Kuntner, M. Biogeographical history of golden orbweavers: Chronology of a global conquest. *J. Biogeogr.* **2020**, *47*, 1333–1344. [CrossRef]
6. Kuntner, M.; Hamilton, C.A.; Cheng, R.-C.; Gregorič, M.; Lupše, N.; Lokovšek, T.; Lemmon, E.M.; Lemmon, A.R.; Agnarsson, I.; Coddington, J.A.; et al. Golden orbweavers ignore biological rules: Phylogenomic and comparative analyses unravel a complex evolution of sexual size dimorphism. *Syst. Biol.* **2019**, *68*, 555–572. [CrossRef] [PubMed]
7. Thorell, T. Studi sui Ragni Malesi e Papuani. I. Ragni di Selebes raccolti nel 1874 dal Dott. O. Beccari. *Ann. del Mus. Civ. Stor. Nat. Genova* **1877**, *10*, 341–637.
8. Kuntner, M. A revision of *Herennia* (Araneae: Nephilidae: Nephilinae), the Australasian "coin spiders". *Invertebr. Syst.* **2005**, *19*, 391–436. [CrossRef]
9. Doleschall, L. Tweede Bijdrage tot de kennis der Arachniden van den Indischen Archipel. *Acta Soc. Sci. Indica-Neerl.* **1859**, *5*, 1–60.
10. Thorell, T. Studi sui Ragni Malesi e Papuani. III. Ragni dell'Austro Malesia e del Capo York, conservati nel Museo civico di storia naturale di Genova. *Ann. Mus. Civ. Stor. Nat. Genova* **1881**, *17*, 7–27.
11. Bell, J.R.; Bohan, D.A.; Shaw, E.M.; Weyman, G.S. Ballooning dispersal using silk: World fauna, phylogenies, genetics and models. *Bull. Entomol. Res.* **2005**, *95*, 69–114. [CrossRef] [PubMed]
12. Lee, V.M.J.; Kuntner, M.; Li, D. Ballooning behavior in the golden orbweb spider *Nephila pilipes* (Araneae: Nephilidae). *Front. Ecol. Evol.* **2015**, *3*, 1–5. [CrossRef]
13. O'Connell, J.F.; Allen, J.; Williams, M.A.J.; Williams, A.N.; Turney, C.S.M.; Spooner, N.A.; Kamminga, J.; Brown, G.; Cooper, A. When did *Homo sapiens* first reach Southeast Asia and Sahul? *Proc. Natl. Acad. Sci. USA* **2018**, *115*, 8482–8490. [CrossRef] [PubMed]
14. Hamilton, C.A.; Lemmon, A.R.; Lemmon, E.M.; Bond, J.E. Expanding Anchored Hybrid Enrichment to resolve both deep and shallow relationships within the spider tree of life. *BMC Evol. Biol.* **2016**, *16*, 212. [CrossRef] [PubMed]
15. Opatova, V.; Hamilton, C.A.; Hedin, M.; De Oca, L.M.; Král, J.; Bond, J.E.; Wiegmann, B. Phylogenetic Systematics and Evolution of the Spider Infraorder Mygalomorphae Using Genomic Scale Data. *Syst. Biol.* **2020**, *69*, 671–707. [CrossRef] [PubMed]
16. Hamilton, C.A.; Hendrixson, B.E.; Bond, J.E. Taxonomic revision of the tarantula genus *Aphonopelma* Pocock, 1901 (Araneae, Mygalomorphae, Theraphosidae) within the United States. *Zookeys* **2016**, *2016*, 1–340. [CrossRef] [PubMed]
17. Chamberland, L.; McHugh, A.; Kechejian, S.; Binford, G.J.; Bond, J.E.; Coddington, J.A.; Dolman, G.; Hamilton, C.A.; Harvey, M.S.; Kuntner, M.; et al. From Gondwana to GAARlandia: Evolutionary history and biogeography of ogre-faced spiders (*Deinopis*). *J. Biogeogr.* **2018**, *45*, 2442–2457. [CrossRef]
18. Maddison, W.P.; Evans, S.C.; Hamilton, C.A.; Bond, J.E.; Lemmon, A.R.; Lemmon, E.M. A genome-wide phylogeny of jumping spiders (Araneae, Salticidae), using Anchored Hybrid Enrichment. *Zookeys* **2017**, *695*, 89–101. [CrossRef]
19. Godwin, R.L.; Opatova, V.; Garrison, N.L.; Hamilton, C.A.; Bond, J.E. Phylogeny of a cosmopolitan family of morphologically conserved trapdoor spiders (Mygalomorphae, Ctenizidae) using Anchored Hybrid Enrichment, with a description of the family, Halonoproctidae Pocock 1901. *Mol. Phylogenet. Evol.* **2018**, *126*, 303–313. [CrossRef]
20. Bond, J.E.; Hamilton, C.A.; Godwin, R.L.; Ledford, J.M.; Starrett, J. Phylogeny, evolution, and biogeography of the North American trapdoor spider family Euctenizidae (Araneae: Mygalomorphae) and the discovery of a new "endangered living fossil" along California's Central Coast. *Insect Syst. Divers.* **2020**, *4*, 1–14. [CrossRef]
21. Hebets, E.A.; Bern, M.; McGinley, R.H.; Roberts, A.; Kershenbaum, A.; Starrett, J.; Bond, J.E. Sister species diverge in modality-specific courtship signal form and function. *Ecol. Evol.* **2021**, *11*, 852–871. [CrossRef] [PubMed]
22. Kuntner, M.; Arnedo, M.A.; Trontelj, P.; Lokovšek, T.; Agnarsson, I. A molecular phylogeny of nephilid spiders: Evolutionary history of a model lineage. *Mol. Phylogenet. Evol.* **2013**, *69*, 961–979. [CrossRef]
23. Bouckaert, R.R.; Heled, J.; Kühnert, D.; Vaughan, T.; Wu, C.H.; Xie, D.; Suchard, M.A.; Rambaut, A.; Drummond, A.J. BEAST 2: A software platform for Bayesian evolutionary analysis. *PLoS Comput. Biol.* **2014**, *10*, e1003537. [CrossRef] [PubMed]
24. Bouckaert, R.R.; Drummond, A.J. bModelTest: Bayesian phylogenetic site model averaging and model comparison. *BMC Evol. Biol.* **2017**, *17*, 42. [CrossRef]

25. Bidegaray-Batista, L.; Arnedo, M.A. Gone with the plate: The opening of the Western Mediterranean basin drove the diversification of ground-dweller spiders. *BMC Evol. Biol.* **2011**, *11*, 317. [CrossRef]
26. Ritchie, A.M.; Lo, N.; Ho, S.Y.W. The impact of the tree prior on molecular dating of data sets containing a mixture of inter- and intraspecies sampling. *Syst. Biol.* **2017**, *66*, 413–425. [CrossRef] [PubMed]
27. Müller, R.D.; Zahirovic, S.; Williams, S.E.; Cannon, J.; Seton, M.; Bower, D.J.; Tetley, M.; Heine, C.; Le Breton, E.; Liu, S.; et al. A global plate model including lithospheric deformation along major rifts and orogens since the Triassic. *Tectonics* **2019**, *38*, 1884–1907. [CrossRef]
28. Müller, R.D.; Cannon, J.; Qin, X.; Watson, R.J.; Gurnis, M.; Williams, S.; Pfaffelmoser, T.; Seton, M.; Russell, S.H.J.; Zahirovic, S. GPlates: Building a virtual Earth through deep time. *Geochem. Geophys. Geosys.* **2018**, *19*, 2243–2261. [CrossRef]
29. Sibuet, J.C.; Hsu, S.K. How was Taiwan created? *Tectonophysics* **2004**, *379*, 159–181. [CrossRef]
30. Voris, H.K. Maps of Pleistocene sea levels in Southeast Asia: Shorelines, river systems and time durations. *J. Biogeogr.* **2000**, *27*, 1153–1167. [CrossRef]
31. Park, Y.; Maffre, P.; Godderis, Y.; MacDonald, F.A.; Anttila, E.S.C.; Swanson-Hysell, N.L. Emergence of the Southeast Asian islands as a driver for Neogene cooling. *Proc. Natl. Acad. Sci. USA* **2020**, *117*, 25319–25326. [CrossRef] [PubMed]
32. Hall, R. The palaeogeography of Sundaland and Wallacea since the Late Jurassic. *J. Limnol.* **2013**, *72*, 1–17. [CrossRef]
33. Moss, S.J.; Wilson, M.E.J. Biogeographic implications of the Tertiary palaeogeographic evolution of Sulawesi and Borneo. In *Biogeography and Geological Evolution of SE Asia*; Hall, R., Holloway, J.D., Eds.; Backhuys Publishers: Leiden, The Netherlands, 1998; pp. 133–163.
34. Yu, Y.; Harris, A.J.; Blair, C.; He, X. RASP (Reconstruct Ancestral State in Phylogenies): A tool for historical biogeography. *Mol. Phylogenet. Evol.* **2015**, *87*, 46–49. [CrossRef] [PubMed]
35. Mirarab, S.; Warnow, T. ASTRAL-II: Coalescent-based species tree estimation with many hundreds of taxa and thousands of genes. *Bioinformatics* **2015**, *31*, i44–i52. [CrossRef] [PubMed]
36. Kass, R.E.; Raftery, A.E. Bayes factors. *J. Am. Stat. Assoc.* **1995**, *90*, 773–795. [CrossRef]
37. Čandek, K.; Agnarsson, I.; Binford, G.J.; Kuntner, M. Biogeography of the Caribbean *Cyrtognatha* spiders. *Sci. Rep.* **2019**, *9*, 1–14. [CrossRef] [PubMed]
38. Gillespie, R.G.; Baldwin, B.G.; Waters, J.M.; Fraser, C.I.; Nikula, R.; Roderick, G.K. Long-distance dispersal: A framework for hypothesis testing. *Trends Ecol. Evol.* **2012**, *27*, 47–56. [CrossRef]

 *diversity*

MDPI

Article

Single-Island Endemism despite Repeated Dispersal in Caribbean *Micrathena* (Araneae: Araneidae): An Updated Phylogeographic Analysis

Lily Shapiro [1,*], Greta J. Binford [2] and Ingi Agnarsson [1,3,*]

1   Department of Biology, College of Arts and Sciences, University of Vermont, 109 Carrigan Drive, Burlington, VT 05401, USA
2   Department of Biology, Lewis and Clark College, 615 S. Palatine Hill Road, Portland, OR 97219, USA; binford@lclark.edu
3   Faculty of Life and Environmental Sciences, University of Iceland, Sturlugata 7, 102 Reykjavik, Iceland
*   Correspondence: lily.shapiro@uvm.edu (L.S.); iagnarsson@gmail.com (I.A.)

Abstract: Island biogeographers have long sought to elucidate the mechanisms behind biodiversity genesis. The Caribbean presents a unique stage on which to analyze the diversification process, due to the geologic diversity among the islands and the rich biotic diversity with high levels of island endemism. The colonization of such islands may reflect geologic heterogeneity through vicariant processes and/ or involve long-distance overwater dispersal. Here, we explore the phylogeography of the Caribbean and proximal mainland spiny orbweavers (*Micrathena*, Araneae), an American spider lineage that is the most diverse in the tropics and is found throughout the Caribbean. We specifically test whether the vicariant colonization via the contested GAARlandia landbridge (putatively emergent 33–35 mya), long-distance dispersal (LDD), or both processes best explain the modern *Micrathena* distribution. We reconstruct the phylogeny and test biogeographic hypotheses using a 'target gene approach' with three molecular markers (CO1, ITS-2, and 16S rRNA). Phylogenetic analyses support the monophyly of the genus but reject the monophyly of Caribbean *Micrathena*. Biogeographical analyses support five independent colonizations of the region via multiple overwater dispersal events, primarily from North/Central America, although the genus is South American in origin. There is no evidence for dispersal to the Greater Antilles during the timespan of GAARlandia. Our phylogeny implies greater species richness in the Caribbean than previously known, with two putative species of *M. forcipata* that are each single-island endemics, as well as deep divergences between the Mexican and Floridian *M. sagittata*. *Micrathena* is an unusual lineage among arachnids, having colonized the Caribbean multiple times via overwater dispersal after the submergence of GAARlandia. On the other hand, single-island endemism and undiscovered diversity are nearly universal among all but the most dispersal-prone arachnid groups in the Caribbean.

Keywords: phylogeny; Caribbean biogeography; GAARlandia; arachnid; araneae; *Micrathena*; vicariance; long distance dispersal

Citation: Shapiro, L.; Binford, G.J.; Agnarsson, I. Single-Island Endemism despite Repeated Dispersal in Caribbean *Micrathena* (Araneae: Araneidae): An Updated Phylogeographic Analysis. *Diversity* **2022**, *14*, 128. https://doi.org/10.3390/d14020128

Academic Editors: Luc Legal and Matjaž Kuntner

Received: 7 October 2021
Accepted: 5 February 2022
Published: 10 February 2022

1. Introduction

Understanding the evolutionary machinery of biodiversity genesis in island systems has long been a focus of fundamental biological research [1–4]. Islands serve as discrete, isolated systems in which to study the generation of biodiversity, resulting from complex patterns of (sometimes) repeated colonization, radiation, and extinction. The isolated nature of islands also allows for the evolution of increased magnitudes of endemic forms; archipelagos facilitate these processes, which are replicated continuously across the entire system [5–7]. Such biodiversity is exemplified within Caribbean archipelagoes and can be observed across taxonomic groups, including arthropods, amphibians, fish, mammals, birds, and plants [7,8]. The proximity of the Caribbean islands to continental blocks has resulted in the production of a unique assemblage of endemic biota, while still being remote enough for the formation of effective oceanic barriers for dispersal [7].

The geologic history of the Caribbean is intrinsically coupled with this biological diversity, and the region itself is composed of islands with varying geologic origins and different regional tectonic influences [9–12]. This complex geology includes old islands such as the Greater Antilles, which have been emergent for at least 40 million years (mid-Eocene) [13] and younger, primarily volcanic islands (e.g., Lesser Antilles) that emerged less than 10 mya (upper Miocene). The distinct geologic history of each island in the Caribbean should be reflected in the modern patterns of organismal diversity, resulting from its colonization via long-distance dispersal and/or vicariant processes, potentially leading to diversification. Newer volcanic islands and isolated limestone/sedimentary oceanic islands, separated from other landmasses by large swaths of ocean, will likely have species assemblages exclusively resulting from long-distance dispersal from the mainland or other island sources. Continental islands, such as the Greater Antilles, are much older island systems with a complex history of islands becoming emergent or submerged, and splintering and rejoining [12,14,15]. Unraveling the role of LDD and vicariance for a specific group depends on the geology of an individual island, in conjunction with the biology of that lineage [14–18]. As these islands are deferentially isolated from continents, the dispersal ability of a selected lineage is especially significant in understanding its historical colonization of the Caribbean [19].

The GAARlandia (Greater Antilles Aves Ridge) landbridge is a hypothetical sub-aerial connection between South America and the Greater Antilles, in which parts of the previously submerged Aves Ridge became exposed as a consequence of dropping sea levels and the Greater Antillean uplift during the Eocene-Oligocene transition (35–33 mya) [20,21]. This ephemeral connection would have permitted direct overland colonization of South American taxa to the Greater Antilles, followed by the subsequent diversification and speciation as organisms filled previously empty niches before the landbridge was re-submerged around 30 mya [20]. The GAARlandia hypothesis, therefore, predicts the simultaneous colonization across diverse taxa to the Greater Antilles within this timespan, a readily testable biological prediction that has recently been evaluated in a variety of Caribbean biogeographic studies across multiple arthropod taxa [14,16,22–36]. While recent chronostratigraphic data suggests the emergence of a landmass between Puerto Rico and the Lesser Antilles in the mid-Eocene, corresponding with crustal shortening and thickening that is consistent with GAARlandia [37], the hypothesis remains contested due to limited [38,39] or conflicting geological and paleo-oceanographic data [40,41]. Ali and Hedges [40], and others cited therein, also emphasize that biogeographic evidence, consistent with the hypothesis, may offer only weak support due to ambiguity in lineage dating. Recent meta-analyses, uniting multiple studies, generally rejected the role of GAARlandia in the biogeography of Caribbean land vertebrates [40], continuing this active debate.

This complex geologic and evolutionary history can be clarified with phylogeographic evidence from densely sampled, regionally-focused clades. Spiders have increasingly been used, in recent years, as biogeographical models not only in the Caribbean but on global and finer scales [23,42–46], as they form a hyperdiverse group with corresponding diversity in dispersal ability and lineage age. While much of the historical research concerning Caribbean biogeography has been vertebrate-based [14,34,47–49], invertebrates, such as arachnids, can provide fine-scale signals of historical dispersal and colonization [16,50]. Recent evidence from these animals have found mixed support for vicariance and LDD, with a large diversity of focal lineages [16,23,26,29,31,32,36,51,52].

Micrathena, the spiny orbweavers (Araneae, Araneidae), are a colorful, highly ornate, and sexually dimorphic group of 119 New World species, distributed from northern Argentina, throughout the Caribbean and Central America, to the New York state, and into southern Ontario [53,54]. Members of the genus reside in forests or woodlands, constructing webs in the understory up to approximately 4 m off the ground [55]. The large, colorful adult females are sedentary and solitary, while the much tinier males wander in search of a mate, preferably a penultimate-instar female (as noted in the case of *Microthena gracilis*) [55]. Ballooning behavior has only been formally observed in the juveniles of *Microthena sagittata* [56] but the biogeographic patterns [36,51,53] suggest that it may have played a role in overwater dispersal in the Caribbean.

About 67 *Microthena* species are South American endemics (most found in Colombia and Brazil), with an additional 25 potentially widespread species that have part of their range in South America [57]. Fourteen species are Central American endemics, and eight are Caribbean endemics. Of the eight Caribbean species, four are known single-island endemics: two from Cuba (*M. banksi* and *M. cubana*), one from Jamaica (*M. rufopuncata*), and one from Hispaniola (*M. similis*). In addition, *Microthena forcipata* from Cuba and Hispaniola, and *Microthena militaris* from Puerto Rico and Hispaniola, have recently been suggested to represent clearly divergent lineages, potentially yielding four additional single-island endemics in the Caribbean [51]. Four species are found in North America (*M. funebris*, *M. gracilis*, *M. mitrata*, and *M. sagittata*), and each of these species is in the Caribbean. A previous phylogeographic analysis of Caribbean *Microthena* by McHugh et al. [51] proposed three Caribbean species-groups (the *militaris* group, the *furcula* group, and the *gracilis* group), in agreement with studies by Magalhães et al. [51,53]. Each of these species groups included members of the North, Central, and South American *Microthena*, indicating that Caribbean *Microthena* are not monophyletic, and that colonization of the Caribbean must have been repetitive [51]. Similar patterns are found in some other members of Araneidae (I. Agnarsson unpublished data).

This paper expands on the work of McHugh et al. [51] with increased taxon sampling of Caribbean *Microthena* and additional North and South American mainland species (Colombia and Florida). These additional taxa allow more refined tests of patterns of single-island endemism and more a rigorous evaluation of factors influencing divergence patterns. McHugh et al. [51] rejected the hypothesis that *Microthena* colonized the Greater Antilles via the GAARlandia landbridge. Here, we explicitly test the dispersal route using our additional data on previously omitted and undersampled species that help clarify patterns and timelines for the Caribbean colonization in the genus. These tests strengthen our understanding of the continental-island interchange and other biogeographic patterns of *Microthena* within the region.

2. Materials and Methods

2.1. Specimen and Taxon Sampling

Microthena specimens were collected in the field from 1997–2015 (Table 1, Figure 1). Specimens were stored at −20 °C in 95% ethanol at the University of Vermont. In this work, we added 50 individuals, representing 14 additional *Microthena* species, to the previous McHugh et al. [51] *Microthena* phylogeography study (*M. duodecimspinosa*, *M. lucasi*, *M. sp* (putative species) *M. mitrata*, *M. beta*, *M. cornuta*, *M. embira*, *M. exlinae*, *M. miles*, *M. perfida*, *M. reimoseri*, *M. spinulata*, *M. triangularispinosa*, and *M. yanomami* (Table 1)). We also added previously represented species from new localities: *M. gracilis* from Florida; *M. horrida* from Jamaica; *M. militaris* from Dominica; *M. sagittata* from Florida and Mexico; *M. schreibersi* from Colombia, Trinidad, and Costa Rica; *M. sexspinosa* from Colombia; and expanded sites of *M. forcipata* from Cuba, which were sampled on CarBio trips from 2012–2015 (Table 1). We used a specimen of *Achaearanea* sp. (Theridiidae) as the primary outgroup, along with five araneid members: two *Argiope* specimens and three *Gasteracantha cancriformis* individuals. The outgroups included some relatively near relatives of *Microthena* [58], along with more distantly related araneid members in *Argiope* [49], with members of Theridiidae being used to root the tree.

Table 1. Taxon sampling table with barcodes, locality data, and GenBank accession numbers. "x" denotes GenBank submission in progress.

Genus	Species	Barcode	Country/Region	Latitude	Longitude	16S	CO1	ITS2
Micrathena	*annulata*	MIC007	Brazil	26.08933S	48.64006W		KJ157272	
Micrathena	*aureola*	MIC009	Brazil	4.904167S	42.79083W		KJ157249	
Micrathena	*banksi*	784750	Cuba	20.05269N	76.50296W	KJ156991	KJ157215	KJ157104
Micrathena	*banksi*	784760	Cuba	20.0107N	76.8843W	KJ156992	KJ157216	
Micrathena	*banksi*	784976	Cuba	20.00939N	76.89402W	KJ156993	KJ157217	KJ157105
Micrathena	*banksi*	785101	Cuba	20.00939N	76.89402W	KJ156994	KJ157220	KJ157106
Micrathena	*banksi*	785175	Cuba	20.33178N	74.56919W	KJ156995	KJ157219	KJ157107
Micrathena	*banksi*	787933	Cuba	20.01742N	76.89781W	KJ156996	KJ157218	KJ157108
Micrathena	*beta*	MIC238	Peru	4.5674444S	73.45925W		KX687306	
Micrathena	*bimucronata*	MIC123	Costa Rica	10.233518N	84.075411W		KJ157236	
Micrathena	*brevipes*	MIC121	Costa Rica	9.552960N	83.112910W		KJ157223	
Micrathena	*cornuta*	MIC199	Peru	12.8088056S	69.30175W		KX687309	
Micrathena	*cubana*	784355	Cuba	20.01309N	76.83400W	KJ156997	KJ157224	KJ157109
Micrathena	*cubana*	784820	Cuba	20.00874N	76.88777W	KJ156998	KJ157225	KJ157110
Micrathena	*cubana*	785048	Cuba	22.65707N	83.70161W	KJ156999	KJ157226	KJ157111
Micrathena	*cubana*	787840	Cuba	20.33178N	74.56919W	KJ157000	KJ157227	
Micrathena	*digitata*	MIC017	Brazil	11.39983S	40.52206W		KJ157238	
Micrathena	*duodecimspinosa*	00004833A	Costa Rica	San Antonio de Escazú			x	x
Micrathena	*embira*	MIC182	Brazil	9.642419S	41.446727W		KX687311	
Micrathena	*exlinae*	MIC147	Brazil	0.99185S	62.15915W		KX687313	
Micrathena	*forcipata*	00002846A	Cuba	Juan Gonzalez, Guamá			x	x
Micrathena	*forcipata*	00002848A	Cuba	20.01309N	76.83400W		x	x
Micrathena	*forcipata*	00002845A	Cuba	20.01309N	76.83400W		x	x
Micrathena	*forcipata*	784425	Cuba	20.00939N	76.89402W	KJ157002	KJ157256	KJ157113
Micrathena	*forcipata*	787842	Cuba	20.33178N	74.56919W	KJ157003	KJ157257	
Micrathena	*forcipata*	782311	Hispaniola	18.355536N	68.61825W	KJ157004	KJ157258	
Micrathena	*forcipata*	782434	Hispaniola	19.34405N	69.46635W	KJ157005	KJ157260	KJ157114
Micrathena	*forcipata*	784362	Hispaniola	18.32902N	68.80995W	KJ157006	KJ157264	KJ157115
Micrathena	*forcipata*	784366	Hispaniola	18.32902N	68.80995W		KJ157271	KJ157116
Micrathena	*forcipata*	784447	Hispaniola	18.2205360N	68.480607W	KJ157007	KJ157261	KJ157117
Micrathena	*forcipata*	785054	Hispaniola	19.746175N	71.257726W	KJ157008	KJ157263	KJ157118
Micrathena	*forcipata*	785282	Hispaniola	18.355536N	68.6185W	KJ157009	KJ157259	KJ157119
Micrathena	*forcipata*	785682	Hispaniola	18.2205360N	68.480607W	KJ157010	KJ157	
Micrathena	*forcipata*	787132	Hispaniola	18.310010 N	71.6000 W		KJ157265	
Micrathena	*forcipata*	787135	Hispaniola	18.310010 N	71.6000 W	KJ157011	KJ157266	
Micrathena	*forcipata*	787150	Hispaniola	18.310010 N	71.6000 W	KJ157012	KJ157267	KJ157121
Micrathena	*forcipata*	787153	Hispaniola	18.310010 N	71.6000 W	KJ157013	KJ157269	KJ157122
Micrathena	*forcipata*	787210	Hispaniola	18.310010 N	71.6000 W	KJ157014	KJ157268	KJ157123
Micrathena	*forcipata*	787243	Hispaniola	18.310010 N	71.6000 W	KJ157015	KJ157270	KJ157124

Table 1. *Cont.*

Genus	Species	Barcode	Country/Region	Latitude	Longitude	16S	CO1	ITS2
Micrathena	*furcata*	MIC037	Brazil	27.66667 S	49.01667W		KJ157242	
Micrathena	*gracilis*	10000619A	FL, USA	29.4776N	82.5627W		x	x
Micrathena	*gracilis*	10000629A	FL, USA	29.62986N	82.29880W		x	
Micrathena	*gracilis*	10000627A	FL, USA	29.62986N	82.29880W		x	
Micrathena	*gracilis*	10000638A	FL, USA	29.63680N	82.23961W		x	x
Micrathena	*gracilis*	10000644A	FL, USA	29.46368N	82.52898W		x	
Micrathena	*gracilis*	10000642A	FL, USA	29.62688N	82.29878W		x	
Micrathena	*gracilis*	10000643A	FL, USA	29.62688N	82.29878W		x	
Micrathena	*gracilis*	00000804A	NC, USA	35.44842N	81.58694W		KJ157250	KJ157188
Micrathena	*gracilis*	00000954A	SC, USA	33.03913N	79.56459W	KJ157084	KJ157252	KJ157192
Micrathena	*gracilis*	00000935A	SC, USA	33.03913N	79.56459W	KJ157083	KJ157254	KJ157191
Micrathena	*gracilis*	00000889A	SC, USA	33.03913N	79.56459W	KJ157082	KJ157251	KJ157190
Micrathena	*gracilis*	00000984A	SC, USA	33.03913N	79.56459W	KJ157086	KJ157253	KJ157194
Micrathena	*gracilis*	00000988A	SC, USA	33.03913N	79.56459W	KJ157087	KJ157255	KJ157195
Micrathena	*gracilis*	00002487A	NY, USA	42.01807N	73.91707W	KJ157088		KJ157196
Micrathena	*gracilis*	00002501A	NY, USA	42.01807N	73.91707W	KJ157089		KJ157197
Micrathena	*gracilis*	00000976A	SC, USA	33.03913N	79.56459W	KJ157085		KJ157193
Micrathena	*horrida*	MIC042	Brazil	16.59553S	41.57925W		KJ157248	
Micrathena	*horrida*	MIC122	Costa Rica	10.233518N	84.075411W		KJ157245	
Micrathena	*horrida*	00003552A	Jamaica	18.1635N	77.39410W		x	x
Micrathena	*horrida*	784351	Cuba	20.00939N	76.89402W	KJ157016	KJ157243	KJ157125
Micrathena	*horrida*	784751	Cuba	20.00939N	76.89402W	KJ157017	KJ157246	KJ157126
Micrathena	*horrida*	787913	Cuba	20.00939N	76.89402W	KJ157018	KJ157247	KJ157127
Micrathena	*horrida*	787919	Cuba	20.00939N	76.89402W	KJ157019	KJ157244	KJ157128
Micrathena	*lucasi*	00004785A	Costa Rica	San Antonio de Escazú				
Micrathena	*macfarlanei*	MIC054	Brazil	19.65000S	42.56667W		KJ157241	
Micrathena	*miles*	MIC142	Peru	3.82975S	73.375333W		KX687317	
Micrathena	*militaris*	10000526A	Dominica	15.32710N	61.3381W		x	x
Micrathena	*militaris*	10000528A	Dominica	15.32710N	61.3381W		x	x
Micrathena	*militaris*	782365	Hispaniola	18.355536N	068.61825W	KJ157020		KJ157129
Micrathena	*militaris*	784338	Hispaniola	18.32902N	068.80995W	KJ157021	KJ157273	
Micrathena	*militaris*	784363	Hispaniola	18.32902N	068.80995W	KJ157022	KJ157293	KJ157130
Micrathena	*militaris*	784403	Hispaniola	18.32902N	068.80995W	KJ157023	KJ157298	KJ157131
Micrathena	*militaris*	784430	Hispaniola	18.32902N	068.80995W	KJ157024		KJ157132
Micrathena	*militaris*	784448	Hispaniola	18.32902N	068.80995W	KJ157025	KJ157294	KJ157133
Micrathena	*militaris*	784458	Hispaniola	18.32902N	068.80995W	KJ157026		KJ157134
Micrathena	*militaris*	784503	Hispaniola	18.3150011N	71.580556W	KJ157027	KJ157300	KJ157135
Micrathena	*militaris*	784531	Hispaniola	18.355536N	068.61825W	KJ157028		KJ157136
Micrathena	*militaris*	784566	Hispaniola	18.32902N	068.80995W	KJ157029	KJ157296	KJ157137
Micrathena	*militaris*	784671	Hispaniola	19.06707N	069.46355W	KJ157030		KJ157138
Micrathena	*militaris*	784721	Hispaniola	18.32902N	068.80995W	KJ157031	KJ157310	KJ157139

Table 1. *Cont.*

Genus	Species	Barcode	Country/Region	Latitude	Longitude	16S	CO1	ITS2
Micrathena	*militaris*	784759	Hispaniola	18.355536N	068.61825W	KJ157032	KJ157277	KJ157140
Micrathena	*militaris*	784762	Hispaniola	18.2205360N	68.4806070W	KJ157033		KJ157141
Micrathena	*militaris*	784772	Hispaniola	18.32902N	068.80995W	KJ157034	KJ157287	KJ157142
Micrathena	*militaris*	784806	Hispaniola			KJ157035		KJ157143
Micrathena	*militaris*	784926	Hispaniola			KJ157036		KJ157144
Micrathena	*militaris*	785066	Hispaniola	19.06707N	069.46355W	KJ157037		KJ157145
Micrathena	*militaris*	785080	Hispaniola	18.32902N	068.80995W	KJ157038	KJ157274	KJ157146
Micrathena	*militaris*	785099	Hispaniola	18.32902N	068.80995W		KJ157313	
Micrathena	*militaris*	785128	Hispaniola	18.355536N	068.61825W	KJ157039		KJ157147
Micrathena	*militaris*	785144	Hispaniola	19.746175N	71.257726W	KJ157040		KJ157148
Micrathena	*militaris*	785169	Hispaniola	18.355536N	068.61825W	KJ157041	KJ157290	KJ157149
Micrathena	*militaris*	785173	Hispaniola	19.06707N	069.46355W	KJ157042	KJ157314	KJ157150
Micrathena	*militaris*	785174	Hispaniola	19.06707N	069.46355W	KJ157043	KJ157292	KJ157151
Micrathena	*militaris*	785194	Hispaniola	18.355536N	068.61825W	KJ157044		
Micrathena	*militaris*	785208	Hispaniola	18.2205360N	68.4806070W	KJ157045	KJ157297	KJ157152
Micrathena	*militaris*	785219	Hispaniola	18.355536N	068.61825W	KJ157046	KJ157286	KJ157153
Micrathena	*militaris*	785263	Hispaniola	18.355536N	068.61825W	KJ157047		KJ157154
Micrathena	*militaris*	785273	Hispaniola	19.432213N	070.371412W	KJ157048	KJ157275	KJ157155
Micrathena	*militaris*	785280	Hispaniola	18.32902N	068.80995W	KJ157049	KJ157315	KJ157156
Micrathena	*militaris*	785312	Hispaniola	19.34405N	069.46635W	KJ157050	KJ157280	KJ157157
Micrathena	*militaris*	785401	Hispaniola	19.06707N	069.46355W	KJ157051	KJ157276	KJ157158
Micrathena	*militaris*	785402	Hispaniola	19.34405N	069.46635W	KJ157052	KJ157285	KJ157159
Micrathena	*militaris*	785423	Hispaniola	18.355536N	068.61825W	KJ157053		KJ157160
Micrathena	*militaris*	785461	Hispaniola	19.06707N	069.46355W	KJ157054	KJ157281	
Micrathena	*militaris*	785502	Hispaniola	19.06707N	069.46355W	KJ157055	KJ157301	KJ157161
Micrathena	*militaris*	785512	Hispaniola	19.06707N	069.46355W	KJ157056	KJ157316	KJ157162
Micrathena	*militaris*	785524	Hispaniola	18.355536N	068.61825W	KJ157057	KJ157311	KJ157163
Micrathena	*militaris*	785527	Hispaniola	19.34405N	069.46635W	KJ157058	KJ157279	KJ157164
Micrathena	*militaris*	785563	Hispaniola	19.06707N	069.46355W	KJ157059	KJ157295	KJ157165
Micrathena	*militaris*	785604	Hispaniola	19.06707N	069.46355W	KJ157060	KJ157288	KJ157166
Micrathena	*militaris*	785706	Hispaniola	19.06707N	069.46355W	KJ157061	KJ157278	KJ157167
Micrathena	*militaris*	785709	Hispaniola	19.06707N	069.46355W		KJ157312	KJ157168
Micrathena	*militaris*	785722	Hispaniola	19.06707N	069.46355W	KJ157062	KJ157283	KJ157169
Micrathena	*militaris*	785729	Hispaniola	19.34405N	069.46635W	KJ157063	KJ157284	KJ157170
Micrathena	*militaris*	785743	Hispaniola	19.06707N	069.46355W	KJ157064	KJ157282	KJ157171
Micrathena	*militaris*	785769	Hispaniola	19.06707N	069.46355W	KJ157065		KJ157172
Micrathena	*militaris*	787068	Hispaniola	18.980122N	70.798425W	KJ157066	KJ157299	KJ157173
Micrathena	*militaris*	787106	Hispaniola	18.980122N	70.798425W	KJ157067	KJ157289	KJ157174
Micrathena	*militaris*	787148	Hispaniola	18.3150011N	71.580556W	KJ157068	KJ157291	KJ157175
Micrathena	*militaris*	787152	Hispaniola	18.3150011N	71.580556W	KJ157069		KJ157176
Micrathena	*militaris*	787166	Hispaniola	18.3150011N	71.580556W	KJ157070		KJ157177
Micrathena	*militaris*	787190	Hispaniola	18.3150011N	71.580556W	KJ157071		KJ157178

Table 1. *Cont.*

Genus	Species	Barcode	Country/Region	Latitude	Longitude	16S	CO1	ITS2
Microthena	*militaris*	787208	Hispaniola	18.3150011N	71.580556W	KJ157072		KJ157179
Microthena	*militaris*	787212	Hispaniola	18.3150011N	71.580556W	KJ157073		KJ157180
Microthena	*militaris*	787214	Hispaniola	18.3150011N	71.580556W	KJ157001		KJ157112
Microthena	*militaris*	392672	Puerto Rico	17.971472N	66.867958W	KJ157074	KJ157302	KJ157181
Microthena	*militaris*	392677	Puerto Rico	17.971472N	66.867958W	KJ157075	KJ157303	KJ157182
Microthena	*militaris*	782048	Puerto Rico	18.414373N	66.728722W	KJ157076	KJ157307	KJ157183
Microthena	*militaris*	782126	Puerto Rico	18.173264N	66.590149W	KJ157077	KJ157308	KJ157184
Microthena	*militaris*	782153	Puerto Rico	18.414373N	66.728722W	KJ157078	KJ157306	KJ157185
Microthena	*militaris*	782174	Puerto Rico	18.414373N	66.728722W	KJ157079	KJ157304	KJ157186
Microthena	*militaris*	782201	Puerto Rico	18.032518N	67.094653W	KJ157080	KJ157305	KJ157187
Microthena	*militaris*	783400	Puerto Rico	18.45226N	66.59711W		KJ157309	
Microthena	*mitrata*	10000679A	Mexico	19.79357N	104.0554W		x	x
Microthena	*mitrata*	00002849A	Mexico	19.79357N	104.0554W		x	x
Microthena	*nigrichelis*	MIC056	Brazil	20.43481S	43.50906W		KJ157239	
Microthena	*perfida*	MIC026	Brazil	24.387111S	47.017583W		KX687318	
Microthena	*plana*	MIC062	Brazil	16.53294S	41.51042W		KJ157240	
Microthena	*reimoseri*	MIC072	Brazil	11.399833S	40.522056W		KX687321	
Microthena	*saccata*	MIC076	Brazil	1.424828S	48.43802W		KJ157237	
Microthena	*sagittata*	10000618A	FL, USA	29.4776N	082.5627W		x	
Microthena	*sagittata*	10000621A	FL, USA	29.63703N	082.23976W		x	
Microthena	*sagittata*	10000631A	FL, USA	29.62986N	082.29880W		x	x
Microthena	*sagittata*	10000633A	FL, USA	29.62986N	082.29880W		x	
Microthena	*sagittata*	10000636A	FL, USA	29.63680N	082.23961W		x	x
Microthena	*sagittata*	10000634A	FL, USA	29.46397N	082.55285W		x	x
Microthena	*sagittata*	10000639A	FL, USA	29.63680N	082.23961W		x	
Microthena	*sagittata*	10000640A	FL, USA	29.62688N	082.29878W		x	
Microthena	*sagittata*	00002847A	Mexico	18.18963N	89.46333W		x	
Microthena	*sagittata*	00000833A	SC, USA	33.03913 N	79.56459W	KJ157081	KJ157221	KJ157189
Microthena	*schreibersi*	00002357A	Colombia	Bucaramanga			x	
Microthena	*schreibersi*	10000650A	Colombia	8.39104N	77.21548W		x	
Microthena	*schreibersi*	10000652A	Colombia	8.39104N	77.21548W		x	
Microthena	*schreibersi*	10000653A	Colombia	8.39104N	77.21548W		x	x
Microthena	*schreibersi*	10000664A	Colombia	8.424N	77.29216W		x	
Microthena	*schreibersi*	10000673A	Colombia	8.39104N	77.21548W		x	
Microthena	*schreibersi*	10000658A	Colombia	8.39104N	77.21548W		x	
Microthena	*schreibersi*	10000651A	Colombia	8.39104N	77.21548W		x	x
Microthena	*schreibersi*	10000663A	Colombia	8.424N	77.29216W		x	
Microthena	*schreibersi*	10000665A	Colombia	8.424N	77.29216W		x	x
Microthena	*schreibersi*	00004787A	Colombia	10.21192N	75.25403W		x	x
Microthena	*schreibersi*	00004818A	Trinidad				x	x
Microthena	*schreibersi*	00002900A	Costa Rica	10.430686N	84.007089W		x	x
Microthena	*schreibersi*	00000936A	Colombia	7.062695N	73.073058W	KJ157090	KJ157318	KJ157198
Microthena	*schreibersi*	00002357A	Colombia	7.062695N	73.073058W	KJ157092	KJ157319	KJ157199

Table 1. *Cont.*

Genus	Species	Barcode	Country/Region	Latitude	Longitude	16S	CO1	ITS2
Micrathena	*sexspinosa*	10000690A	Colombia	8.35249N	77.22118W		x	
Micrathena	*sexspinosa*	10000659A	Colombia	8.35249N	77.22118W		x	
Micrathena	*sexspinosa*	10000674A	Colombia	8.35249N	77.22118W		x	x
Micrathena	*sexspinosa*	10000677A	Colombia	11.120083N	74.082805W		x	
Micrathena	*sexspinosa*	10000683A	Colombia	11.120083N	74.082805W		x	
Micrathena	*sexspinosa*	10000669A	Colombia	8.39104N	77.21548W		x	x
Micrathena	*sexspinosa*	10000670A	Colombia	8.39104N	77.21548W		x	x
Micrathena	*sexspinosa*	10000681A	Colombia	8.35249N	77.22118W		x	
Micrathena	*sexspinosa*	10000678A	Colombia	8.35249N	77.22118W		x	
Micrathena	*sexspinosa*	00000987A	Colombia	7.062695N	73.073058W	KJ157091	KJ157222	
Micrathena	*similis*	785024	Hispaniola	19.34405N	69.46635W	KJ157093	KJ157228	KJ157200
Micrathena	*similis*	785496	Hispaniola	19.34405N	69.46635W	KJ157094	KJ157232	KJ157201
Micrathena	*similis*	787265	Hispaniola	19.05116N	70.88866W	KJ157095	KJ157233	KJ157202
Micrathena	*similis*	787297	Hispaniola	19.05116N	70.88866W	KJ157096		KJ157203
Micrathena	*similis*	787308	Hispaniola	19.03627N	70.54337W	KJ157097	KJ157229	KJ157204
Micrathena	*similis*	787309	Hispaniola	19.05116N	70.88866W	KJ157098		KJ157205
Micrathena	*similis*	787311	Hispaniola	19.05116N	70.88866W		KJ157235	KJ157206
Micrathena	*similis*	787318	Hispaniola	19.03627N	70.54337W	KJ157099	KJ157234	KJ157207
Micrathena	*similis*	787320	Hispaniola	19.05116N	70.88866W	KJ157100	KJ157230	KJ157208
Micrathena	*similis*	787322	Hispaniola	19.05116N	70.88866W	KJ157101	KJ157231	KJ157209
Micrathena	sp.	10000656A	Colombia	11.120083N	74.082805W		x	
Micrathena	sp.	10000671A	Colombia	11.120083N	74.082805W		x	x
Micrathena	sp.	00006693A	Colombia	11.120083N	74.082805W		x	x
Micrathena	*spinulata*	MIC205	Mexico	19.1381667N	97.2045W		KX687324	
Micrathena	*triangularispinosa*	MIC156	Brazil	0.97799S	62.10292W		KX687327	
Micrathena	*yanomami*	MIC193	Peru	13.055639S	71.546194W		KX687332	
Outgroups								
Achaearanea	sp.	784841	Cuba	21.59166N	77.78822W		KJ157211	
Argiope	*lobata*	Arg0160	Spain	Missing GPS data		KJ156988		KJ157103
Gasteracantha	*cancriformis*	787198	Hispaniola	18.3150011N	71.580556W	KJ156989	KJ157212	
Gasteracantha	*cancriformis*	784515	Hispaniola	18.2205260N	68.480607W		KJ157213	
Gasteracantha	*cancriformis*	782149	Puerto Rico	18. 172979N	66.491798W	KJ156990	KJ157214	

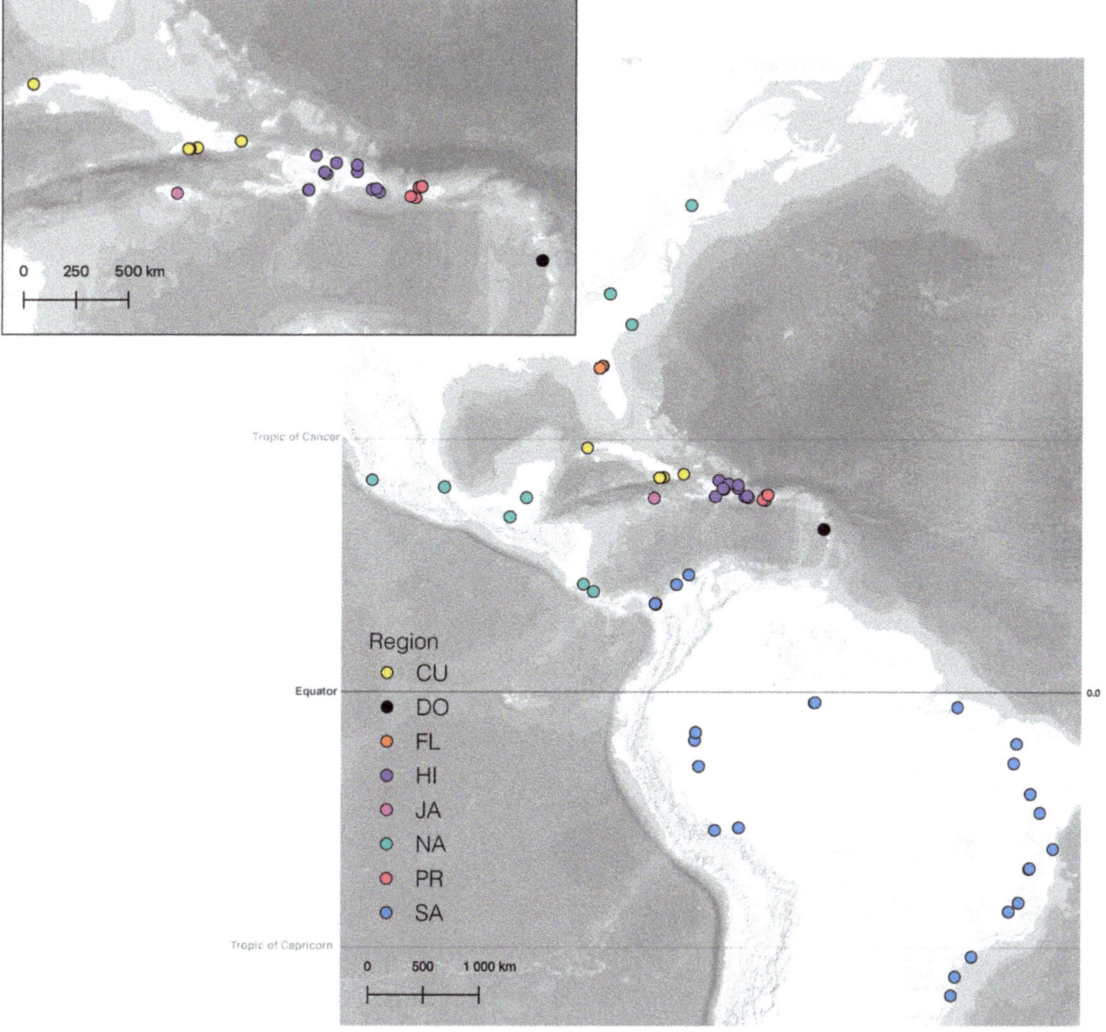

Figure 1. Map of collection localities of all specimens included in analysis. Points are colored by biogeographic area assigned for BioGeoBEARS analysis (see supporting material).

2.2. Tissue Extraction and PCR

Tissue samples were taken from the right legs, and DNA was isolated using the QIA-GEN DNeasy Tissue Kit (Qiagen, Inc., Valencia, CA, USA). Fragments of one mitochondrial locus (CO1: cytochrome c oxidase subunit 1) and one nuclear locus (ITS-2: internal transcribed spacer 2) were sequenced. The 16S data, along with the previous ITS-2 and CO1 data, were retrieved from McHugh et al. [51]. Both ITS-2 and CO1 have demonstrated utility in illuminating relationships between species-level and low-level taxonomic clades in previous arachnid phylogenetics studies [59,60]. The CO1 locus was amplified using the primers *Jerry* [61] and C1-N-2776 [62] for the majority of specimens ($n = 43$), while a select number were amplified using LCO1490 [63] and C1-N-2776 ($n = 7$), which resulted in a higher success rate of amplification within this group. The ITS2 locus was amplified using the primers ITS5.8S and ITS4S [64]. The conditions for each PCR are listed in Table 2. Sanger sequencing was conducted by the University of Vermont Cancer Center DNA Analysis Facility within the Vermont Integrative Genomics Resource (VIGR) facility. Additional sequences used to inform deficiencies in our South American *Microthena* collection were retrieved from GenBank. All novel sequences have been submitted to GenBank (in progress).

Table 2. Polymerase chain reaction (PCR) conditions for ITS-2 and CO1. Conditions were split for CO1, given that two sets of primers were used.

	Polymerase Chain Reaction (PCR) Conditions			
Gene	**Forward Primer**	**Reverse Primer**	**Annealing Temp. (°C)**	**Fragment Length (bp)**
Internal transcribed spacer 2 (ITS-2)	ITS4	ITS5.8	47	350–500
	Jerry	C1-N-2776	46	~1250
Cytochrome oxidase subunit 1 (CO1)	LCO11490	C1-N-2776	48	~1250

2.3. Alignment and Phylogeny Building

Phred and Phrap [65,66] were used to compile sequence chromatograms. Chromatograms were inspected and sequences were edited using the Chromaseq module [67] within the program Mesquite 3.61 [68]. Sequences were aligned using the MAFFT online service [69] with gaps treated as missing characters and all other settings set to default. The substitution models and partitioning schemes for a Bayesian analysis were selected with PartitionFinder 2.1 [70], using AIC (Akaike's information criterion) [71] amongst the 24 available models in MrBayes [72]. Sequence data were partitioned by gene, and additionally by codon, for CO1 as input for PartitionFinder. We ran a Bayesian inference using the CIPRES online portal [73] on a concatenated matrix where each locus was separately partitioned using MrBayes 3.2.7.a [72]. The Markov Chain Monte Carlo (MCMC) algorithm was run with four chains for 30,000,000 generations, sampling every 1000 generations. Tracer 1.71 [74] was used to verify the proper mixing of chains, to confirm that stationarity had been achieved, and to determine the adequate burn-in.

2.4. Divergence Time Estimation and Biogeographic Modeling

To estimate node ages among *Microthena*, we used BEAST 2.60 [75] under a relaxed clock model. Because the South American species only had CO1 sequence data available, we used only this locus in the BEAST analysis. Terminal taxa were pruned for redundancy so that one representative of each critical species remained. BEAST analyses for CO1 were run with both an alignment partitioned by codon, using the best-fit models extracted from PartitionFinder [70] (GTR + I + Γ for position 1, TVM + I + Γ for position 2, and TRN + Γ for position 3), along with an unpartitioned analysis, which was run using the best-fit model for CO1 overall (GTR + I + Γ). Both analyses returned identical results. The analyses in *BEAST* were run for 30,000,000 generations, sampling every 1000 generations with a Yule Tree prior. *Microthena*, along with closely related lineages, lack a fossil record, so the phylogeny was calibrated using the estimated age of Araneidae and the most recent common ancestor (MRCA), including Theridiidae and Araneidae derived from a recent fossil calibrated study by Kuntner et al. [76]. The minimum age of Araneidae was set as a normal prior with a mean of 70 million years and a standard deviation of 3. The minimum age of Theridiidae + Araneidae was also set as a normal prior with a mean of 100 million years and a standard deviation of 9; both prior distributions covered the 95% confidence intervals derived from Kuntner et al. [76]. Based on the estimated substitution rates of CO1 that have been found to be consistent across spider lineages [76,77], the mitochondrial substitution rate parameter (ucld.mean) mean value was set to 0.0112 and the s.d. was set to 0.001. We confined the monophyly of *Microthena* based on the results of our Bayesian analyses. *Tracer 1.7* [74] was again utilized to visualize the results of our node age estimation analysis, to determine burn-in and to check for stationarity.

An ancestral range analysis was conducted using the BioGeoBEARS v.1.1.2 package in R [78]. The maximum range was constrained to three areas, due to the widespread distribution of some focal taxa. In this analysis, we employed our CO1 dated phylogeny with terminals pruned to represent single species or genetically distinct single-island endemics based on our Bayesian tree. We defined seven geographic areas: North America (NA), South America (SA), Florida (FL), Cuba (CU), Hispaniola (HI), Jamaica (JA), and Puerto Rico (PR) (see Supplementary File S1). Mexico, and all of Central America north of Panama, were included as part of North America, given that the edge of the Maya Block in southern Mexico corresponds to the southernmost boundary of the North American Tectonic Plate and that the Chorotega and Chortís blocks of Central America were associated with North America as a geologic entity for our focal time period [79–81]. Florida was coded as a separate entity from North America, as the land was unavailable until about 5 mya [82].

We tested a GAARlandia model and a no-GAARlandia model (the distribution was explained by overwater dispersal) by applying probabilities to paleogeographical-based time slices coded on the emergence or submergence of the defined areas at a given period, following Chamberland et al. [46] and Tong et al. [31] (see Supplementary Material). GAARlandia was modeled as the connections between islands making up the Greater Antilles, along with their connection to South America from 35–30 mya [20,21]. We also modeled the geologic splits among the Greater Antillean islands in both the GAARlandia and no-GAARlandia models, specifically the opening of the Mona Passage between Hispaniola and Puerto Rico at 23 mya, and the opening of the Windward Passage, separating Cuba and Hispaniola, at 15 mya [20]. In addition, we encoded for the fluctuating emergence of Jamaica at various periods, and on the timing of the appearance and distance of Central America to other landmasses within the region [20]. In *BioGeoBEARS* and within *R*, we applied the dispersal-extinction-cladogenesis (DEC) and DEC + J models, the latter of which accounts for founder-event speciation. It should be mentioned that the DEC + J model has been criticized as a poor explanator of geographic range evolution due to its parameterization of the speciation mode, as opposed to speciation rate [83]. Here, we tested DEC and DEC + J under the no-GAARlandia and GAARlandia models. The Akaike information criterion (AIC) [71] and relative likelihoods were used to assess model probabilities, given the data. We compared the likelihood scores obtained from each run to test for significance (ΔAICc of 2 was considered significant) [84].

2.5. Specimen Photography

Specimen photographs, depicting morphological variation between the populations or species, were taken using a Canon 5D camera with a 65 mm macro 5× zoom lens attached to the Visionary Digital BK laboratory system rig (Dun Inc., Palmyra, VA, USA). Specimens were placed in a dish filled with alcohol-based hand sanitizer (65% ethanol), and covered with a thin film of 95% ethanol to in order to produce a clear image. Multiple image slices were stacked using the Helicon Focus [85] and were refined in Adobe Photoshop 22.1, where dust and other residues were removed from the background and the image was fine-tuned to adjust for contrast and sharpness. Scale measurements for each specimen were also added via Photoshop. Figures were generated and edited using Adobe Illustrator and exported as PDFs.

3. Results

3.1. Sequence Alignment

A total of 76 sequences were generated from the CO1 and ITS2 fragments of the *Microthena* sample set ($n_{CO1} = 50$, $n_{ITS2} = 26$). These were combined with sequences retrieved from data generated by McHugh et al. [51] to form a combined dataset of 405 sequences ($n_{CO1} = 164$, $n_{ITS2} = 131$, $n_{16S} = 110$), representing 189 individuals. The additional 24 CO1 sequences, representing unaccounted-for species, were retrieved from GenBank. Alignment lengths were CO1-1162 bp, 16S-458 bp, and ITS2-554 bp for a total of 2174 base pairs.

3.2. Phylogenetics

Relationships based on the Bayesian inference were robustly supported, with posterior probability values of most nodes >0.95 (Figure 2). Relationships within *Microthena militaris* showed considerably lower support than the other nodes along the tree, as did some of the other fine-scale relationships highlighted in this analysis (mostly individual specimens representing tree tips) (Figures 2–5). However, support for major clade divisions and deep-rooted nodes remained consistently robust throughout the concatenated phylogeny (Figure 2).

Our results support the monophyly of *Microthena*, but reject the monophyly of Caribbean *Microthena* (Figures 2–5). All named *Microthena* species were monophyletic. Caribbean taxa are distributed among three species groups, previously defined by Magalhães and Santos [53] (Figure 3). We identified Caribbean *Microthena* to belong to the nominal *militaris*-group, including *M. sexspinosa*, *M. militaris*, *M. sagittata*, and *M. banksi* (Figure 3). In addition, we substantiated the *furcula*-group, containing *M. cubana* and *M. similis*.

The *gracilis*-group, including *M. gracilis* and *M. horrida*, was additionally delineated but did not include *M. forcipata* in our multillocus analysis (Figure 3). Instead, we found that *Microthena forcipata* was located as a sister to *M. schreibersi*, together forming the sister group to the *furcula* group. However, the topology of our CO1 trees indicated that the positionality of the *furcula* group (*M. cubana and M. similis*) and *M. schreibersi* were unstable. In our CO1 analysis, *M. schreibersi* is sister to the *gracilis*-group, instead of *M. forcipata*, while both *M. schreibersi* and the *gracilis*-group were, together, sisters to *M. forcipata* (Figure 4).

Our analysis also produced evidence in support of single-island endemism and island monophyly of *Microthena forcipata*. High levels of island genetic structuring and relatively deep divergences were observed between *M. forcipata* from Cuba and *M. forcipata* from Hispaniola (Figures 2–5). At a finer scale, *M. forcipata* groups from Hispaniola further demonstrated intra-island structuring (Figure 2).

A Puerto Rican *M. militaris* clade was nested within Hispaniolan *M. militaris*; thus, it is not a single-island endemic (Figure 2). *Microthena horrida* from Cuba, Jamaica, and Central America were not found to be genetically distinct from one another, but were distinct from South American *M. horrida* (Figures 2–5). Furthermore, *M. sagittata* from Mexico, North America (South Carolina), and Florida were genetically distinct from one another, and may represent isolated, morphologically similar, but distinguishable species (Figures 2 and 3, L. Shapiro unpublished data). A putative new species, sister to *M. nigrichelis*, was additionally delineated, here denoted as *M.* sp. (Figure 2). In the Bayesian analysis two South American *Microthena*: *M. perfida* and *M. beta* were used as outgroups, as they were found to be sister to the least inclusive clade containing Caribbean *Microthena* (Figure 2).

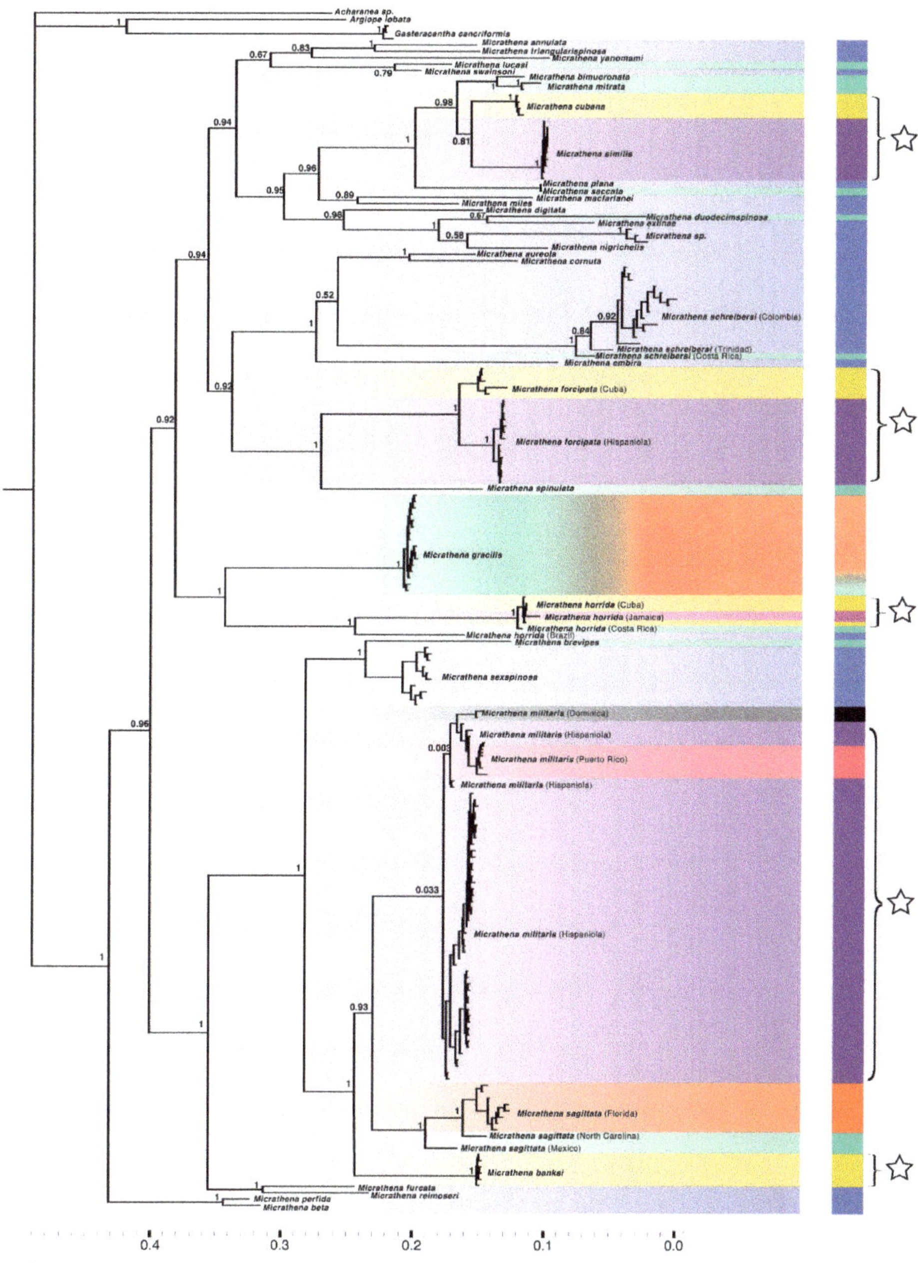

Figure 2. Complete consensus tree from MrBayes concatenated analysis depicting relationships among all sampled *Micrathena* species. Outgroups are located at the top of the phylogeny. Here, terminal individual labels have been replaced with species names along with locality. Overlaying colors are in accordance with color-coded map areas. *M. gracilis* was sampled from both North America and Florida and, therefore, is shaded with an analogous gradient. Stars represent the placement of Caribbean groups within the phylogeny. Posterior probability values are indicated.

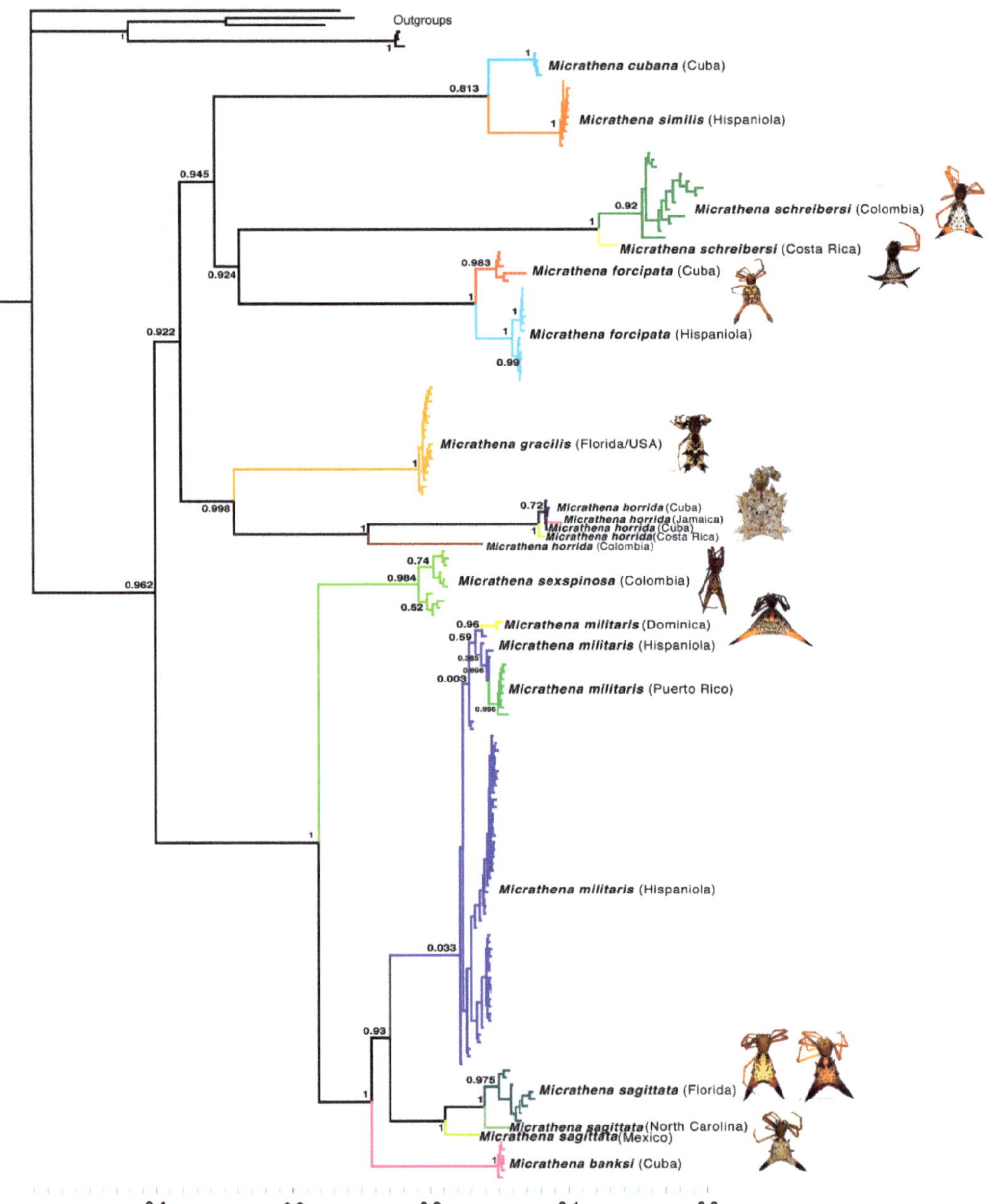

Figure 3. Pruned Bayesian inference tree depicting relationships among Caribbean species groups with associated posterior probability values. Branches are colored by species and individual taxa and have been replaced by species names at tips, but full clade structure is preserved. *Micrathena* dorsal habitus images represent adjacently located taxa. Branches are proportional to evolutionary distances.

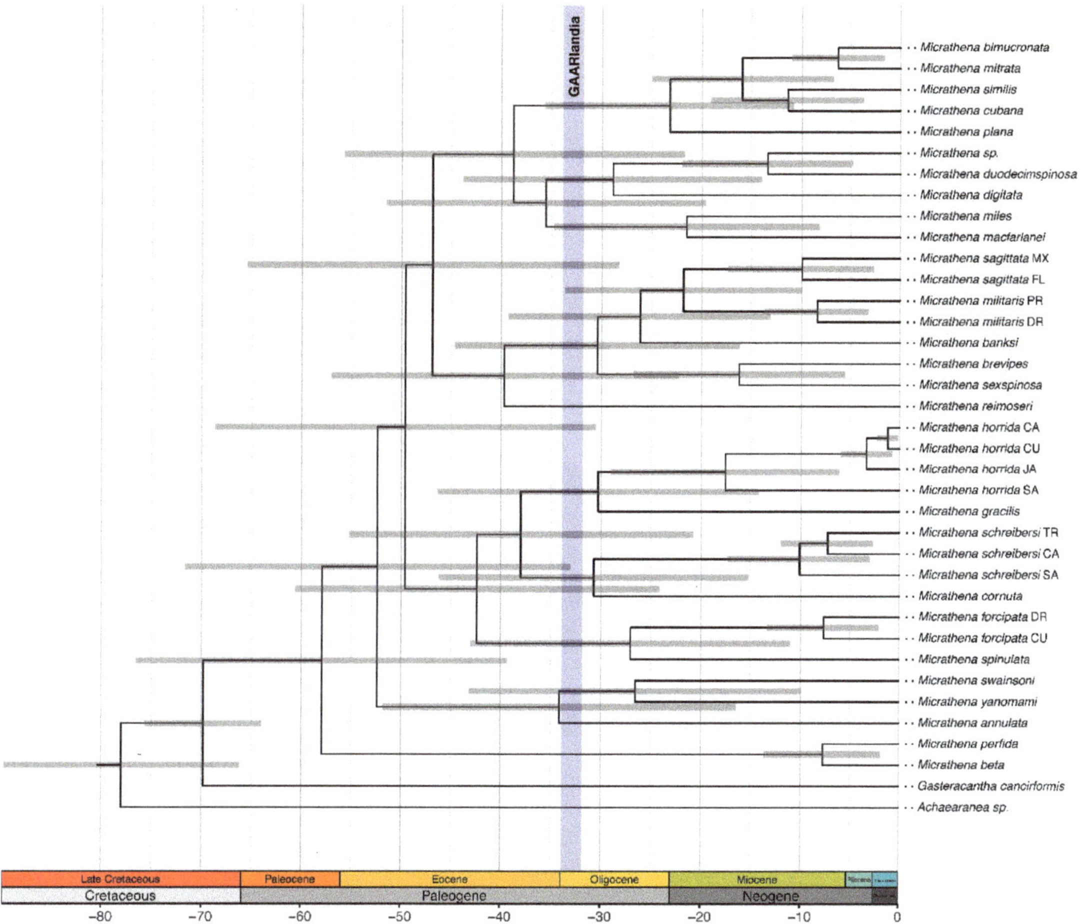

Figure 4. BEAST divergence time estimations of pruned taxa from CO1 data. Grey error bars show error margins around splits calculated in BEAST. Bottom scale is in millions of years and indicates associated geologic time units (periods on lower scale, epochs on upper scale). The timing of the GAARlandia landbridge is also shown from 33–35 Ma. Regional codes associated with taxon names are as follows: CA = Central America, CU = Cuba, DR = Dominican Republic, FL = Florida, JA = Jamaica, MX = Mexico, PR = Puerto Rico, TR = Trinidad.

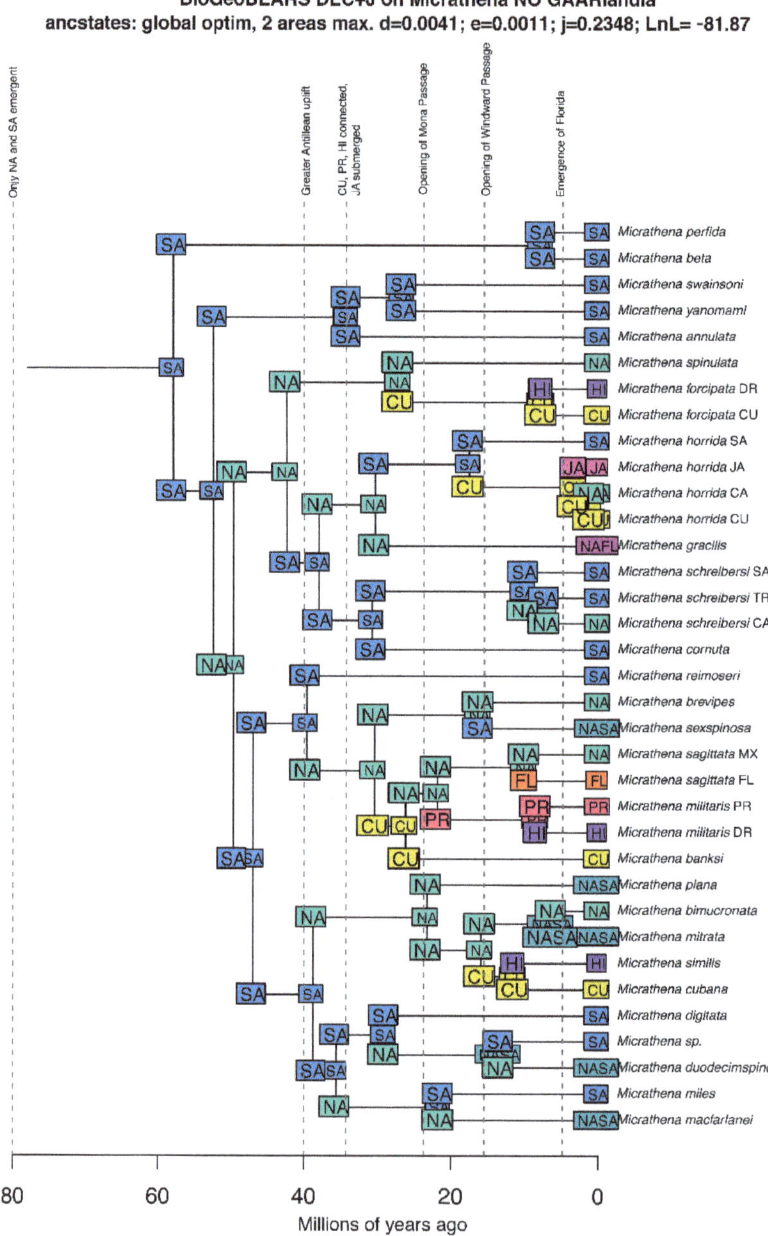

Figure 5. Ancestral range estimation output from BioGeoBEARS on the DEC + J no-GAARlandia model. Colored nodes indicate the most probable range of the MRCA (most recent common ancestor); SA = South America, NA = North America + Central America, CU = Cuba, PR = Puerto Rico, HI = Hispaniola, FL = Florida, JA = Jamaica. Some boxes indicate multiple probable ranges. Boxes are colored by species area labels (See Figure 1). Relevant geologic events corresponding with BioGeoBEARS time slice inputs (see Supplementary Material) are indicated by dotted lines.

3.3. Divergence Times

Only CO1 data were used to build our dated phylogeny, as sequences were available for various South American taxa for which data on other loci were absent. BEAST analyses indicated that the age of Araneidae was estimated at 70 my (64–76), while the age of the Araneidae–Theridiidae split was placed at 78 my (67–91) (Figure 4). The age of *Micrathena* was estimated to be around 58 my (33–71) (Paleocene, Thanetian, supported by Garrison et al. [86]), corroborating that they are representative of a relatively old New World araneid lineage and were present in the Caribbean region within the timing of the GAARlandia landbridge (Figure 4). Caribbean lineages diverged from mainland groups at variable geologic timepoints, with the oldest split dating back to around 30 mya between Cuba and North America and, additionally, implied five possible colonizations of the Caribbean (Figure 4). More recent Caribbean taxa, exemplified by *M. cubana* and *M. similis*, split from their Mexican and Central American relatives (*M. mitrata* and *M. bimucronata*) at approximately 16 mya (Figure 4). The Caribbean and Central American lineages of *M. horrida* split from South American *M. horrida* at around 17 mya (Figure 4). Deep divergences between Mexican and Floridian *M. sagittata* were also suggested, with a split occurring approximately 10 mya (Figures 2–4). Caribbean *Micrathena* were ostensibly polyphyletic (Figures 2–5).

For further detail on topological comparisons between the Bayesian and CO1 BEAST trees, see Supplementary File S3.

3.4. Biogeographic Patterns

3.4.1. Overview

The ancestral range reconstruction in *BioGeoBEARS* suggested five independent colonizations of the Caribbean by *Micrathena* (the *similis/cubana* clade, *banksi* clade, *militaris* clade, *horrida* clade, and *forcipata* clade) (Figure 5). The larger *banksi/militaris* group is considered a Caribbean clade, but *M. banksi* and *M. militaris* from Hispaniola and Puerto Rico each arrived to the Greater Antilles independently (Figure 6). *Micrathena* originated in South America; an early branching South American lineage is sister to a lineage represented by another South American clade that is then, in turn, sister to the rest of the genus, including further South American members and those found in North and Central America and the Caribbean (Figure 5). There existed an early split between South and North American *Micrathena* 52 million years ago and, subsequently, multiple bifurcations between North/Central and South American *Micrathena* occurred thereafter (Figure 5). These results indicated that a fraction of *Micrathena*, other than the *swainsoni* and *perfida* clades, were indeed North American/Central American in origin, the ancestor having split from South America at this 52 mya timepoint, and this clade originating in North America 50 million years ago (Figure 5).

Four of the five clades containing Greater Antillean taxa are North American/Central American in origin (Figure 5). *M. horrida* is the exception, with South America denoted as ancestral, originating about 17 ma (Figure 5). However the common ancestor of *M. horrida* and *M. gracilis* appears to be North American (30 Ma) (Figure 5). While Cuba is resolved as ancestral to the entirety of the sagitatta/militaris clade (including *M. banksi*), North America is the origin of *M. militaris* from both Puerto Rico and Hispaniola (its pre-dispersal to Puerto Rico was approximately 21 ma) (Figure 5). After colonization from South America, *M. horrida* appears to have diversified to form the Central American, Jamaican, and Cuban clades. Jamaican *M. horrida* split off from this group first at 3.3 Ma, with North/Central American *M. horrida* and Cuban *M. horrida* subsequently bifurcating at 1.18 Ma (Figure 5).

Figure 6. High-resolution composite photographs of female *M. sagittata* specimens from Florida and Mexico depicting morphological variation between populations. Images are of dorsal and ventral habitus of each specimen. Scale bars are associated with each photograph (all lines are 1 mm in length). Habitus shape, along with posterior spine proportion and form, differ between the two groups, although spine number is consistent. Posterior spines of *M. sagittata* from Mexico appear more rounded and wider-set than Floridian *M. sagittata*. Obvious differences in coloration are apparent, with Mexican *M. sagittata* lacking the bright red and yellow pigmentation of Floridian *M. sagittata* on dorsal and ventral sides. Further sampling of Mexican *M. sagittata* is necessary to ensure within-population morphology is consistently distinct from Floridian *M. sagittata*.

Cuba was the first of the Greater Antillean islands to be colonized by South and North/Central American ancestors among all Caribbean groups in our analyses, preceding dispersal to other Caribbean islands (Puerto Rico, Hispaniola, or Jamaica (or mainland sources in select aforementioned cases)) (Figure 5). The initial splits between mainland and Cuban taxa occur at 27 Ma (in the *M. spinulata*/*M. forcipata* group), 17 Ma (amongst *M. horrida*), 30 Ma (in the *M. militaris* clade), and 16 Ma (within the *M. simils*/*M.cubana*/*M. mitrata* clade) (Figure 5).

We additionally observed multiple inter-island colonization events within the Greater Antilles; this included movement from Puerto Rico to Hispaniola at 8 mya within *M. militaris*, and two Cuba–Hispaniola splits at 7 and 11 mya within *M. forcipata* and between *M. cubana* and *M similis* (Figure 5).

3.4.2. Vicariance vs. Long Distance Dispersal

The DEC + J no-GAARlandia hypothesis demonstrated the best statistical fit, given our input phylogeny, applied time-slices, and affiliated chrono-geographical probabilities (Table 3). The model comparison using AICc also distinguished the BAYAREALIKE + J as significant (Table 3). The top three models determined by AICc were all representative of no-GAARlandia hypotheses (Table 3) with mixed support for lower-ranked models, although none are of statistical significance (Table 3). Both the model ranking and Bio-GeoBEARS results are in agreement that colonization events are not tied to dispersal via the GAARlandia landbridge.

Table 3. BioGeoBEARS model probabilities and rankings. Six models were used in our analysis (DEC, DEC + J, BAYAREALIKE, BAYAREALIKE + J, DIVALIKE, DIVALIKE + J) to test data in the presence or absence of GAARlandia (GAARlandia and no-GAARlandia models). *LnL* is log likelihood, *d* is dispersal rate, *e* is extinction rate, *j* is the relative probability of founder event speciation at cladogenesis, *AICc* is Akaike's information criterion (with correction for smaller sample sizes), *AICc weight* is the normalized relative model likelihood, and $\Delta AICc$ is AIC—min(AIC).

Model	*LnL*	Number of Parameters	*d*	*e*	*j*	*AICc*	*AICc Weight*	$\Delta AICc$
DEC + J no-GAARlandia	−81.87	3	0.0041	0.0011	0.2	170.5	0.56	0
BAYAREALIKE + J no-GAARlandia	−82.46	3	0.0019	0.01	0.2	171.7	0.31	1.2
DIVALIKE + J no-GAARlandia	−83.53	3	0.0048	0.001	0.2	173.8	0.11	3.3
BAYAREALIKE + J GAARlandia	−85.26	3	0.023	0.011	0.8	177.3	0.019	6.8
DIVALIKE no-GAARlandia	−95.23	2	0.013	0.0033	0	194.8	2.9×10^{-6}	24.3
DEC + J GAARlandia	−94.48	3	0.025	1.00×10^{-12}	2.4	195.7	1.90×10^{-6}	25.2
DIVALIKE + J GAARlandia	−97.42	3	0.027	1.00×10^{-12}	1.7	201.6	9.90×10^{-8}	31.1
DEC no-GAARlandia	−99.69	2	0.013	0.0063	0	203.8	3.40×10^{-8}	33.3
BAYAREALIKE no-GAARlandia	−107.9	2	0.017	0.025	0	220.2	8.90×10^{-12}	49.7
BAYAREALIKE GAARlandia	−112	2	0.24	0.025	0	228.4	1.50×10^{-13}	57.9
DIVALIKE GAARlandia	−112.8	2	0.11	0.0058	0	230	6.90×10^{-14}	59.5
DEC GAARlandia	−112.9	2	0.16	0.01	0	230.2	6.00×10^{-14}	59.7

4. Discussion

Molecular analyses, with the expanded taxon sampling of *Micrathena*, resolved the genus as monophyletic with polyphyletic Caribbean taxa (Figures 2–5), consistent with the findings of McHugh et al. [51], Crews and Esposito [36], and Magalhães and Santos [53] (Figures 2–5). We detected five independent colonization events to the Caribbean from varying mainland sources (Figure 5). While South America was the ancestral *Micrathena* range, four of the five Caribbean groups were actually North American/Central American in origin (Figure 5), corroborating evidence by other authors [36]. Crews and Esposito [36] found evidence that *Micrathena* had repeatedly dispersed to the Caribbean (six times) and suggested that GAARlandia likely played some role in this dispersal. We did not find evidence for the latter hypothesis [36,51]. Rather, the BioGeoBEARS results and the biogeographic model ranking indicated that *Micrathena* colonized the Caribbean multiple times, but each time outside of the timespan of the proposed GAARlandia landbridge.

In addition to the dispersal from continental sources, we found evidence for movement among islands, as well as the reverse colonization of North America from Cuba (Figure 5). The phenomenon of movement from island-to-continent has been documented in other spider lineages, including *Deinopis* [46] and *Tetragnatha* [87], adding to the growing frequency of this pattern observed in arachnids, even across groups with variable dispersal strategies [87]. Movement among the Greater Antillean islands reflected both long-distance dispersal and the dispersal to nearby islands (e.g., two pairs of HI-CU sister taxa and the *M. militaris* groups from PR and HI) (Figures 2–5).

Independent dispersals at various geologic timepoints (Figure 5) suggested that stochastic events, such as extreme weather events (e.g., hurricanes) or ocean currents, could have played a role in transporting *Micrathena* across the Caribbean, as proposed for other arthropod groups [88–90]. Given that the Caribbean lineages of *Micrathena* have a North/Central American origin, the loop current, wrapping around the Gulf of Mexico, entering by the Yucatán peninsula, and exiting via the straights of Florida [91], may be of particular import as it brushes close to Greater Antillean islands. The long-distance dispersal, via rafting in arachnids, has been documented in *Moggridgea* mygalomorphs in Australia [92] and in *Amaurobioides* [93]. Paleocurrent directionality in the Caribbean, which most likely mirrors that of the Holocene (although a thruway between the Atlantic and Pacific existed before the closure of the Panama isthmus at 3.5 Ma) [94–96], and it can be hypothesized that the dispersal routes that allowed *Micrathena* to colonize the Caribbean reflect modern and paleooceanographic dynamics. Future investigations may consider integrating paleowind and paleocurrent data to better explain fine-scale dispersal routes of Caribbean colonization that criss-cross the region. While such analyses have been undertaken for Caribbean mammals in terms of utilizing "floating islands" [97], these data have not been applied to biogeographic investigations of spiders. However, hurricanes (with modern directionality) have been shown to be a mechanism important in arthropod dispersal [90] and the dispersal effects have also been empirically noted [89]. The habitat choice in *Micrathena*, often occupying the center of wide-open spaces in forests where the web and animal are readily exposed to weather conditions reaching inside the forest, could render them relatively prone to weather-related involuntary aerial dispersal.

This study adds to the growing composite of data suggesting manifold Caribbean dispersals in *Micrathena* and indicates that, although they are considered relatively poor dispersers due to their apparent bulkiness and elaborate spine coverage, *Micrathena* may actually be relatively proficient dispersers. We would predict this dispersal would mostly occur as juveniles, when they are less heavily ornamented. Other large araneids, including *Nephila* [98] and various *Argiope* and *Araneus* species, do balloon [56]. Not much is known about the physical capacity for dispersal in *Micrathena*, and biogeographic investigations may benefit from increased physiological and behavioral analyses of the genus.

We recovered four distinct *Micrathena* clades containing Caribbean taxa, which roughly correspond to the species-groups defined by Magalhães and Santos [53] and are corroborated by McHugh et al. [51]: the *militaris*-group, the *gracilis*-group, and the *furcula*-group + *M. forcipata*

(Figure 3, Table 4). Like McHugh et al. [51], our analyses do not place *M. forcipata* within the *gracilis* group. However, the placement of *M. forcipata* differs from McHugh et al. [51] and is influenced by taxon sampling and phylogenetic methods (Table 4). It is likely that gaps in taxon sampling are responsible for the instability of *M. schreibersi* and the *furcula* group, that is noted between the multilocus and the CO1 analyses.

Table 4. Comparisons between species-group delineations for three *Micrathena* phylogenetic analyses performed by Magalhaẽs et al. [53], McHugh et al. [51], and this investigation (multilocus datset, Figures 1 and 2). Caribbean species groups are listed along with species belonging to that group in each study. Additional notes on the differing position of *M. schreibersi*, as it relates to these groups, the study by McHugh et al. [51], and this analysis, are listed as footnotes.

Species-Group	Magalhaẽs et al., 2012	McHugh et al., 2014	Current *Micrathena* Study
furcula	*M. cubana, M. similis*	*M. cubana, M. similis*	*M. cubana, M. similis*
militaris	*M. banksi, M. militaris, M. sagittata, M. sexspinosa*	*M. banksi, M. militaris, M. sagittata, M. sexspinosa*	*M. banksi, M. militaris, M. sagittata, M. sexspinosa*
gracilis	*M. horrida, M. gracilis, M. forcipata*	*M. horrida, M. gracilis* [1]	*M. horrida, M. gracilis* [2]

[1] *M. schreibersi* is the sister to the *gracilis* group; *M. forcipata* is the sister to the *furcula* group. [2] *M. schreibersi* is the sister to *M. forcipata*, and both are sisters to the *furcula* group.

Our analyses indicated deep divergences within 'widespread taxa', suggesting that such taxa would be better characterized as multiple single-island endemics. For example, *M. forcipata* from Cuba and Hispaniola are genetically distinct from one another, as indicated by deep branching separating the two on the phylogeny. These taxa may also be distinguishable based on morphology (Figure 3 and L. Shapiro's unpublished data). The divergence among these similar taxa is likely due to the segregation of these two islands by the Windward Passage, acting as a geographic barrier post-dispersal (Figures 2–5). While McHugh et al. [51] also determined that the *M. militaris* groups represent single-island endemics from Puerto Rico and Hispaniola, we found that, although *M. militaris* from Puerto Rico are monophyletic, they are nested within the Hispaniolan members of the species, hence rejecting a model of purely single-island endemics in this genus (Figure 2).

Genetic divergences between *M. sagittata* from North America (North Carolina), Florida, and Mexico were also noted in our analyses, where the Mexican *M. sagittata* is the sister to the North American group (Figures 2 and 3). Morphological distinctions between Mexican *M. sagittata*, in comparison to our *M. sagittata* sample from Florida, can be clearly observed (Figure 6). An additional putative, currently undescribed sister species to *M. nigrichelis* was identified in the phylogeny, *Micrathena* sp. The preliminary habitus photographs of *M.* sp. are displayed in Figure 7. Integrative genetic and morphological analyses are currently underway to solidify evidence for the species delimitations of new clades and divergent species uncovered in this study.

Our work, combined with previous biogeographic analyses, substantiates *Micrathena* spiders as an excellent model for Caribbean biogeography of a dispersal-prone lineage. The additional depth in taxon sampling of *Micrathena* and the related genera, especially across Central and South America, as well as expanded data with next-generation sequencing and the greater availability of fossil evidence for calibration, will add to the resolution of factors influencing biodiversity in this region.

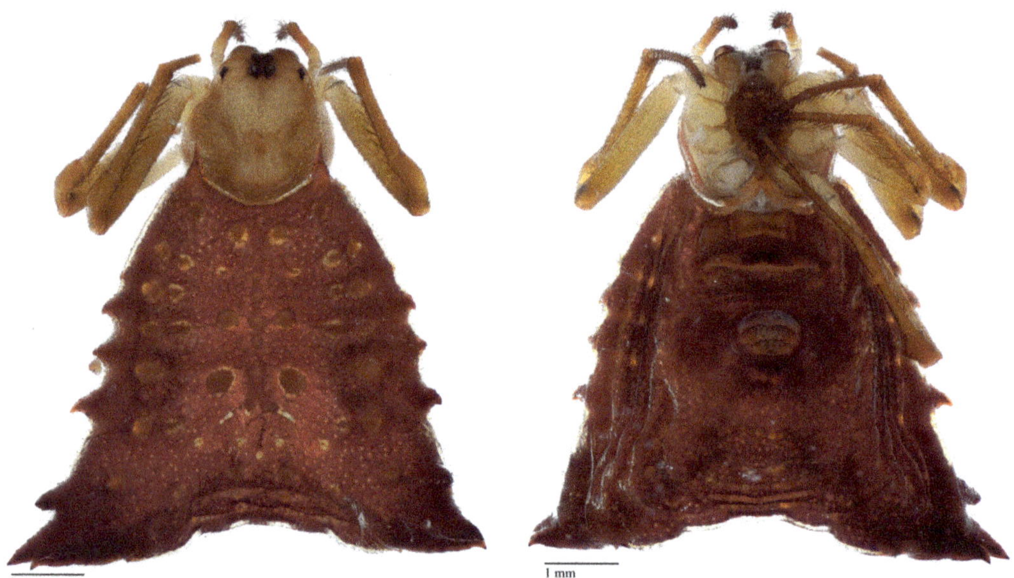

Figure 7. High-resolution composite photographs of putative new species *M.* sp. from Colombia. Photographs depict dorsal and ventral habitus of a female specimen. Future studies will hopefully provide more data detailing important morphological characters. Scale is depicted at the bottom of each photograph.

5. Conclusions

We present a detailed molecular phylogenetic and biogeographic analysis of *Micrathena*, demonstrating that the group likely colonized the Caribbean region multiple times independently during the last 30 million years, and that diversification was likely a result of multiple overwater dispersal events and not GAARlandia vicariance. This finding suggests that *Micrathena*, while potentially dispersal-limited due to its size and morphology, have nevertheless been carried across oceanic barriers to colonize Caribbean islands five times in 30 million years, perhaps as juveniles. We found interesting evidence for single-island endemics in *M. forcipata* and have unveiled the cryptic diversity in *M. sagittata* and within the genus altogether. Further studies will focus on taxonomic examinations of potential species uncovered in this phylogeny.

Supplementary Materials: The following are available online at https://www.mdpi.com/article/10.3390/d14020128/s1, File S1: Dispersal probabilities and geography input for BioGeoBEARS, File S2: List of *Micrathena* species in study, File S3: Comparison of concatenated Bayesian and BEAST phylogenies, File S4: Raw BEAST.xml output file.

Author Contributions: Conceptualization, L.S. and I.A.; methodology L.S. and I.A.; software, L.S. and I.A.; formal analysis, L.S. and I.A.; investigation, L.S. and I.A.; resources, I.A.; data curation, L.S. and I.A.; writing—original draft preparation, L.S.; writing—review and editing, L.S., I.A. and G.J.B.; visualization, L.S. and I.A.; supervision, I.A. and G.J.B.; project administration, I.A.; funding acquisition, I.A. and G.J.B. All authors have read and agreed to the published version of the manuscript.

Funding: This research was funded by the National Science Foundation, grants numbered DEB-1314749 and DEB-1050253 awarded to G. Binford and I. Agnarsson, and by a grant from the National Geographic Society (WW-203R017) to I. Agnarsson.

Institutional Review Board Statement: All material was collected under appropriate collection permits and approved guidelines.

Informed Consent Statement: Not applicable.

Data Availability Statement: Code can be found at https://github.com/lkshapir/Micrathena_paper_scripts (accessed on 6 October 2021).

Acknowledgments: We would like to thank all members of the CarBio team who were involved in collecting and cataloguing specimens used in this study. We thank members of the Agnarsson laboratory-specifically Lisa Chamberland and Laura Caicedo-Quiroga for their guidance and advice in developing this project, and Matjaz Gregoric and Ren-Chun Cheng of the Kuntner lab in Slovenia for providing outgroup sequence data on *Argiope*. Special thanks to Anne McHugh who initiated this research project and published a paper on earlier findings.

Conflicts of Interest: The authors declare no conflict of interest.

References

1. Wilson, E.O.; MacArthur, R.H. *The Theory of Island Biogeography*; Princeton University Press: Princeston, NJ, USA, 1967.
2. Carlquist, S.J. *Island Biology*; Columbia University Press: New York, NY, USA, 1974.
3. Whittaker, R.J. *Island Biogeography*; Oxford University Press: Oxford, UK, 1998.
4. Whittaker, R.J.; Fernández-Palacios, J.M. *Island Biogeography: Ecology, Evolution, and Conservation*; Oxford University Press: Oxford, UK, 2007.
5. Gillespie, R. Community Assembly Through Adaptive Radiation in Hawaiian Spiders. *Science* **2004**, *303*, 356–359. [CrossRef]
6. Craig, D.A.; Currie, D.C.; Joy, D.A. Geographical history of the central-western Pacific black fly subgenus *Inseliellum* (Diptera: Simuliidae: *Simulium*) based on a reconstructed phylogeny of the species, hot-spot archipelagoes and hydrological considerations. *J. Biogeogr.* **2001**, *28*, 1101–1127. [CrossRef]
7. Ricklefs, R.; Bermingham, E. The West Indies as a laboratory of biogeography and evolution. *Philos. Trans. R. Soc. B Biol. Sci.* **2008**, *363*, 2393–2413. [CrossRef]
8. Myers, N.; Mittermeler, R.A.; Mittermeler, C.G.; Da Fonseca, G.A.B.; Kent, J. Biodiversity hotspots for conservation priorities. *Nature* **2000**, *403*, 853–858. [CrossRef] [PubMed]
9. Rosen, D.E. A Vicariance Model of Caribbean Biogeography. *Syst. Biol.* **1975**, *24*, 431–464. [CrossRef]
10. Pindell, J.L.; Barrett, S.F. Geologic evolution of the Caribbean region: A plate-tectonic persepective. In *Decade of American Geology*; Case, J.E., Dengo, G., Eds.; Geological Society of America: Boulder, CO, USA, 1990; The Geology of North America v. H; pp. 405–432.
11. Pindell, J.L. Evolution of the Gulf of Mexico and the Caribbean. In *Caribbean Geology, an Introduction*; Donovan, S.K., Jackson, T.A., Eds.; The University of West Indies Publishers' Association: Kingston, Jamaica, 1994; pp. 13–39.
12. Burke, K. Tectonic evolution of the Caribbean. *Annu. Rev. Earth Planet. Sci.* **1988**, *16*, 201–230. [CrossRef]
13. Donnelly, T.W. Geologic constraints on Caribbean biogeography. In *Zoogeography of Caribbean Insects*; Liebherr, J.K., Ed.; Cornell University Press: Ithaca, NY, USA, 1988; pp. 15–37.
14. Alonso, R.; Crawford, A.J.; Bermingham, E. Molecular phylogeny of an endemic radiation of Cuban toads (Bufonidae: Peltophryne) based on mitochondrial and nuclear genes. *J. Biogeogr.* **2012**, *39*, 434–451. [CrossRef]
15. Keppel, G.; Lowe, A.J.; Possingham, H.P. Changing Perspectives on the Biogeography of the Tropical South Pacific: Influences of Dispersal, Vicariance and Extinction. *J. Biogeogr.* **2009**, *36*, 1035–1054. [CrossRef]
16. Crews, S.C.; Gillespie, R.G. Molecular systematics of *Selenops* spiders (Araneae: Selenopidae) from North and Central America: Implications for Caribbean biogeography. *Biol. J. Linn. Soc.* **2010**, *101*, 288–322. [CrossRef]
17. Heinicke, M.P.; Duellman, W.E.; Hedges, S.B. Major Caribbean and Central American frog faunas originated by ancient oceanic dispersal. *Proc. Natl. Acad. Sci. USA* **2007**, *104*, 10092–10097. [CrossRef]
18. Toussaint, E.F.A.; Balke, M. Historical biogeography of *Polyura* butterflies in the oriental Palaeotropics: Trans-archipelagic routes and South Pacific island hopping. *J. Biogeogr.* **2016**, *43*, 1560–1572. [CrossRef]
19. Agnarsson, I.; Cheng, R.-C.; Kuntner, M. A multi-clade test supports the intermediate dispersal model of biogeography. *PLoS ONE* **2014**, *9*, e86780. [CrossRef]
20. Iturralde-Vinent, M.A. Meso-Cenozoic Caribbean paleogeography: Implications for the historical biogeography of the region. *Int. Geol. Rev.* **2006**, *48*, 791–827. [CrossRef]
21. Iturralde-Vinent, M.A.; MacPhee, R.D.E. Paleogeography of the Caribbean region: Implications for Cenozoic biogeography. *Bull. Am. Museum Nat. Hist.* **1999**, *238*, 1–95.
22. Fabre, P.-H.; Vilstrup, J.T.; Raghavan, M.; Der Sarkissian, C.; Willerslev, E.; Douzery, E.J.P.; Orlando, L. Rodents of the Caribbean: Origin and diversification of hutias unravelled by next-generation museomics. *Biol. Lett.* **2014**, *10*, 20140266. [CrossRef] [PubMed]
23. Klemen, Č.; Ingi, A.; Binford, G.J.; Matjaž, K. Biogeography of the Caribbean *Cyrtognatha* spiders. *Sci. Reports (Nature Publ. Group)* **2019**, *9*, 1–14.
24. Esposito, L.A.; Bloom, T.; Caicedo-Quiroga, L.; Alicea-Serrano, A.M.; Sánchez-Ruíz, J.A.; May-Collado, L.J.; Binford, G.J.; Agnarsson, I. Islands within islands: Diversification of tailless whip spiders (Amblypygi, *Phrynus*) in Caribbean caves. *Mol. Phylogenet. Evol.* **2015**, *93*, 107–117. [CrossRef]

25. Condamine, F.L.; Silva-Brandão, K.L.; Kergoat, G.J.; Sperling, F.A.H. Biogeographic and diversification patterns of Neotropical Troidini butterflies (Papilionidae) support a museum model of diversity dynamics for Amazonia. *BMC Evol. Biol.* **2012**, *12*, 82. [CrossRef]

26. Dziki, A.; Binford, G.J.; Coddington, J.A.; Agnarsson, I. *Spintharus flavidus* in the Caribbean—A 30 million year biogeographical history and radiation of a 'widespread species'. *PeerJ* **2015**, *3*, e1422. [CrossRef]

27. Pfingstl, T.; Lienhard, A.; Baumann, J.; Koblmüller, S. A taxonomist's nightmare–Cryptic diversity in Caribbean intertidal arthropods (Arachnida, Acari, Oribatida). *Mol. Phylogenet. Evol.* **2021**, *163*, 107240. [CrossRef]

28. Esposito, L.A.; Prendini, L. Island ancestors and New World biogeography: A case study from the scorpions (Buthidae: Centruroidinae). *Sci. Rep.* **2019**, *9*, 1–11. [CrossRef]

29. Chamberland, L.; Salgado-Roa, F.C.; Basco, A.; Crastz-Flores, A.; Binford, G.J.; Agnarsson, I. Phylogeography of the widespread Caribbean spiny orb weaver *Gasteracantha cancriformis*. *PeerJ* **2020**, *8*, e8976. [CrossRef]

30. Říčan, O.; Piálek, L.; Zardoya, R.; Doadrio, I.; Zrzavý, J. Biogeography of the Mesoamerican Cichlidae (Teleostei: Heroini): Colonization through the GAARlandia land bridge and early diversification. *J. Biogeogr.* **2013**, *40*, 579–593. [CrossRef]

31. Tong, Y.; Binford, G.; Rheims, C.A.; Kuntner, M.; Liu, J.; Agnarsson, I. Huntsmen of the Caribbean: Multiple tests of the GAARlandia hypothesis. *Mol. Phylogenet. Evol.* **2019**, *130*, 259–268. [CrossRef]

32. Agnarsson, I.; LeQuier, S.M.; Kuntner, M.; Cheng, R.-C.; Coddington, J.A.; Binford, G. Phylogeography of a good Caribbean disperser: *Argiope argentata* (Araneae, Araneidae) and a new 'cryptic' species from Cuba. *Zookeys* **2016**, 25. [CrossRef] [PubMed]

33. Nieto-Blázquez, M.E.; Antonelli, A.; Roncal, J. Historical Biogeography of endemic seed plant genera in the Caribbean: Did GAAR landia play a role? *Ecol. Evol.* **2017**, *7*, 10158–10174. [CrossRef] [PubMed]

34. Dávalos, L.M. Phylogeny and biogeography of Caribbean mammals. *Biol. J. Linn. Soc.* **2004**, *81*, 373–394. [CrossRef]

35. Dávalos, L.M.; Turvey, S.T. West Indian mammals: The old, the new, and the recently extinct. In *Bones, Clones, and Biomes: The History and Geography of Recent Neotropical Mammals*; Patterson, B.D., Costa, L.P., Eds.; University of Chicago Press: Chicago, IL, USA, 2012; pp. 157–202.

36. Crews, S.C.; Esposito, L.A. Towards a synthesis of the Caribbean biogeography of terrestrial arthropods. *BMC Evol. Biol.* **2020**, *20*, 12. [CrossRef] [PubMed]

37. Philippon, M.; Cornée, J.-J.; Münch, P.; Van Hinsbergen, D.J.J.; Boudagher-Fadel, M.; Gailler, L.; Boschman, L.M.; Quillevere, F.; Montheil, L.; Gay, A.; et al. Eocene intra-plate shortening responsible for the rise of a faunal pathway in the northeastern Caribbean realm. *PLoS ONE* **2020**, *15*, e0241000.

38. Ali, J.R. Colonizing the Caribbean: Is the GAARlandia land-bridge hypothesis gaining a foothold? *J. Biogeogr.* **2012**, *39*, 431–433. [CrossRef]

39. Hedges, S.B. Paleogeography of the Antilles and Origin of West Indian Terrestrial Vertebrates. *Ann. Missouri Bot. Gard.* **2006**, *93*, 231–244. [CrossRef]

40. Ali, J.R.; Hedges, S.B. Colonizing the Caribbean: New geological data and an updated land-vertebrate colonization record challenge the GAARlandia land-bridge hypothesis. *J. Biogeogr.* **2021**, *48*, 2699–2707. [CrossRef]

41. Garrocq, C.; Lallemand, S.; Marcaillou, B.; Lebrun, J.; Padron, C.; Klingelhoefer, F.; Laigle, M.; Münch, P.; Gay, A.; Schenini, L.; et al. Genetic Relations Between the Aves Ridge and the Grenada Back-Arc Basin, East Caribbean Sea. *J. Geophys. Res. Solid Earth* **2021**, *126*, e2020JB020466. [CrossRef]

42. Bond, J.E.; Hamilton, C.A.; Godwin, R.L.; Ledford, J.M.; Starrett, J. Phylogeny, evolution, and biogeography of the North American trapdoor spider family Euctenizidae (Araneae: Mygalomorphae) and the discovery of a new 'endangered living fossil' along California's central coast. *Insect Syst. Divers.* **2020**, *4*, 2. [CrossRef]

43. Chousou-Polydouri, N.; Carmichael, A.; Szűts, T.; Saucedo, A.; Gillespie, R.; Griswold, C.; Wood, H.M. Giant Goblins above the waves at the southern end of the world: The biogeography of the spider family Orsolobidae (Araneae, Dysderoidea). *J. Biogeogr.* **2019**, *46*, 332–342. [CrossRef]

44. Rix, M.G.; Wilson, J.D.; Huey, J.A.; Hillyer, M.J.; Gruber, K.; Harvey, M.S. Diversification of the mygalomorph spider genus *Aname* (Araneae: Anamidae) across the Australian arid zone: Tracing the evolution and biogeography of a continent-wide radiation. *Mol. Phylogenet. Evol.* **2021**, *160*, 107127. [CrossRef]

45. Richardson, B.J. Evolutionary biogeography of Australian jumping spider genera (Araneae: Salticidae). *Aust. J. Zool.* **2020**, *67*, 162–172. [CrossRef]

46. Chamberland, L.; McHugh, A.; Kechejian, S.; Binford, G.J.; Bond, J.E.; Coddington, J.; Dolman, G.; Hamilton, C.A.; Harvey, M.S.; Kuntner, M.; et al. From Gondwana to GAARlandia: Evolutionary history and biogeography of ogre-faced spiders (*Deinopis*). *J. Biogeogr.* **2018**, *45*, 2442–2457. [CrossRef]

47. Hedges, S.B.; Hass, C.A.; Maxson, L.R. Caribbean biogeography: Molecular evidence for dispersal in West Indian terrestrial vertebrates. *Proc. Natl. Acad. Sci. USA* **1992**, *89*, 1909–1913. [CrossRef]

48. Betancur-R, R.; Hines, A.; Acero, P.A.; Ortí, G.; Wilbur, A.E.; Freshwater, D.W. Reconstructing the lionfish invasion: Insights into Greater Caribbean biogeography. *J. Biogeogr.* **2011**, *38*, 1281–1293. [CrossRef]

49. Klein, N.K.; Burns, K.J.; Hackett, S.J.; Griffiths, C.S. Molecular phylogenetic relationships among the wood warblers (Parulidae) and historical biogeography in the Caribbean basin. *J. Caribb. Ornithol.* **2004**, *17*, 3–17.

50. Ferrier, S.; Powell, G.V.N.; Richardson, K.S.; Manion, G.; Overton, J.M.; Allnutt, T.F.; Cameron, S.E.; Mantle, K.; Burgess, N.D.; Faith, D.P.; et al. Mapping more of terrestrial biodiversity for global conservation assessment. *Bioscience* **2004**, *54*, 1101–1109. [CrossRef]

51. McHugh, A.; Yablonsky, C.; Binford, G.; Agnarsson, I. Molecular phylogenetics of Caribbean *Micrathena* (Araneae:Araneidae) suggests multiple colonisation events and single island endemism. *Invertebr. Syst.* **2014**, *28*, 337–349. [CrossRef]

52. Zhang, J.-X.; Maddison, W.P. Molecular phylogeny, divergence times and biogeography of spiders of the subfamily Euophryinae (Araneae: Salticidae). *Mol. Phylogenet. Evol.* **2013**, *68*, 81–92. [CrossRef] [PubMed]

53. Magalhães, I.L.F.; Santos, A.J. Phylogenetic analysis of *Micrathena* and *Chaetacis* spiders (Araneae: Araneidae) reveals multiple origins of extreme sexual size dimorphism and long abdominal spines. *Zool. J. Linn. Soc.* **2012**, *166*, 14–53. [CrossRef]

54. Levi, H.W. The spiny orb-weaver genera *Micrathena* and *Chaetacis* (Araneae: Araneidae). Los géneros de tejedoras de esferas espinosas *Micrathena* y *Chaetacis* (Araneae: Araneidae). *Bull. Museum Comp. Zool.* **1985**, *150*, 429–618.

55. Bukowski, T.C.; Christenson, T.E. Natural history and copulatory behavior of the spiny orbweaving spider *Micrathena gracilis* (Araneae, Araneidae). *J. Arachnol.* **1997**, 307–320.

56. Bell, J.R.; Bohan, D.A.; Shaw, E.M.; Weyman, G.S. Ballooning dispersal using silk: World fauna, phylogenies, genetics and models. *Bull. Entomol. Res.* **2005**, *95*, 69–114. [CrossRef]

57. World Spider Catalog. World Spider Catalog. Version 18.5. National History Museum Bern. Available online: http://ws.nmbe.ch (accessed on 1 March 2018).

58. Scharff, N.; Coddington, J.A.; Blackledge, T.A.; Agnarsson, I.; Framenau, V.W.; Szűts, T.; Hayashi, C.Y.; Dimitrov, D. Phylogeny of the orb-weaving spider family Araneidae (Araneae: Araneoidea). *Cladistics* **2020**, *36*, 1–21. [CrossRef]

59. Agnarsson, I.; Maddison, W.P.; Avilés, L. The phylogeny of the social *Anelosimus* spiders (Araneae: Theridiidae) inferred from six molecular loci and morphology. *Mol. Phylogenet. Evol.* **2007**, *43*, 833–851. [CrossRef]

60. Kuntner, M.; Agnarsson, I. Biogeography and diversification of hermit spiders on Indian Ocean islands (Nephilidae: *Nephilengys*). *Mol. Phylogenet. Evol.* **2011**, *59*, 477–488. [CrossRef]

61. Simon, C.; Frati, F.; Beckenbach, A.; Crespi, B.; Liu, H.; Flook, P. Evolution, weighting, and phylogenetic utility of mitochondrial gene sequences and a compilation of conserved polymerase chain reaction primers. *Ann. Entomol. Soc. Am.* **1994**, *87*, 651–701. [CrossRef]

62. Hedin, M.C.; Maddison, W.P. A combined molecular approach to phylogeny of the jumping spider subfamily Dendryphantinae (Araneae: Salticidae). *Mol. Phylogenet. Evol.* **2001**, *18*, 386–403. [CrossRef] [PubMed]

63. Folmer, O.; Black, M.; Hoeh, W.; Lutz, R.; Vrijenhoek, R. DNA primers for amplification of mitochondrial cytochrome c oxidase subunit I from diverse metazoan invertebrates. *Mol. Mar. Biol. Biotechnol.* **1994**, *3*, 294–299. [PubMed]

64. White, T.J.; Bruns, T.; Lee, S.; Taylor, J. Amplification and direct sequencing of fungal ribosomal RNA genes for phylogenetics. *PCR Protoc. Guid. Methods Appl.* **1990**, *18*, 315–322.

65. Green, P. Phrap. Version 1.090518. Available online: http://phrap.org (accessed on 19 June 2019).

66. Green, P.; Ewing, B. Phred. Version 0.020425c. Available online: http://phrap.org/othersoftware.html (accessed on 19 June 2019).

67. Maddison, D.R.; Maddison, W.P. Chromaseq: A Mesquite Package for Analyzing Sequence Chromatograms. Version 1.0 2011. Available online: http://mesquiteproject.org/packages/chromaseq (accessed on 19 June 2019).

68. Maddison, W.P.; Maddison, D.R. Mesquite: A Modular System for Evolutionary Analysis. Version 3.6. Available online: http://mesquiteproject.org (accessed on 19 June 2019).

69. Katoh, K.; Rozewicki, J.; Yamada, K.D. MAFFT online service: Multiple sequence alignment, interactive sequence choice and visualization. *Brief. Bioinform.* **2019**, *20*, 1160–1166. [CrossRef]

70. Lanfear, R.; Frandsen, P.B.; Wright, A.M.; Senfeld, T.; Calcott, B. PartitionFinder 2: New methods for selecting partitioned models of evolution for molecular and morphological phylogenetic analyses. *Mol. Biol. Evol.* **2017**, *34*, 772–773. [CrossRef]

71. Akaike, H. Information theory and an extension of the maximum likelihood principle. In *Selected Papers of Hirotugu Akaike*; Springer: Berlin/Heidelberg, Germany, 1998; pp. 199–213.

72. Ronquist, F.; Teslenko, M.; Van Der Mark, P.; Ayres, D.L.; Darling, A.; Höhna, S.; Larget, B.; Liu, L.; Suchard, M.A.; Huelsenbeck, J.P. MrBayes 3.2: Efficient Bayesian phylogenetic inference and model choice across a large model space. *Syst. Biol.* **2012**, *61*, 539–542. [CrossRef]

73. Miller, M.A.; Pfeiffer, W.; Schwartz, T. The CIPRES science gateway: A community resource for phylogenetic analyses. In Proceedings of the 2011 TeraGrid Conference: Extreme Digital Discovery, Salt Lake City, UT, USA, 18–21 July 2011; pp. 1–8.

74. Rambaut, A.; Drummond, A.J.; Xie, D.; Baele, G.; Suchard, M.A. Posterior summarization in Bayesian phylogenetics using Tracer 1.7. *Syst. Biol.* **2018**, *67*, 901. [CrossRef]

75. Bouckaert, R.; Vaughan, T.G.; Barido-Sottani, J.; Duchêne, S.; Fourment, M.; Gavryushkina, A.; Heled, J.; Jones, G.; Kühnert, D.; De Maio, N.; et al. BEAST 2.5: An advanced software platform for Bayesian evolutionary analysis. *PLoS Comput. Biol.* **2019**, *15*, e1006650. [CrossRef]

76. Kuntner, M.; Arnedo, M.A.; Trontelj, P.; Lokovšek, T.; Agnarsson, I. A molecular phylogeny of nephilid spiders: Evolutionary history of a model lineage. *Mol. Phylogenet. Evol.* **2013**, *69*, 961–979. [CrossRef] [PubMed]

77. Bidegaray-Batista, L.; Arnedo, M.A. Gone with the plate: The opening of the Western Mediterranean basin drove the diversification of ground-dweller spiders. *BMC Evol. Biol.* **2011**, *11*, 317. [CrossRef] [PubMed]

78. Matzke, N.J. BioGeoBEARS: BioGeography with Bayesian (and likelihood) evolutionary analysis in R Scripts. *R Packag.* **2013**, *1*, 2013.

79. Bartolini, C.; Buffler, R.T.; Blickwede, J.F. *The Circum-Gulf of Mexico and the Caribbean: Hydrocarbon Habitats, Basin Formation, and Plate Tectonics, AAPG Memoir 79*; American Association of Petroleum Geologists: Tulsa, OK, USA, 2003.

80. Marshall, J.S. The geomorphology and physiographic provinces of Central America. *Cent. Am. Geol. Resour. Hazards* **2007**, *1*, 75–121.
81. Mann, P. Overview of the tectonic history of northern Central America. *Geol. Soc. Am. Spec. Pap.* **2007**, *428*, 1–19.
82. Randazzo, A.F.; Jones, D.S. *The Geology of Florida*; University Press of Florida: Gainesville, FL, USA, 1997.
83. Ree, R.H.; Sanmartín, I. Conceptual and statistical problems with the DEC+ J model of founder-event speciation and its comparison with DEC via model selection. *J. Biogeogr.* **2018**, *45*, 741–749. [CrossRef]
84. Ree, R.H.; Smith, S.A. Maximum Likelihood Inference of Geographic Range Evolution by Dispersal, Local Extinction, and Cladogenesis. *Syst. Biol.* **2008**, *57*, 4–14. [CrossRef]
85. Kozub, D.; Khmelik, V.; Shapoval, Y.; Chentsov, V.; Yatsenko, S.; Litovchenko, B.; Starykh, V. Helicon Focus Software. Available online: http://heliconsoft.com (accessed on 26 January 2017).
86. Garrison, N.L.; Rodriguez, J.; Agnarsson, I.; Coddington, J.A.; Griswold, C.E.; Hamilton, C.A.; Hedin, M.; Kocot, K.M.; Ledford, J.M.; Bond, J.E. Spider phylogenomics: Untangling the Spider Tree of Life. *PeerJ* **2016**, *4*, e1719. [CrossRef]
87. Čandek, K.; Agnarsson, I.; Binford, G.J.; Kuntner, M. Global biogeography of *Tetragnatha* spiders reveals multiple colonization of the Caribbean. *BioRxiv* **2018**. [CrossRef]
88. Peck, S.B. Aerial dispersal of insects between and to islands in the Galapagos Archipelago, Ecuador. *Ann. Entomol. Soc. Am.* **1994**, *87*, 218–224. [CrossRef]
89. Aldrich, J.R. Others Dispersal of the southern green stink bug, *Nezara viridula* (L.) (Heteroptera: Pentatomidae), by hurricane Hugo. *Proc. Entomol. Soc. Washingt.* **1990**, *92*, 757–759.
90. Zimmerman, E.C. *Insects of Hawaii*; University of Hawaii Press: Honolulu, HI, USA, 1948; Volume 7.
91. Oey, L.; Ezer, T.; Lee, H. Loop Current, rings and related circulation in the Gulf of Mexico: A review of numerical models and future challenges. *Geophys. Monogr. Geophys. Union* **2005**, *161*, 31.
92. Harrison, S.E.; Harvey, M.S.; Cooper, S.J.B.; Austin, A.D.; Rix, M.G. Across the Indian Ocean: A remarkable example of trans-oceanic dispersal in an austral mygalomorph spider. *PLoS ONE* **2017**, *12*, e0180139. [CrossRef]
93. Ceccarelli, F.S.; Opell, B.D.; Haddad, C.R.; Raven, R.J.; Soto, E.M.; Ramírez, M.J. Around the world in eight million years: Historical biogeography and evolution of the spray zone spider *Amaurobioides* (Araneae: Anyphaenidae). *PLoS ONE* **2016**, *11*, e0163740. [CrossRef] [PubMed]
94. Duque-Caro, H. Neogene stratigraphy, paleoceanography and paleobiogeography in northwest South America and the evolution of the Panama Seaway. *Palaeogeogr. Palaeoclimatol. Palaeoecol.* **1990**, *77*, 203–234. [CrossRef]
95. Coates, A.G.; Jackson, J.B.C.; Collins, L.S.; Cronin, T.M.; Dowsett, H.J.; Bybell, L.M.; Jung, P.; Obando, J.A. Closure of the Isthmus of Panama: The near-shore marine record of Costa Rica and western Panama. *Geol. Soc. Am. Bull.* **1992**, *104*, 814–828. [CrossRef]
96. Holcombe, T.L.; Moore, W.S. Paleocurrents in the eastern Caribbean: Geologic evidence and implications. *Mar. Geol.* **1977**, *23*, 35–56. [CrossRef]
97. Houle, A. Floating islands: A mode of long-distance dispersal for small and medium-sized terrestrial vertebrates. *Divers. Distrib.* **1998**, *4*, 201–216.
98. Lee, V.M.J.; Kuntner, M.; Li, D. Ballooning behavior in the golden orbweb spider *Nephila pilipes* (Araneae: Nephilidae). *Front. Ecol. Evol.* **2015**, *3*, 2. [CrossRef]

diversity

Article

Biogeography of Long-Jawed Spiders Reveals Multiple Colonization of the Caribbean

Klemen Čandek [1,2,3,*], Ingi Agnarsson [4,5,6], Greta J. Binford [7] and Matjaž Kuntner [1,2,5,8]

1   Department of Organisms and Ecosystems Research, National Institute of Biology, 1000 Ljubljana, Slovenia; matjaz.kuntner@nib.si
2   Jovan Hadži Institute of Biology, Research Centre of the Slovenian Academy of the Sciences and Arts, 1000 Ljubljana, Slovenia
3   Department of Biology, Biotechnical Faculty, University of Ljubljana, 1000 Ljubljana, Slovenia
4   Department of Biology, University of Vermont, Burlington, VT 05405, USA; iagnarsson@gmail.com
5   Department of Entomology, National Museum of Natural History, Smithsonian Institution, Washington, DC 20560, USA
6   Faculty of Life and Environmental Sciences, University of Iceland, Sturlugata 7, 102 Reykjavik, Iceland
7   Department of Biology, Lewis and Clark College, Portland, OR 97219, USA; binford@lclark.edu
8   Centre for Behavioural Ecology and Evolution, College of Life Sciences, Hubei University, Wuhan 430011, China
*   Correspondence: Klemen.candek@nib.si or klemen.candek@gmail.com

Abstract: Dispersal ability can affect levels of gene flow thereby shaping species distributions and richness patterns. The intermediate dispersal model of biogeography (IDM) predicts that in island systems, species diversity of those lineages with an intermediate dispersal potential is the highest. Here, we tested this prediction on long-jawed spiders (*Tetragnatha*) of the Caribbean archipelago using phylogenies from a total of 318 individuals delineated into 54 putative species. Our results support a *Tetragnatha* monophyly (within our sampling) but reject the monophyly of the Caribbean lineages, where we found low endemism yet high diversity. The reconstructed biogeographic history detects a potential early overwater colonization of the Caribbean, refuting an ancient vicariant origin of the Caribbean *Tetragnatha* as well as the GAARlandia land-bridge scenario. Instead, the results imply multiple colonization events to and from the Caribbean from the mid-Eocene to late-Miocene. Among arachnids, *Tetragnatha* uniquely comprises both excellently and poorly dispersing species. A direct test of the IDM would require consideration of three categories of dispersers; however, long-jawed spiders do not fit one of these three a priori definitions, but rather represent a more complex combination of attributes. A taxon such as *Tetragnatha*, one that readily undergoes evolutionary changes in dispersal propensity, can be referred to as a 'dynamic disperser'.

Keywords: *Tetragnatha*; dynamic disperser; intermediate dispersal model of biogeography; GAARlandia; Tetragnathidae

Citation: Čandek, K.; Agnarsson, I.; Binford, G.J.; Kuntner, M. Biogeography of Long-Jawed Spiders Reveals Multiple Colonization of the Caribbean. *Diversity* **2021**, *13*, 622. https://doi.org/10.3390/d13120622

Academic Editors: Michael Wink and Martín J. Ramírez

Received: 4 October 2021
Accepted: 16 November 2021
Published: 26 November 2021

Publisher's Note: MDPI stays neutral with regard to jurisdictional claims in published maps and institutional affiliations.

1. Introduction

Species distributions and species richness can vastly vary among taxonomic units of comparable ranks. Evolutionary biology aims to understand which factors contribute to such variation [1,2]. On the one hand, abiotic factors such as habitable area size, climate conditions or the presence of barriers may all contribute [3,4]. On the other hand, biological attributes such as species generation time [5], clade age [6] and species dispersal ability [7,8] may be equally important. Organismal dispersal ability, in particular, has the potential to directly affect levels of gene flow among populations and, consequently, affect species' potential to reach new habitats.

A low dispersal potential of a taxon can limit its colonization success and gene flow among populations, while a high dispersal potential enables the colonization of remote areas and maintains higher levels of gene flow. In theory, both of these extreme cases (low and high dispersal) constrain the number of speciation events [9,10]. The intermediate dispersal model (IDM) [9,10] predicts that in island systems, the species diversity of those lineages with an intermediate dispersal potential is the highest.

Taxa that contain species with poor dispersal abilities are associated with high levels of endemism, which is often associated with biogeographic patterns of multiple single island endemics across archipelagos [11,12] insofar as extinction rates do not overwhelm their diversification. It is the taxa that contain species with excellent dispersal potential that are usually widely, even globally distributed [13–15]. Studying and understanding the biogeography of species and higher taxonomic ranks on local as well as on global scales could illuminate a clade's intrinsic propensities to disperse. Island systems present compelling biogeographic laboratories [16,17]. Islands are discrete units with different degrees of barriers to gene flow for terrestrial organisms leading to different degrees of adaptation and diversification [18,19]. Extreme forms of morphological adaptations or even secondary loss of dispersal abilities in island taxa are not unusual [20–22]. While some species undergo rapid local adaptation and diversification on islands, others readily disperse among islands and form wide distributions across archipelagos.

Comprising over 700 islands and being a genuine biodiversity hotspot, the Caribbean archipelago is among the most suitable natural biogeographic laboratories [19,23]. Although the Caribbean region encompasses adjacent parts of the mainland Americas, the Caribbean islands are classified into three distinct units: the Greater Antilles comprise the large and older islands of Cuba, Hispaniola, Jamaica, and Puerto Rico; the Lesser Antilles are geologically younger, volcanic islands to the east; and the Lucayan Archipelago is the group of tiny Atlantic islands north of the Greater Antilles, mostly encompassing the Bahamas. The Caribbean archipelago has likely been isolated from the mainland since its emergence (see [24] for an overview of the Caribbean geology from a biogeographical perspective); however, a hypothetical land-bridge, GAARlandia, might have connected the greater Antilles with South America between 33 and 35 million years ago (MYA) [25,26] but see [27]. If this land-bridge indeed existed, it would have enabled the colonization of the Caribbean for various organisms, regardless of their dispersal potential.

Spiders are good models in biogeographic research due to the breath of their taxonomic, genetic, evolutionary, and biological diversity [15,28–33]. Spiders represent an ancient lineage with nearly 50 thousand species from 129 families [34] and inhabit most terrestrial ecosystems. Interestingly, various taxonomic groups of comparable ranks within spiders exhibit highly variable dispersal abilities, degrees of endemism, species richness and species distributions [35,36]

Our study focuses on long-jawed spiders (genus *Tetragnatha*, family Tetragnathidae), with special emphasis on the Caribbean archipelago. This diverse genus includes 323 described species [34] (and probably numerous undescribed ones) and has been extensively used in biogeographic research, notably in Hawaii and other Pacific archipelagos [37–40]. While some species are single island endemics, others show extremely wide, even cosmopolitan distributions like *T. nitens* [34]. *Tetragnatha* is generally considered to have an extraordinary dispersal potential and is able to quickly reach even the most remote, newly formed islands [37]. This assumption, reinforced by a study of Okuma and Kisimoto (1981) [41], found that 96% of the aerial plankton collected 400 km off the Chinese shore were *Tetragnatha* spiders, passively dispersing through ballooning behavior [35,42].

Here, we report on our study of a rich original collection of *Tetragnatha* from the Caribbean archipelago. With species delimitation methods we estimate 25 putative species in our original dataset. We then reconstruct the evolutionary history of long-jawed spiders of the Caribbean using a mitochondrial COI gene and a nuclear 28S gene fragment. We place the Caribbean phylogeny into a global context by adding a single sequence for numerous published *Tetragnatha* species on GenBank, increasing the number of putative species to 54 using a Bayesian analysis. Moreover, we estimate the number, the timing, and the directionality of all Caribbean colonization events by *Tetragnatha*, then look for potential agreement with common biogeographic scenarios on the Caribbean such as colonization by overwater dispersal or vicariant origins [43]. As does the paper on *Micrathena* spiders [44], we also test for a biogeographical pattern in *Tetragnatha* that would support the hypothetical GAARlandia land-bridge scenario. We contrast the biogeography of *Tetragnatha* with those of other spider lineages in the Caribbean, including a close relative *Cyrtognatha*, and broadly estimate the dispersal abilities of *Tetragnatha* in the context of the IDM.

2. Materials and Methods

2.1. Material Acquisition

We collected the material for our research as a part of a large-scale Caribbean biogeography (CarBio) project. We used standard methods for collecting the spiders [45,46], namely, day- and night-time beating and a visual aerial search. We fixed the collected material in 96% ethanol at the site of field work and stored it at $-20/-80$ °C. We then used light microscopy to verify the genus and to identify the species, where possible.

2.2. Molecular Procedures

We isolated the DNA using a QIAGEN DNeasy Tissue Kit (Qiagen, Inc., Valencia, CA, USA) at the University of Vermont (Vermont, USA), or an Autogenprep965 automated phenol chloroform extraction at the Smithsonian Institution (Washington, DC, USA), or a robotic DNA extraction with a Mag MAX™ Express magnetic particle processor Type 700 with DNA Multisample kit (Applied Biosystems, Foster City, CA, USA), following modified protocols [47] at the EZ Lab (Ljubljana, Slovenia).

We targeted two genetic markers, a mitochondrial (COI) and a nuclear one (28S rRNA). We used the forward LCO1490 (GGTCAACAAATCATAAAGATATTGG) [48] and the reverse C1-N-2776 (GGATAATCAGAATATCGTCGAGG) [49] primers for COI amplification. A 25 µL reaction volume contained the mixture of: 5 µL Promega's GoTaq Flexi Buffer, 0.15 µL GoTaq Flexi Polymerase, 0.5 µL dNTP's (2 mM each, BioTools), 2.3 µL MgCl$_2$ (25 mM, Promega), 0.5 µL of each primer (20 µM), 0.15 µL BSA (10 mg/mL; Promega), 2 µL DNA template and sterile distilled water for the remaining volume. We set the PCR cycling protocol as follows: initial denaturation (5 min at 94 °C), 20 repeats (60 s at 94 °C, 90 s at 44 °C while increasing 0.5° per repeat, 1 min at 72 °C), 15 repeats (90 s at 94 °C, 90 s at 53.5 °C, 60 s at 72 °C), and a final elongation (7 min at 72 °C).

We used the forward 28Sa (also known as 28S-D3A; GACCCGTCTTGAAACA CGGA) [50] and the reverse 28S-rD5b (CCACAGCGCCAGTTCTGCTTAC) [51] primers for 28S amplification. The 35 µL reaction volume contained the mixture of: 7.1 µL Promega's GoTaq Flexi Buffer, 0.2 µL GoTaq Flexi Polymerase, 2.9 µL dNTP's (2 mM each, BioTools), 3.2 µL MgCl$_2$ (25 mM, Promega), 0.7 µL of each primer (20 µM), 0.2 µL BSA (10 mg/mL; Promega), 1 µL DNA template and sterile distilled water for the remaining volume. We set the PCR cycling protocol as follows: initial denaturation (7 min at 94 °C), 20 repeats (45 s at 96 °C, 45 s at 62 °C while decreasing 0.5° per repeat, 1 min at 72 °C), 15 repeats (45 s at 96 °C, 45 s at 52 °C, 60 s at 72 °C), and a final elongation (10 min at 72 °C).

We used Geneious v. 5.6.7 [52] for the de-novo sequence assembly. We used a combination of MEGA [53] and Mesquite [54] for basic sequence analysis, and for the renaming and concatenating matrices of both genetic markers. We then used the online version of MAFFT [55] for sequence alignment.

We obtained original COI sequences for all 254 specimens and 54 original 28S sequences (based on the preliminary species delimitation results; described below) of *Tetragnatha* spiders. We incorporated an additional 45 *Tetragnatha* COI sequences from GenBank. These were chosen to be of sufficient quality (over 70% of overlap with our sequences) and to represent either those species not sampled in the Caribbean, or if sampled there, with clearly outlying populations (such as *T. pallescens* from Canada). For the outgroups, we used nine COI and seven 28S originally generated sequences as well as ten COI and five 28S sequences mined from GenBank, representing species of *Arkys*, *Linyphia*, *Meta*, *Metellina*, *Pachygnatha*, *Chrysometa*, *Cyrtognatha* and *Leucauge* (see Supplementary Table S1 for details). Altogether, our broadest dataset comprised 318 COI sequences and 66 28S sequences. Our specimen selection focused on the Caribbean, but our broad taxon sampling ensured a global representation of *Tetragnatha*. The concatenated matrix for COI and 28S genes was 1199 nucleotides long, with 649 bp for the COI and 550 bp for the 28S. Relevant specimen details and GenBank accession codes are presented in the Supplementary Table S1.

2.3. Species Delimitation

To estimate the number of molecular operational taxonomic units (MOTUs), we analyzed all COI sequences in the dataset using the Automatic Barcode Gap Discovery (ABGD) [56], setting Pmin to 0.001 and Pmax to 0.2 with 30 steps between those values. We set the X (relative gap width) to different values from 1.5 to 3 to check for the consistency of the results.

2.4. Phylogenetic Analyses

2.4.1. Two Gene, Species Level Phylogeny

Based on the species delimitation analysis, we selected a subset of the data to create a concatenated matrix (COI and 28S) for a Bayesian phylogenetic reconstruction. We used 54 of the *Tetragnatha* individuals collected for this work that represent 20 MOTUs and added 12 sequences as outgroups (Supplementary Table S1). We used MrBayes v3.2 [57] with two independent runs, each with four MCMC chains, for 30 million generations. We partitioned the dataset per genetic marker, set a sampling frequency of 2000 and set a relative burn-in to 25%. As the nucleotide substitution model, we used the generalized time-reversible model with gamma distribution and invariant sites (GTR+G+I), as per AIC and BIC criteria derived in jModelTest2 [58]. The starting tree was random. We examined the statistical parameters and MCMC chains convergence with sump command within MrBayes and with Tracer v1.7.1 [59]. We visualized the trees with FigTree v1.4.3 [60].

2.4.2. All-Terminal, Single Gene Phylogeny

Using MrBayes, we reconstructed two Bayesian phylogenies using all originally sampled *Tetragnatha* (N = 254) and a single COI sequence for 45 other *Tetragnatha* species. We added 19 specimens as outgroups (Supplementary Table S1). We included multiple Hawaiian *Tetragnatha* species belonging to the "spiny-leg" clade, to serve as a check, since their monophyly is well established [22,40] and should thus be recovered in our own phylogenetic reconstructions.

The first all-terminal phylogenetic reconstruction used an unconstrained approach while we enforced the *Tetragnatha* monophyly for the second one. The settings for both analyses were as in the above species level analysis except with the number of MCMC generations increased to 100 million. Additionally, we set the parameter "contype" within the "sumt" command to "allcompat" to obtain a fully resolved tree.

We then tested whether the unconstrained or the constrained model better fit our data by comparing marginal likelihood scores and by Tracer model comparison analysis. We examined the statistical parameters and MCMC chains convergence with a sump command within MrBayes and with Tracer.

2.4.3. Caribbean *Tetragnatha* Monophyly Testing

To test the Caribbean *Tetragnatha* monophyly, we ran two additional Bayesian analyses. We created a subset of COI sequences, and a single for each MOTU/recognized species. We then ran the unconstrained analysis and compared it to the analysis with an enforced monophyly of the Caribbean endemic *Tetragnatha* species. To confirm or reject the monophyly of *Tetragnatha* species present exclusively on the Caribbean islands, we compared the marginal likelihood scores between the models and ran a Tracer model comparison analysis. The general settings of both (constrained and unconstrained) phylogenetic analyses were as above except with the number of MCMC generations set to 50 million.

2.4.4. Time-Calibrated Phylogenetic Reconstruction

We used BEAST v2.5.1 [60] for a time-calibrated phylogeny on a subsample of the total taxon selection (Table S1). For the substitution model selection, we used a bModelTest [61] expansion in BEAST, following model comparison approaches described by Bidegaray-Batista and Arnedo (2011) [62]. We selected a relaxed log-normal molecular clock and set the priors accordingly: ucld.mean with normal distribution, mean value of 0.0112 and standard deviation 0.001; ucld.stdev with exponential distribution and a mean value of 0.666. We set the number of MCMC generations to 30 million with a sampling frequency of 1000. We constrained the topology to comply with the all-terminal COI phylogeny and left BEAST to only estimate branch lengths and associated timing. In the absence of relevant fossils, we used biogeographic and secondary calibrations of the chronogram. The first calibration point was the well-supported appearance of the Hawaiian Islands at 5.1 MYA, therefore the diversification of the Hawaiian clade was constrained with a uniform prior with the upper bound at 5.1 MYA and the lower bound at 0. We used the second calibration point as the time of the appearance of the Lesser Antillean islands, which are presumably not older than 11 million years [63]. Therefore, the diversification of the clade uniting two putatively Lesser Antillean endemic species (SP5 and SP9) was constrained using a uniform prior with the upper bound at 11 MYA and the lower bound at 0. The final calibration point originated from the estimated ages for Tetragnathidae at 100 (44–160) [64] and 99 (64–133) MYA [65]. Therefore, we set the MRCA prior on the node *Tetragnatha + Arkys* (the latter outgroup in Arkyidae) [66] with the following settings: exponential distribution, mean value of 31.5 and 44 offset, corresponding to the soft upper bound at 160 MYA and the hard lower bound at 44.8 MYA.

We used Tracer to determine burn-in, to check for the MCMC convergence and for other statistics. We used TreeAnnotator [60] to summarize trees with a 10% burn-in and node heights set as median heights. We used FigTree for the maximum clade credibility tree visualization.

2.5. Ancestral Area Estimation and Biogeographic Stochastic Mapping

For the biogeographic analyses we employed BioGeoBEARS v0.2.1 [67] implemented in R version 3.5.0 [68] using the ultrametric tree from BEAST with outgroups removed. We conducted the analysis with the remaining 54 species/MOTUs and created a geographic data list with two possible areas: the Caribbean (C) and "Other" (O), meaning there were four possible ancestral states at each node (C, O, CO and 'null'). Although the Caribbean is a complex island archipelago, we simplified its definition here in order to increase the resolution for estimating biogeographic events between the Caribbean and non-Caribbean areas. We tested all six possible biogeographical models implemented in BioGeoBEARS: DEC (+J), DIVALIKE (+J) and BAYAREALIKE (+J), testing their suitability for our data with Akaike information criterion (AIC) and sample-size corrected AIC (AICc).

To estimate the number and types of biogeographic events in the Caribbean, we performed biogeographical stochastic mapping (BSM) analysis expansion in BioGeoBEARS. We used the most suitable (BAYAREALIKE + J) model, as discovered in the results from previous analysis. We simulated 100 exact biogeographic histories, extracted the estimated number and types of events and presented them as histograms.

3. Results

3.1. Species Delimitation

We collected 254 individuals from Cuba, Jamaica, Lesser Antilles, Hispaniola, Puerto Rico, South-East USA and Central America (Figure 1, Supplementary Table S1). We confirmed that all individuals were morphologically *Tetragnatha* and identified eight described species. Using ABGD, we additionally estimated another 17 MOTUs (labeled SP and number; Supplementary Note S1). ABGD was consistent in estimating the number of MOTUs for all tested X (relative gap width) values. Moreover, ABGD detected a 3.8%

wide barcoding gap, measured in K2P percent distance [69], closely matching that of the barcoding gap width representative of the family Tetragnathidae [70].

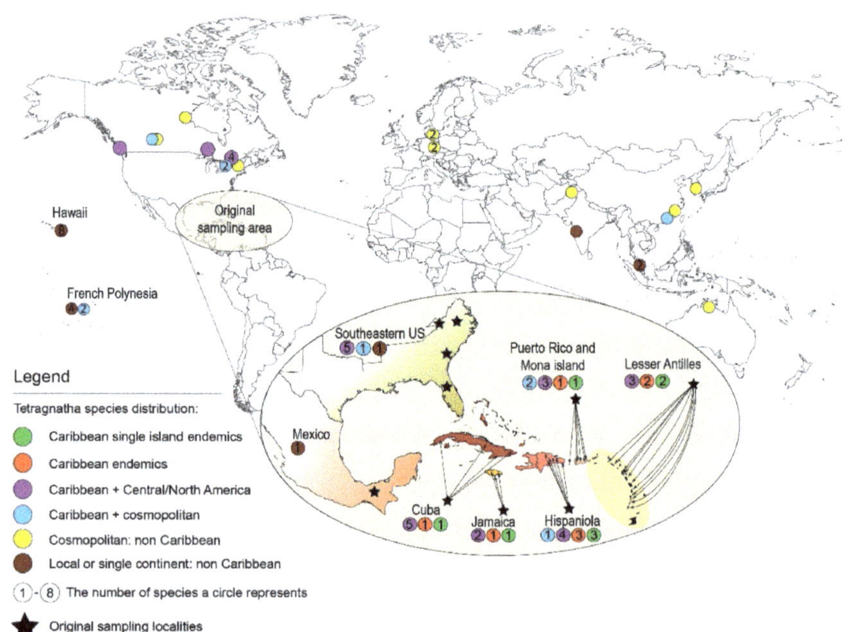

Figure 1. Map of our original sampling localities and origins of GenBank sequences. Circles on the world map and stars in the Caribbean inset indicate where the specimens were collected. Color coding refers to the distribution of a species.

ABGD accurately delimited all species from GenBank with the exception of *T. praedonia* and *T. nigrita*. These two species clustered together with a mere 1.1% sequence divergence between them, suggesting that one of them has been misidentified. Similarly, *T. nitens* was separated into two species, one of them clustering with *T. moua* (Supplementary Note S1). This suggests that *T. nitens* is a species complex, or that it represented another case of GenBank misidentification. Altogether, our originally collected material comprised 25 putative species while together with GenBank sequences our dataset contained 54 putative species. Of those, 11 were Caribbean endemics while 15 species were found both in the Caribbean, as well as elsewhere (Figure 1). In our dataset, 28 species were from outside of the Caribbean.

3.2. Molecular Phylogeny

Our two gene, species level phylogeny (Figure 2) supported the *Tetragnatha* monophyly. A well-supported basal *Tetragnatha* node contained some, but not all *T. shoshone* exemplars, refuting the validity of that MOTU. The remaining species/MOTUs generally grouped together with high support. Except for a single node, all internal nodes in this phylogeny were resolved (Figure 2).

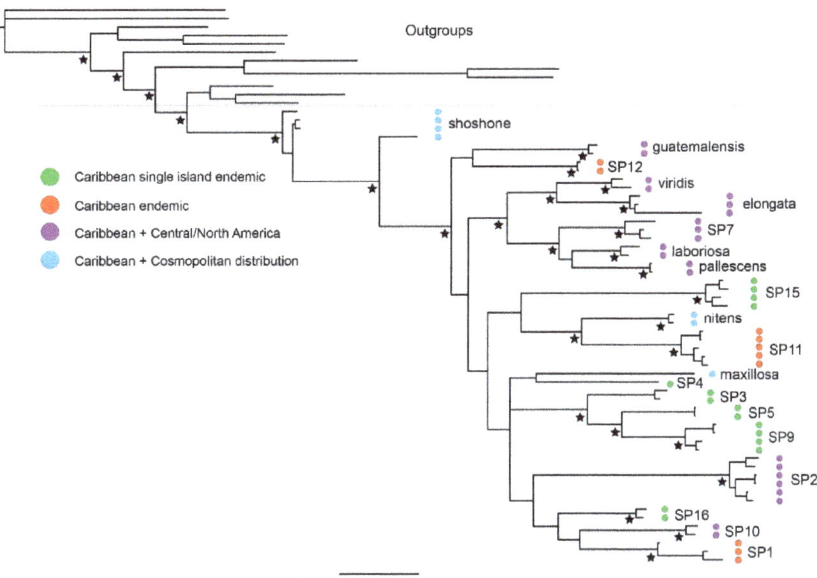

Figure 2. Species level Bayesian phylogeny of the Caribbean *Tetragnatha* based on COI and 28S. These species relationships support the *Tetragnatha* monophyly. Stars at nodes indicate Bayesian posterior probability values > 0.95.

The COI phylogenetic reconstruction in Figure 3 represents the relationships of all originally collected, as well as all data-mined *Tetragnatha* exemplars. Because the unconstrained all-terminal phylogeny (Figure S1) recovered a paraphyletic *Tetragnatha* that also included *Pachygnatha*, we constrained the *Tetragnatha* monophyly as per the species level results (Figure 3). Occasional inaccuracies at deeper levels of the mitochondrial phylogenies are to be expected due to high saturation levels in COI [71]. Our approach to constrain the all-terminal COI phylogeny was justified by the marginal likelihood test with a Tracer model comparison (logBF = 710.4 ± 0.061 from 500 bootstrap replicates favoring the constrained model). The Hawaiian *Tetragnatha* always formed a well-supported clade, a result lending credibility to our COI only analyses.

This phylogenetic pattern (Figure 3) reveals eight MOTUs as Caribbean single island endemics and an additional three that are Caribbean endemics. The remaining putative species have more widespread distributions. Although the phylogenetic pattern in itself already hinted at the Caribbean *Tetragnatha* not being monophyletic, its monophyly was further rejected by the marginal likelihood scores testing with a Tracer model analysis (logBF = 163.6 ± 0.039 SE from 500 bootstrap replicates favoring unconstrained model). The only multispecies *Tetragnatha* clade recovered as purely Caribbean was the group uniting SP3, 5, 9, and 13. All analyses converged and stabilized with ESS > 300.

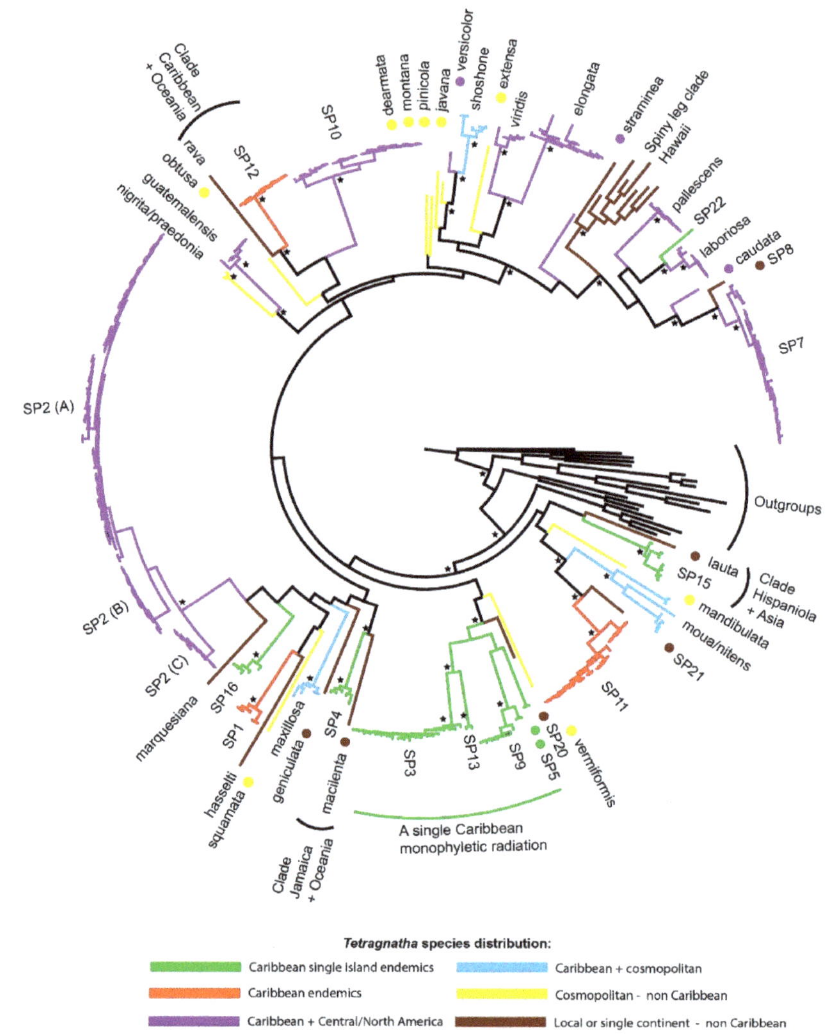

Figure 3. The all-terminal mitochondrial Bayesian phylogeny of the Caribbean and global *Tetragnatha* representatives. The scattered phylogenetic pattern of single island endemic, Caribbean endemic and cosmopolitan species, as well as highly supported clades of geographically distant relatives reveal a profile of a "dynamic disperser". Note the non-monophyly of the Caribbean *Tetragnatha*. Stars at nodes indicate Bayesian posterior probability values > 0.90.

3.3. Time Calibrated Phylogeny

The BEAST chronogram suggested that tetragnathids had diverged from their sister family *Arkyidae* some 58 MYA (million years ago) (95% HPD 44.5–81.9) (Figure 4). We estimated that the diversification of those *Tetragnatha* species represented in our dataset had begun some 46 MYA (44–53.3). The Caribbean endemic clade (with species SP3, 5, 9, 13) appears to have diverged from its sister species from Mexico ca. 20 MYA (12.8–29 MYA). Time estimates of the species divergences generally fit the estimated geologic ages of the Caribbean islands.

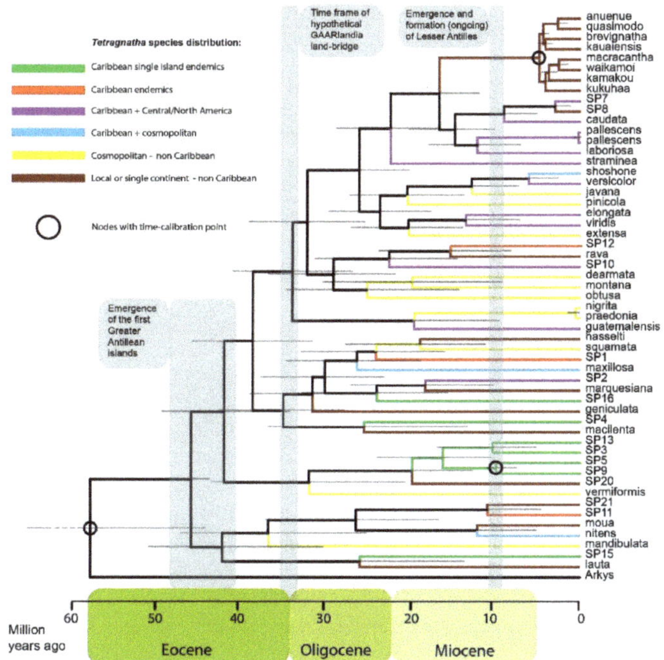

Figure 4. A time-calibrated phylogeny of *Tetragnatha*. This chronogram from BEAST is consistent with an early Caribbean colonization by *Tetragnatha*, with multiple overwater dispersal events over the GAARlandia, or with ancient vicariance scenarios. We constrained the topology based on the results from the all-terminal COI phylogeny, therefore, node supports are not shown. Nodal bars represent 95% HPD intervals.

The examination of log files (all EES > 266) revealed that this BEAST phylogeny was well supported, and additional examination with bModelAnalyzer revealed the most used nucleotide substitution model was a version of TVM, followed by versions of TN93 and GTR (Figure S2) (for details on bModelTest see [61]).

3.4. Biogeographic Analyses

BioGeoBEARS model comparison recovered the BAYAREALIKE+J model as the most suitable for our data, regardless of the scoring criterion (Supplementary Table S2). A founder effect within the model (+J parameter) showed a significantly better fit for our data than the model without this parameter ($p < 0.01$). The ancestral area estimation (Figure 5) hinted at multiple independent origins of the Caribbean *Tetragnatha*.

We simulated 100 biogeographic stochastic mapping repeats of exact biogeographic histories (Supplementary Animation S1). This analysis estimated that on average 21.88 biogeographic events took place between the delimited areas. Of those, 11.72 were anagenetic events (all of which were range expansions), and 10.16 were cladogenetic events (Figure 6). Of the latter, all were dispersals with a founder event, and none were vicariant events. The results of narrow sympatry should not be considered in our case because of the experimental design with only two areas. Being extremely wide, the area classified as "Other" would produce a false positive score under narrow sympatry. According to BSM analyses, 17.86 dispersal events must have taken place from outside sources to the Caribbean (9.51 range expansions, 8.35 founder events) while 4.02 dispersal events took place in the opposite direction (2.21 range expansions, 1.81 founder events).

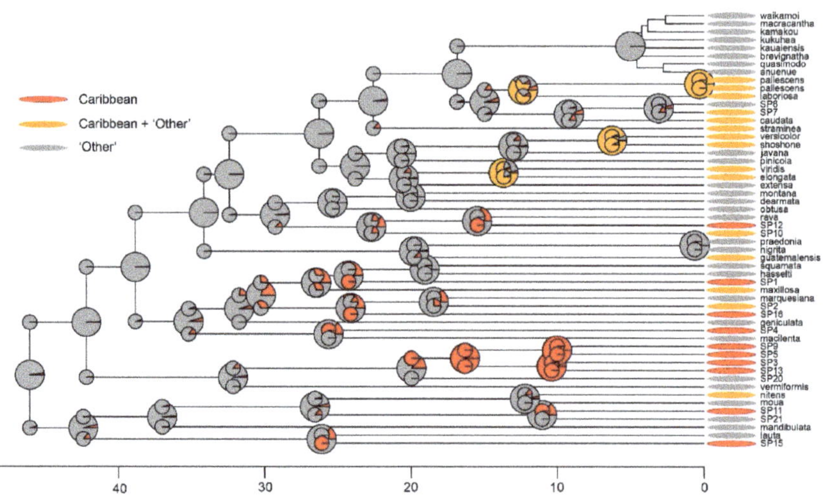

Figure 5. Ancestral area estimation of *Tetragnatha*. The biogeographical analysis in BioGeoBEARS using the most suitable model for our data (BAYAREALIKE + J, max_range_size = 2), reveals multiple origins of the Caribbean taxa from the outside sources. A single subclade (SP 13, 3, 5, 9) has a well-supported Caribbean ancestral range. Evidently, multiple founder events and range expansion took place throughout *Tetragnatha*'s biogeographic history in the Caribbean.

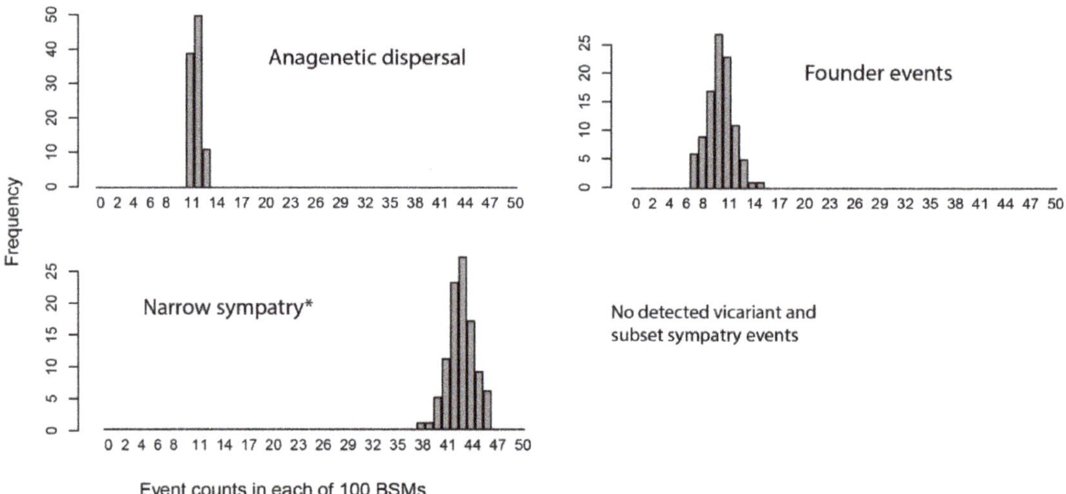

Figure 6. Histogram of event counts from biogeographic stochastic mapping in BioGeoBEARS. Note that * should be interpreted with caution due to experimental design with two areas. The area classified as "Other" is extremely wide and would produce a false positive score under narrow sympatry.

4. Discussion

We reconstructed multiple *Tetragnatha* phylogenies from over 300 individuals of this diverse genus, with a focus on the wider Caribbean region. Our results support the monophyly of the long-jawed spiders but reject the monophyly of the Caribbean *Tetragnatha*. In the Caribbean, we found low levels of endemism yet a high diversity within *Tetragnatha*, an unusual pattern considering other spider biogeographic research in the Caribbean [24,30–33,72–76]. For example, another orb-weaver, *Micrathena*, shows a different pattern in the Caribbean, that of a pronounced single island endemism in spite of rampant dispersal history in that clade [44]. The time calibrated phylogenetic reconstruction is consistent with an early overwater colonization of the Caribbean by long-jawed spiders. Moreover, the combination of chronogram and biogeographic history reconstruction refute the possibility of ancient vicariant origins of Caribbean *Tetragnatha* while favoring overwater dispersals. This result lends no support for the GAARlandia land-bridge hypothesis and lends credibility to the recently reinforced lack of evidence for this land-bridge to have even existed [27]. Biogeographical stochastic mapping recovers multiple colonization events to the Caribbean and evidence of 'reverse colonization' from islands to continents, from the mid-Eocene to late-Miocene. The genus *Tetragnatha* is exceptionally species rich with well over 300 species worldwide. Of these, our study "only" included 54 species. From the point of view of taxonomic completeness, our biogeographic reconstructions need to be seen as somewhat preliminary; however, the patterns we discuss are likely to be representative of the genus. These results, when compared with other lineages with known biogeographic histories in the Caribbean, suggest a unique dispersal history of *Tetragnatha*, combining excellent dispersal ability of the lineage as a whole with subsequent reduction or loss of that trait in individual clades through evolutionary history, as also seen in the Hawaiian *Tetragnatha* [22,77]. As we discuss below, this mixed dispersal biology within the genus *Tetragnatha* complicates our direct testing of the IDM.

4.1. Caribbean Tetragnatha Are Not Monophyletic

Our phylogeny (Figure 2) supports *Tetragnatha* monophyly; however, combining our results with previously collected data is challenging due to the lack of overlapping 28S data available on GenBank. The comparison of this "local" phylogenetic reconstruction, involving exclusively Caribbean species, with the more taxon-rich, global phylogeny, highlights the importance of providing a global context to studies of Caribbean biogeography. While "local" relationships from Figure 2, based upon our sampling, are well supported and appear to tell a clear story, the picture changes substantially with the additional, globally distributed, species. Namely, the all-terminal phylogeny (Figure 3) recovers a complex and mixed pattern in which single island Caribbean endemic species are phylogenetically scattered with other species, some of which are geographically distant. The reasons for such patterns are opaque but could relate to the overall mobility of certain clades that become more global in reach than some others.

Although the phylogeny recovered a single, small-scale, radiation of Caribbean *Tetragnatha* (the clade with SP3, 5, 9 and 13), overall Caribbean *Tetragnatha* are not monophyletic. This combined pattern strongly hints at multiple colonization events of *Tetragnatha* to the archipelago. It further suggests that at least some *Tetragnatha* species maintain relatively high levels of gene flow within and among the islands, as well as between the archipelago and the continents that serve as source populations. Other species, however, seem to have secondarily lost this dispersal potential and form narrow range endemics. Similarly, research detects the absence of large monophyletic radiations of birds on the Caribbean [19], that are otherwise well documented in bird lineages from more remote archipelagos such as Darwin's finches on the Galapagos [78] and Hawaiian honeycreepers [79]; however, focusing on less mobile organisms, the Caribbean does harbor its own exemplary radiation of *Anolis* lizards [80–82]. It seems that clade-specific dispersal abilities must coincide with isolation of an island or archipelago to provide the conditions that lead to adaptive radiations [19].

4.2. Biogeographic History of Caribbean Tetragnatha

Three hypothetical scenarios of Caribbean biotic colonization are commonly reported. The first scenario, an ancient vicariant hypothesis, assuming the colonization of proto-Antilles as early as 70 MYA [83], is refuted by our chronogram (Figure 4). We estimate that *Tetragnatha* had not appeared on the Caribbean before 46 MYA, with more recent estimates even likelier (Figures 4 and 5). Therefore, the ancestral vicariance hypothesis receives no support by our data, because the scenario would predate our time estimates by roughly 30 million years, or more.

The second scenario assumes that non-marine organisms could have used the exposed land-bridge called GAARlandia to reach the Caribbean. GAARlandia may have connected the Greater Antilles with South America between 33 and 35 MYA [25,26], although its very existence is recently questioned [27]. The estimated colonization times of the Caribbean for diverse groups of lineages does seem to coincide with this time period thereby at least indirectly corroborating the possibility of the GAARlandia land-bridge, see fishes [84], frogs [85], mammals [86], invertebrates [32,75,87,88] and plants [89,90], but also see [91]. Our results on *Tetragnatha* do not support the GAARlandia land-bridge scenario. Specifically, the scattered phylogenetic pattern that suggests repeated colonization of the archipelago by various *Tetragnatha* species and lineages, refutes a single colonization scenario. A study on *Cyrtognatha*, another tetragnathid spider genus, also refuted the GAARlandia scenario, but in that case the appearance of the lineage on the archipelago was decisively too recent [24].

The third scenario involves overwater dispersal by terrestrial organisms to reach the Caribbean islands [19]. According to our time estimates (Figure 4) and biogeographic history reconstruction (Figure 5), this scenario best explains our data. *Tetragnatha* (SP15) could have colonized the Caribbean as early as the mid-Eocene, soon after the emergence of the first Caribbean islands between 40–49 MYA [26,92,93]. Our phylogenetic (Figures 3 and 4) and biogeographic (Figure 5) history reconstructions suggest that *Tetragnatha* repeatedly, and independently, colonized the Caribbean until mid-Miocene. Moreover, the *Tetragnatha* biogeographic pattern within the context of the geological history of the Caribbean islands does not support a so-called 'progression rule', a pattern where successive colonization of younger islands is correlated with cladogenesis, e.g., *Tetragnatha* on Hawaii [94].

Our biogeographical stochastic mapping (BSM) analysis (Supplementary Animation S1, Figure 6) provides strong evidence for multiple colonization events of the Caribbean. Simulating biogeographic histories with BSM is a suitable approach to estimating the average number and directionality of biogeographic events in a studied area [95,96]. Within the scope of our analysis, we estimated that, on average, over eight founder events may have taken place in the Caribbean *Tetragnatha*. The estimation of anagenetic events of range expansion further suggest that *Tetragnatha* species have expanded their ranges to the Caribbean without subsequent speciation at least nine times. Caribbean *Tetragnatha* also show evidence of reverse-colonization from islands to the mainland, likely with two range expansions and up to two founder events (Supplementary Table S3).

The reconstructed biogeographic history of the Caribbean *Tetragnatha* is distinct from biogeographic patterns of *Tetragnatha* on other, well studied archipelagos. The Marquesas islands were probably colonized once [97], while the Society islands have been colonized at least twice [97]. On these remote Pacific archipelagos, *Tetragnatha* underwent monophyletic adaptive radiation although on a much smaller scale than *Tetragnatha* on the Hawaiian archipelago [22,98]. *Tetragnatha* is thought to have colonized Hawaii between two and four times and with two possible reverse-colonization events, and the subsequent adaptive radiation(s) have produced at least 38 species [40]. The Indian Ocean Mascarene islands were colonized three times without undergoing an adaptive radiation [40]. Considering the above examples, the Caribbean *Tetragnatha* biogeographic pattern reveals (a) exceptionally high rates of colonization (and reverse-colonization), (b) relatively low levels of endemism, (c) a generally more complex, phylogenetically scattered, species composition, and (d) a very high species richness compared to *Tetragnatha* from other archipelagos that is only surpassed by Hawaii. We therefore conclude that the Caribbean archipelago offers a unique evolutionary arena.

4.3. Tetragnatha Dispersal Abilities

The all-terminal phylogeny (Figure 3) represents a global picture of the evolutionary history of *Tetragnatha*. It unveils repeated cases of sister relationships of Caribbean species and geographically distant taxa. Examples are *Tetragnatha* SP15 from Hispaniola + *Tetragnatha lauta* from Asia; *T.* SP4 from Jamaica + *T. macilenta* from Oceania; *T.* SP12 from Hispaniola and Puerto Rico + *T. rava* from Oceania. Our taxon sampling may have omitted numerous extinct or intermediate species that may in fact group within these small but wide-ranging clades. Nonetheless, these well-supported nodes may hold. Similar to our case, Gillespie [97] found that *Tetragnatha* from the Pacific archipelagos were phylogenetically quite distant relatives.

The extreme geographic distances (over 10,000 km) between pairs of closely related species in our phylogeny (Figure 3), as well as often wide to cosmopolitan distributions of *Tetragnatha* species, together imply that *Tetragnatha* must contain numerous species with extraordinary dispersal abilities. The recurring pattern of single island endemism in *Tetragnatha* hints at evolutionary changes in this dispersal potential where certain species or clades within *Tetragnatha* have become secondarily dispersal-limited. The genus *Tetragnatha* exhibits high dispersal abilities and at the same time a high intrinsic property to quickly adapt and diversify, something that can be labeled as "super-speciator" attributes [99,100].

To construct a more general picture of *Tetragnatha* dispersal ability in the theoretical context of the IDM [9,10], we compare biogeographic patterns of the Caribbean *Tetragnatha* with those of other Caribbean lineages of spiders. In all cases, the local (Caribbean) as well as global species richness of *Tetragnatha* is greater than in other genera with putatively poor dispersers such as: *Deinopis* [32], *Micrathena* [44,76], *Loxosceles* and *Sicarius* [87], *Spintharus* [30,75], and *Selenops* [101]. Moreover, attributes associated with putatively poor dispersers such as high levels of single island endemism and low numbers of colonization events are not reconstructed in the case of the Caribbean *Tetragnatha*. Perhaps as the most relevant comparison, the species richness of the Caribbean *Tetragnatha* is much higher than that of another member of Tetragnathidae, *Cyrtognatha* (~28 vs. 11–14 species; [24]). Given that these genera are closely related, and co-distributed in parts of the region, the differences in their species richness, as well as their biogeographic patterns may, at least in part, be attributed to differences in their dispersal abilities.

Caribbean *Tetragnatha* also show strikingly different patterns compared with some of the spider genera with putatively excellent dispersal abilities, such as *Trichonephila* [14,33] and *Argiope* [31,102]. *Trichonephila* is distributed across the whole Caribbean archipelago but is represented by a single extant species, *T. clavipes*. Although several species of *Argiope* occupy the Caribbean, the most widespread is *A. argentata* (a recently discovered cryptic species *A. butchko* used to be *A. argentata* in part) [31]. This comparison, in addition to the previous comparison with putatively poor dispersers, reveals a much higher local (Caribbean) and global species richness in *Tetragnatha* than in *Trichonephila* or *Argiope*. Moreover, the a priori expected biogeographic pattern with the combination of rare founder events, low number of widely distributed species, and extremely low endemism, all purportedly associated with excellent dispersal ability, is not reconstructed in the Caribbean *Tetragnatha*. *Tetragnatha* is an excellent disperser but appears to readily respond to natural selection upon colonizing islands, which may render individual species to change their dispersal behavior and become endemic. While a direct test of the IDM would require consideration of three, more or less discrete, categories of dispersers (excellent, poor, and intermediate), long-jawed spiders do not readily fit one of these three a priori definitions. Instead, they represent a more complex combination of attributes of a 'dynamic disperser'.

5. Conclusions

Our phylogenetic research of the Caribbean *Tetragnatha* reveals a complex biogeographic history with multiple colonization events to and from the Caribbean. A combination of geographically scattered, closely related species, single island endemics, cosmopolitan species, multiple dispersal events, as well as the high species richness of the Caribbean *Tetragnatha*, presents a unique model for comparative biogeographic research.

Considering the theoretical framework of the IDM [9,10], one cannot classify *Tetragnatha* into one of the three a priori definitions of dispersers (poor vs. intermediate vs. excellent). Instead, they represent a more complex combination of attributes, something we refer to as a 'dynamic disperser'. This refers to a taxon that readily undergoes evolutionary changes in dispersal propensity. In line with the predictions of the IDM, some highly dispersive *Tetragnatha* species fail to speciate, even on more remote islands; however, some lineages have apparently secondarily lost their dispersal ability, thus forming narrow endemic species or groups. While preliminary evidence points towards the general validity of the IDM, more stringent testing will require a less simplistic a priori binning of the dispersal categories of organisms. Nonetheless, our findings demonstrate the very dynamic relationship between dispersal ability and diversity and how changes in dispersal propensity may link to endemism.

Supplementary Materials: The following are available online at https://www.mdpi.com/article/10.3390/d13120622/s1, Table S1: specimen details and GenBank codes, Table S2: BioGeoBEARS model comparison results, Table S3: BioGeoBEARS stochastic mapping results, Supplementary Animation S1. Note S1: ABGD analyses results, Figure S1: unconstrained all-terminal COI *Tetragnatha* phylogeny and Figure S2: detailed results of bModelTest model selection.

Author Contributions: Conceptualization, all authors; methodology, K.Č.; formal analysis, K.Č.; writing—original draft preparation, K.Č. and M.K.; writing—review and editing, all authors; visualization, K.Č.; supervision, M.K.; project administration, M.K., I.A. and G.J.B.; funding acquisition, M.K., I.A. and G.J.B. All authors have read and agreed to the published version of the manuscript.

Funding: This work was supported by grants from the National Science Foundation (DEB-1314749, DEB-1050253), and the Slovenian Research Agency (J1-6729, J1-9163, P1-0255, P1-0236, BI-US/17-18-011).

Institutional Review Board Statement: Not applicable.

Informed Consent Statement: Not applicable.

Data Availability Statement: All results generated in this study and protocols needed to replicate them are included in this published article, public repositories (GenBank) and supplementary material files.

Acknowledgments: We thank the CarBio team (http://www.islandbiogeography.org/participants.html, accessed on 15 November 2021) for collecting the material across the Caribbean. Moreover, we thank Lisa Chamberland and other members of the Agnarsson lab (http://www.theridiidae.com/lab-members.html, accessed on 15 November 2021) for the help with material sorting and molecular procedures. We thank three anonymous reviewers for their constructive comments that improved this paper.

Conflicts of Interest: The authors declare no conflict of interest.

References

1. Kozak, K.H.; Wiens, J.J. What explains patterns of species richness? The relative importance of climatic-niche evolution, morphological evolution, and ecological limits in salamanders. *Ecol. Evol.* **2016**, *6*, 5940–5949. [CrossRef]
2. Kozak, K.H.; Wiens, J.J. Testing the Relationships between Diversification, Species Richness, and Trait Evolution. *Syst. Biol.* **2016**, *65*, 975–988. [CrossRef] [PubMed]
3. Rabosky, D.L.; Hurlbert, A.H. Species richness at continental scales is dominated by ecological limits. *Am. Nat.* **2015**, *185*, 572–583. [CrossRef] [PubMed]
4. Schluter, D.; Pennell, M.W. Speciation gradients and the distribution of biodiversity. *Nature* **2017**, *546*, 48–55. [CrossRef]
5. Rajakaruna, H.; Lewis, M. Do yearly temperature cycles reduce species richness? Insights from calanoid copepods. *Theor. Ecol.* **2018**, *11*, 39–53. [CrossRef]
6. Marin, J.; Hedges, S.B. Time best explains global variation in species richness of amphibians, birds and mammals. *J. Biogeogr.* **2016**, *43*, 1069–1079. [CrossRef]
7. Borda-de-Água, L.; Whittaker, R.J.; Cardoso, P.; Rigal, F.; Santos, A.M.C.C.; Amorim, I.R.; Parmakelis, A.; Triantis, K.A.; Pereira, H.M.; Borges, P.A.V. Dispersal ability determines the scaling properties of species abundance distributions: A case study using arthropods from the Azores. *Sci. Rep.* **2017**, *7*, 3899. [CrossRef] [PubMed]

8. Lenoir, J.; Virtanen, R.; Oksanen, J.; Oksanen, L.; Luoto, M.; Grytnes, J.A.; Svenning, J.C. Dispersal ability links to cross-scale species diversity patterns across the Eurasian Arctic tundra. *Glob. Ecol. Biogeogr.* **2012**, *21*, 851–860. [CrossRef]

9. Claramunt, S.; Derryberry, E.P.; Remsen, J.V.; Brumfield, R.T. High dispersal ability inhibits speciation in a continental radiation of passerine birds. *Proc. R. Soc. B Biol. Sci.* **2012**, *279*, 1567–1574. [CrossRef] [PubMed]

10. Agnarsson, I.; Cheng, R.-C.; Kuntner, M. A multi-clade test supports the intermediate dispersal model of biogeography. *PLoS ONE* **2014**, *9*, e86780. [CrossRef]

11. Opatova, V.; Arnedo, M.A. Spiders on a Hot Volcanic Roof: Colonisation Pathways and Phylogeography of the Canary Islands Endemic Trap-Door Spider Titanidiops canariensis (Araneae, Idiopidae). *PLoS ONE* **2014**, *9*, e115078. [CrossRef] [PubMed]

12. Pitta, E.; Kassara, C.; Tzanatos, E.; Giokas, S.; Sfenthourakis, S. Between-island compositional dissimilarity of avian communities. *Ecol. Res.* **2014**, *29*, 835–841. [CrossRef]

13. Essl, F.; Steinbauer, K.; Dullinger, S.; Mang, T.; Moser, D. Telling a different story: A global assessment of bryophyte invasions. *Biol. Invasions* **2013**, *15*, 1933–1946. [CrossRef]

14. Kuntner, M.; Agnarsson, I. Phylogeography of a successful aerial disperser: The golden orb spider *Nephila* on Indian Ocean islands. *BMC Evol. Biol.* **2011**, *11*, 119. [CrossRef]

15. Turk, E.; Čandek, K.; Kralj-Fišer, S.; Kuntner, M. Biogeographical history of golden orbweavers: Chronology of a global conquest. *J. Biogeogr.* **2020**, *47*, 1333–1344. [CrossRef]

16. Darwin, C. *On the Origin of the Species*; John Murray: London, UK, 1859; Volume 5, ISBN 1546622497.

17. Warren, B.H.; Simberloff, D.; Ricklefs, R.E.; Aguilée, R.; Condamine, F.L.; Gravel, D.; Morlon, H.; Mouquet, N.; Rosindell, J.; Casquet, J.; et al. Islands as model systems in ecology and evolution: Prospects fifty years after MacArthur-Wilson. *Ecol. Lett.* **2015**, *18*, 200–217. [CrossRef] [PubMed]

18. Reynolds, R.G.; Collar, D.C.; Pasachnik, S.A.; Niemiller, M.L.; Puente-Rolón, A.R.; Revell, L.J. Ecological specialization and morphological diversification in Greater Antillean boas. *Evolution* **2016**, *70*, 1882–1895. [CrossRef]

19. Ricklefs, R.E.; Bermingham, E. The West Indies as a laboratory of biogeography and evolution. *Philos. Trans. R. Soc. B Biol. Sci.* **2008**, *363*, 2393–2413. [CrossRef]

20. Frankham, R. Inbreeding and Extinction: Island Populations. *Conserv. Biol.* **2008**, *12*, 665–675. [CrossRef]

21. Meiri, S. Oceanic island biogeography: Nomothetic science of the anecdotal. *Front. Biogeogr.* **2017**, *9*, e32081. [CrossRef]

22. Gillespie, R.G.; Croom, H.B.; Hasty, G.L. Phylogenetic Relationships and Adaptive Shifts among Major Clades of *Tetragnatha* Spiders (Araneae: Tetragnathidae) in Hawaí. *Pacific Sci.* **1997**, *51*, 380–394.

23. Crews, S.C.; Esposito, L.A. Towards a synthesis of the Caribbean biogeography of terrestrial arthropods. *BMC Evol. Biol.* **2020**, *20*, 12. [CrossRef]

24. Čandek, K.; Agnarsson, I.; Binford, G.J.; Kuntner, M. Biogeography of the Caribbean *Cyrtognatha* spiders. *Sci. Rep.* **2019**, *9*, 397. [CrossRef]

25. Ali, J.R. Colonizing the Caribbean: Is the GAARlandia land-bridge hypothesis gaining a foothold? *J. Biogeogr.* **2012**, *39*, 431–433. [CrossRef]

26. Iturralde-Vinent, M.A.; MacPhee, R.D.E. Paleogeography of the Caribbean Region: Implications for Cenozoic Biogeography. *Bull. Am. Museum Nat. Hist.* **1999**, *238*, 791. [CrossRef]

27. Ali, J.R.; Hedges, S.B. Colonizing the Caribbean: New geological data and an updated land-vertebrate colonization record challenge the GAARlandia land-bridge hypothesis. *J. Biogeogr.* **2021**, *48*, jbi.14234. [CrossRef]

28. Gillespie, R.G.; Benjamin, S.P.; Brewer, M.S.; Rivera, M.A.J.; Roderick, G.K. Repeated Diversification of Ecomorphs in Hawaiian Stick Spiders. *Curr. Biol.* **2018**, *28*, 941–947.e3. [CrossRef]

29. Cotoras, D.D.; Bi, K.; Brewer, M.S.; Lindberg, D.R.; Prost, S.; Gillespie, R.G. Co-occurrence of ecologically similar species of Hawaiian spiders reveals critical early phase of adaptive radiation. *BMC Evol. Biol.* **2018**, *18*, 100. [CrossRef] [PubMed]

30. Agnarsson, I.; van Patten, C.; Sargeant, L.; Chomitz, B.; Dziki, A.; Binford, G.J. A radiation of the ornate Caribbean 'smiley-faced spiders', with descriptions of 15 new species (Araneae: Theridiidae, Spintharus). *Zool. J. Linn. Soc.* **2018**, *182*, 758–790. [CrossRef]

31. Agnarsson, I.; LeQuier, S.M.; Kuntner, M.; Cheng, R.-C.; Coddington, J.A.; Binford, G.J. Phylogeography of a good Caribbean disperser: *Argiope argentata* (Araneae, Araneidae) and a new 'cryptic' species from Cuba. *Zookeys* **2016**, *2016*, 25–44. [CrossRef] [PubMed]

32. Chamberland, L.; McHugh, A.; Kechejian, S.; Binford, G.J.; Bond, J.E.; Coddington, J.; Dolman, G.; Hamilton, C.A.; Harvey, M.S.; Kuntner, M.; et al. From Gondwana to GAARlandia: Evolutionary history and biogeography of ogre-faced spiders (*Deinopis*). *J. Biogeogr.* **2018**, *45*, 2442–2457. [CrossRef]

33. Čandek, K.; Agnarsson, I.; Binford, G.J.; Kuntner, M. Caribbean golden orbweaving spiders maintain gene flow with North America. *Zool. Scr.* **2020**, *49*, 210–221. [CrossRef]

34. World Spider Catalog World Spider Catalog. 2021. Available online: https://wsc.nmbe.ch/ (accessed on 10 October 2021).

35. Foelix, R. *Biology of Spiders*, 3rd ed.; Oxford University Press: Oxford, UK, 2011; ISBN 9780199734825.

36. Čandek, K.; Pristovšek Čandek, U.; Kuntner, M. Machine learning approaches identify male body size as the most accurate predictor of species richness. *BMC Biol.* **2020**, *18*, 105. [CrossRef] [PubMed]

37. Cotoras, D.D.; Murray, G.G.R.; Kapp, J.; Gillespie, R.G.; Griswold, C.E.; Simison, W.B.; Green, R.E.; Shapiro, B. Ancient DNA resolves the history of *Tetragnatha* (Araneae, Tetragnathidae) spiders on Rapa Nui. *Genes* **2017**, *8*, 403. [CrossRef]

38. Gillespie, R.G. Spiders of the Genus *Tetragnatha*) (Araneae, Tetragnathidae) in the Society Islands. *J. Arachnol.* **2003**, *31*, 157–172. [CrossRef]

39. Gillespie, R.G. Geographical context of speciation in a radiation of Hawaiian *Tetragnatha* spiders (Araneae, Tetragnathidae). *J. Arachnol.* **2005**, *33*, 313–322. [CrossRef]

40. Casquet, J.; Bourgeois, Y.X.C.; Cruaud, C.; Gavory, F.; Gillespie, R.G.; Thébaud, C. Community assembly on remote islands: A comparison of Hawaiian and Mascarene spiders. *J. Biogeogr.* **2015**, *42*, 39–50. [CrossRef]

41. Okuma, C.; Kisimoto, R. Airborne spiders collected over the East China Sea. *Jpn. J. Appl. Entomol. Zool.* **1981**, *25*, 296–308. [CrossRef]

42. Bell, J.R.; Bohan, D.A.; Shaw, E.M.; Weyman, G.S. Ballooning dispersal using silk: World fauna, phylogenies, genetics and models. *Bull. Entomol. Res.* **2005**, *95*, 69–114. [CrossRef]

43. Hedges, S.B. Paleogeography of the Antilles and Origin of West Indian Terrestrial Vertebrates. *Ann. Missouri Bot. Gard.* **2006**, *93*, 231–244. [CrossRef]

44. Shapiro, L.; Binford, G.J.; Agnarsson, I. Single island endemism despite repeated dispersal in Caribbean *Microthena* (Araneae: Araneidae), an updated phylogeographic analysis. *Diversity* **2021**, *13*, 1–27, in press.

45. Agnarsson, I.; Coddington, J.A.; Kuntner, M. Systematics: Progress in the study of spider diversity and evolution. In *Spider Research in the 21st Century: Trends and Perspectives*; Penney, D., Ed.; Siri Scientific Press: Manchester, UK, 2013; pp. 55–111. ISBN 0957453019.

46. Coddington, J.A.; Griswold, C.E.; Silva, D.; Peqaranda, E.; Larcher, S.F.; Penaranda, E.; Larcher, S.F. Designing and testing sampling protocols to estimate biodiversity in tropical ecosystems. In *The Unity of Evolutionary Biology: Proceedings of the Fourth International Congress of Systematic and Evolutionary Biology*; Dioscorides Press: Portland, OR, USA, 1991; pp. 44–60.

47. Vidergar, N.; Toplak, N.; Kuntner, M. Streamlining DNA barcoding protocols: Automated DNA extraction and a new cox1 primer in arachnid systematics. *PLoS ONE* **2014**, *9*, e113030. [CrossRef]

48. Folmer, O.; Black, M.; Hoeh, W.; Lutz, R.; Vrijenhoek, R. DNA primers for amplification of mitochondrial cytochrome c oxidase subunit I from diverse metazoan invertebrates. *Mol. Mar. Biol. Biotechnol.* **1994**, *3*, 294–299. [CrossRef]

49. Hedin, M.C.; Maddison, W.P. A Combined Molecular Approach to Phylogeny of the Jumping Spider Subfamily Dendryphantinae (Araneae: Salticidae). *Mol. Phylogenet. Evol.* **2001**, *18*, 386–403. [CrossRef] [PubMed]

50. Nunn, G.B.; Theisen, B.F.; Christensen, B.; Arctander, P. Simplicity-correlated size growth of the nuclear 28S ribosomal RNA D3 expansion segment in the crustacean order isopoda. *J. Mol. Evol.* **1996**, *42*, 211–223. [CrossRef] [PubMed]

51. Whiting, M.F. Mecoptera is paraphyletic: Multiple genes and phylogeny of Mecoptera and Siphonaptera. *Zool. Scr.* **2002**, *31*, 93–104. [CrossRef]

52. Kearse, M.; Moir, R.; Wilson, A.; Stones-Havas, S.; Cheung, M.; Sturrock, S.; Buxton, S.; Cooper, A.; Markowitz, S.; Duran, C.; et al. Geneious Basic: An integrated and extendable desktop software platform for the organization and analysis of sequence data. *Bioinformatics* **2012**, *28*, 1647–1649. [CrossRef] [PubMed]

53. Tamura, K.; Stecher, G.; Peterson, D.; Filipski, A.; Kumar, S. MEGA6: Molecular Evolutionary Genetics Analysis Version 6.0. *Mol. Biol. Evol.* **2013**, *30*, 2725–2729. [CrossRef]

54. Maddison, W.P.; Maddison, D.R. Mesquite: A modular system for evolutionary analysis. *Evolution* **2008**, *62*, 1103–1118.

55. Katoh, K.; Standley, D.M. MAFFT Multiple Sequence Alignment Software Version 7: Improvements in Performance and Usability. *Mol. Biol. Evol.* **2013**, *30*, 772–780. [CrossRef]

56. Puillandre, N.; Lambert, A.; Brouillet, S.; Achaz, G. ABGD, Automatic Barcode Gap Discovery for primary species delimitation. *Mol. Ecol.* **2012**, *21*, 1864–1877. [CrossRef]

57. Huelsenbeck, J.P.; Ronquist, F. MrBayes: Bayesian inference of phylogenetic trees. *Bioinformatics* **2001**, *17*, 754–755. [CrossRef]

58. Darriba, D.; Taboada, G.L.; Doallo, R.; Posada, D. jModelTest 2: More models, new heuristics and parallel computing. *Nat. Methods* **2012**, *9*, 772. [CrossRef]

59. Rambaut, A.; Drummond, A.J.; Xie, D.; Baele, G.; Suchard, M.A. Posterior summarisation in Bayesian phylogenetics using Tracer 1.7. *Syst. Biol.* **2018**, *67*, 901. [CrossRef]

60. Bouckaert, R.; Heled, J.; Kühnert, D.; Vaughan, T.; Wu, C.-H.; Xie, D.; Suchard, M.A.; Rambaut, A.; Drummond, A.J. BEAST 2: A software platform for Bayesian evolutionary analysis. *PLoS Comput. Biol.* **2014**, *10*, e1003537. [CrossRef]

61. Bouckaert, R.; Drummond, A.J. bModelTest: Bayesian phylogenetic site model averaging and model comparison. *BMC Evol. Biol.* **2017**, *17*, 42. [CrossRef]

62. Bidegaray-Batista, L.; Arnedo, M.A. Gone with the plate: The opening of the Western Mediterranean basin drove the diversification of ground-dweller spiders. *BMC Evol. Biol.* **2011**, *11*, 317. [CrossRef]

63. Peck, S.B. Diversity and distribution of beetles (Insecta: Coleoptera) of the northern Leeward Islands. *Insecta Mundi* **2011**, *678*, 159.

64. Garrison, N.L.; Rodriguez, J.; Agnarsson, I.; Coddington, J.A.; Griswold, C.E.; Hamilton, C.A.; Hedin, M.C.; Kocot, K.M.; Ledford, J.M.; Bond, J.E. Spider phylogenomics: Untangling the spider tree of life. *PeerJ* **2016**, *4*, e1719. [CrossRef]

65. Bond, J.E.; Garrison, N.L.; Hamilton, C.A.; Godwin, R.L.; Hedin, M.C.; Agnarsson, I. Phylogenomics resolves a spider backbone phylogeny and rejects a prevailing paradigm for orb web evolution. *Curr. Biol.* **2014**, *24*, 1765–1771. [CrossRef]

66. Dimitrov, D.; Benavides, L.R.; Arnedo, M.A.; Giribet, G.; Griswold, C.E.; Scharff, N.; Hormiga, G. Rounding up the usual suspects: A standard target-gene approach for resolving the interfamilial phylogenetic relationships of ecribellate orb-weaving spiders with a new family-rank classification (Araneae, Araneoidea). *Cladistics* **2017**, *33*, 221–250. [CrossRef]

67. Matzke, N.J. BioGeoBEARS: BioGeography with Bayesian (and Likelihood) Evolutionary Analysis in R Scripts. R Packag. Version 0.2 2013, 1. Available online: https://CRAN.R-project.org/package=BioGeoBEARS (accessed on 10 September 2021).

68. R Core Team. R: A Language and Environment for Statistical Computing. 2018. Available online: https://www.r-project.org/ (accessed on 15 November 2021).

69. Kimura, M. A simple method for estimating evolutionary rates of base substitutions through comparative studies of nucleotide sequences. *J. Mol. Evol.* **1980**, *16*, 111–120. [CrossRef]

70. Čandek, K.; Kuntner, M. DNA barcoding gap: Reliable species identification over morphological and geographical scales. *Mol. Ecol. Resour.* **2015**, *15*, 268–277. [CrossRef]

71. Brandley, M.C.; Wang, Y.; Guo, X.; Montes de Oca, A.N.; Fería-Ortíz, M.; Hikida, T.; Ota, H. Accommodating heterogenous rates of evolution in molecular divergence dating methods: An example using intercontinental dispersal of Plestiodon (Eumeces) lizards. *Syst. Biol.* **2011**, *60*, 3–15. [CrossRef]

72. Sánchez-Ruiz, A.; Brescovit, A.D.; Alayón, G. Four new caponiids species (Araneae, Caponiidae) from the West Indies and redescription of *Nops blandus* (Bryant). *Zootaxa* **2015**, *3972*, 43–64. [CrossRef]

73. Zhang, J.; Maddison, W.P. New euophryine jumping spiders from the Dominican Republic and Puerto Rico (Araneae: Salticidae: Euophryinae). *Zootaxa* **2012**, *3476*, 1–54. [CrossRef]

74. Crews, S.C.; Puente-Rolón, A.R.; Rutstein, E.; Gillespie, R.G. A comparison of populations of island and adjacent mainland species of Caribbean Selenops (Araneae: Selenopidae) spiders. *Mol. Phylogenet. Evol.* **2010**, *54*, 970–983. [CrossRef]

75. Dziki, A.; Binford, G.J.; Coddington, J.A.; Agnarsson, I. Spintharus flavidus in the Caribbean—A 30 million year biogeographical history and radiation of a 'widespread species'. *PeerJ* **2015**, *3*, e1422. [CrossRef]

76. McHugh, A.; Yablonsky, C.; Binford, G.J.; Agnarsson, I. Molecular phylogenetics of Caribbean *Micrathena* (Araneae: Araneidae) suggests multiple colonisation events and single island endemism. *Invertebr. Syst.* **2014**, *28*, 337–349. [CrossRef]

77. Gillespie, R.G.; Baldwin, B.G.; Waters, J.M.; Fraser, C.I.; Nikula, R.; Roderick, G.K. Long-distance dispersal: A framework for hypothesis testing. *Trends Ecol. Evol.* **2012**, *27*, 47–55. [CrossRef] [PubMed]

78. Grant, P.R.; Grant, B.R. Adaptive radiation of Darwin's finches. *Am. Sci.* **2002**, *90*, 130. [CrossRef]

79. Lovette, I.J.; Bermingham, E.; Ricklefs, R.E. Clade-specific morphological diversification and adaptive radiation in Hawaiian songbirds. *Proc. R. Soc. B Biol. Sci.* **2002**, *269*, 37–42. [CrossRef] [PubMed]

80. Hass, C.A.; Hedges, S.B.; Maxson, L.R. Molecular insights into the relationships and biogeography of West Indian anoline lizards. *Biochem. Syst. Ecol.* **1993**, *21*, 97–114. [CrossRef]

81. Glor, R.E.; Gifford, M.E.; Larson, A.; Losos, J.B.; Schettino, L.R.; Lara, A.R.C.; Jackman, T.R. Partial island submergence and speciation in an adaptive radiation: A multilocus analysis of the Cuban green anoles. *Proc. R. Soc. B Biol. Sci.* **2004**, *271*, 2257–2265. [CrossRef] [PubMed]

82. Losos, J.B.; Schluter, D. Analysis of an evolutionary species-area relationship. *Nature* **2000**, *408*, 847–850. [CrossRef] [PubMed]

83. Hedges, S.B. Biogeography of the West Indies: An overview. In *Biogeography of the West Indies: Patterns and Perspectives*; Woods, C.A., Sergile, F.E., Eds.; CRC Press: Boca Raton, FL, USA, 2001; pp. 15–33. ISBN 0849320011101.

84. Říčan, O.; Piálek, L.; Zardoya, R.; Doadrio, I.; Zrzavý, J. Biogeography of the Mesoamerican Cichlidae (Teleostei: Heroini): Colonization through the GAARlandia land bridge and early diversification. *J. Biogeogr.* **2013**, *40*, 579–593. [CrossRef]

85. Alonso, R.; Crawford, A.J.; Bermingham, E. Molecular phylogeny of an endemic radiation of Cuban toads (Bufonidae: Peltophryne) based on mitochondrial and nuclear genes. *J. Biogeogr.* **2012**, *39*, 434–451. [CrossRef]

86. Dávalos, L.M. Phylogeny and biogeography of Caribbean mammals. *Biol. J. Linn. Soc.* **2004**, *81*, 373–394. [CrossRef]

87. Binford, G.J.; Callahan, M.S.; Bodner, M.R.; Rynerson, M.R.; Núñez, P.B.; Ellison, C.E.; Duncan, R.P. Phylogenetic relationships of Loxosceles and Sicarius spiders are consistent with Western Gondwanan vicariance. *Mol. Phylogenet. Evol.* **2008**, *49*, 538–553. [CrossRef]

88. Matos-Maraví, P.; Núñez Águila, R.; Peña, C.; Miller, J.Y.; Sourakov, A.; Wahlberg, N. Causes of endemic radiation in the Caribbean: Evidence from the historical biogeography and diversification of the butterfly genus *Calisto* (Nymphalidae: Satyrinae: Satyrini). *BMC Evol. Biol.* **2014**, *14*, 199. [CrossRef]

89. Fritsch, P.W. Multiple Geographic Origins of Antillean Styrax. *Syst. Bot.* **2003**, *28*, 421–430. [CrossRef]

90. van Ee, B.W.; Berry, P.E.; Riina, R.; Gutiérrez Amaro, J.E. Molecular Phylogenetics and Biogeography of the Caribbean-Centered Croton Subgenus *Moacroton* (Euphorbiaceae s.s.). *Bot. Rev.* **2008**, *74*, 132–165. [CrossRef]

91. Nieto-Blázquez, M.E.; Antonelli, A.; Roncal, J. Historical Biogeography of endemic seed plant genera in the Caribbean: Did GAARlandia play a role? *Ecol. Evol.* **2017**, *7*, 10158–10174. [CrossRef]

92. Graham, A. Geohistory models and Cenozoic paleoenvironments of the Caribbean region. *Syst. Bot.* **2003**, *28*, 378–386. [CrossRef]

93. MacPhee, R.D.E.; Grimaldi, D.A. Mammal bones in Dominican amber. *Nature* **1996**, *380*, 489–490. [CrossRef]

94. Shaw, K.L.; Gillespie, R.G. Comparative phylogeography of oceanic archipelagos: Hotspots for inferences of evolutionary process. *Proc. Natl. Acad. Sci. USA* **2016**, *113*, 7986–7993. [CrossRef]

95. Dupin, J.; Matzke, N.J.; Särkinen, T.; Knapp, S.; Olmstead, R.G.; Bohs, L.; Smith, S.D. Bayesian estimation of the global biogeographical history of the Solanaceae. *J. Biogeogr.* **2017**, *44*, 887–899. [CrossRef]

96. Magalhaes, I.L.F.; Santos, A.J.; Ramírez, M.J. Incorporating Topological and Age Uncertainty into Event-Based Biogeography of Sand Spiders Supports Paleo-Islands in Galapagos and Ancient Connections among Neotropical Dry Forests. *Diversity* **2021**, *13*, 418. [CrossRef]

97. Gillespie, R.G. Biogeography of spiders on remote oceanic islands of the Pacific: Archipelagoes as stepping stones? *J. Biogeogr.* **2002**, *29*, 655–662. [CrossRef]

98. Gillespie, R.G.; Croom, H.B.; Palumbi, S.R. Multiple origins of a spider radiation in Hawaii. *Proc. Natl. Acad. Sci. USA* **1994**, *91*, 2290–2294. [CrossRef] [PubMed]

99. Linck, E.; Schaack, S.; Dumbacher, J.P. Genetic differentiation within a widespread "supertramp" taxon: Molecular phylogenetics of the Louisiade White-eye (*Zosterops griseotinctus*). *Mol. Phylogenet. Evol.* **2016**, *94*, 113–121. [CrossRef] [PubMed]

100. Pedersen, M.P.; Irestedt, M.; Joseph, L.; Rahbek, C.; Jønsson, K.A. Phylogeography of a 'great speciator' (Aves: *Edolisoma tenuirostre*) reveals complex dispersal and diversification dynamics across the Indo-Pacific. *J. Biogeogr.* **2018**, *45*, 826–837. [CrossRef]

101. Crews, S.C.; Gillespie, R.G. Molecular systematics of Selenops spiders (Araneae: Selenopidae) from North and Central America: Implications for Caribbean biogeography. *Biol. J. Linn. Soc.* **2010**, *101*, 288–322. [CrossRef]

102. Abel, C.; Schneider, J.M.; Kuntner, M.; Harms, D. Phylogeography of the 'cosmopolitan' orb-weaver Argiope trifasciata (Araneae: Araneidae). *Biol. J. Linn. Soc.* **2020**, *131*, 61–75. [CrossRef]

diversity

Article

Island–to–Island Vicariance, Founder–Events and within–Area Speciation: The Biogeographic History of the *Antillattus* Clade (Salticidae: Euophryini)

Franklyn Cala-Riquelme [1,2,*], Patrick Wiencek [3], Eduardo Florez-Daza [2], Greta J. Binford [4] and Ingi Agnarsson [3,5,*]

[1] Institute for Biodiversity Science & Sustainability, California Academy of Sciences, 55 Music Concourse Drive, San Francisco, CA 94118, USA

[2] Grupo de Estudios de Arácnidos y Miriápodos, Instituto de Ciencias Naturales, Universidad Nacional de Colombia, Bogotá 111321, Colombia; aeflorezd@unal.edu.co

[3] Department of Biology, College of Arts and Sciences, University of Vermont, 109 Carrigan Drive, Burlington, VT 05401, USA; wiencek.patrick@gmail.com

[4] Department of Biology, 615 S Palatine Hill Rd. Lewis and Clark College, Portland, OR 97219, USA; binford@lclark.edu

[5] Faculty of Life and Environmental Sciences, University of Iceland, Sturlugata 7, 102 Reykjavik, Iceland

[*] Correspondence: fcalariquelme@calacademia.org (F.C.-R.); iagnarsson@gmail.com (I.A.)

Abstract: The Caribbean Archipelago is a biodiversity hotspot that plays a key role in developing our understanding of how dispersal ability affects species formation. In island systems, species with intermediate dispersal abilities tend to exhibit greater diversity, as may be the case for many of the salticid lineages of the insular Caribbean. Here, we use molecular phylogenetic analyses to infer patterns of relationships and biogeographic history of the Caribbean endemic *Antillattus* clade (*Antillattus*, *Truncattus*, and *Petemethis*). We test if the timing of origin of the *Antillatus* clade in the Greater Antilles is congruent with GAARlandia and infer patterns of diversification within the *Antillattus* clade among Cuba, Hispaniola, and Puerto Rico. Specifically, we evaluate the relative roles of dispersal over land connections, and overwater dispersal events in diversification within the Greater Antilles. Time tree analysis and model-based inference of ancestral ranges estimated the ancestor of the *Antillattus* clade to be c. 25 Mya, and the best model suggests dispersal via GAARlandia from northern South America to Hispaniola. Hispaniola seems to be the nucleus from which ancestral populations dispersed into Cuba and Puerto Rico via land connections prior to the opening of the Mona Passage and the Windward Passage. Divergences between taxa of the *Antillattus* clade from Cuban, Hispaniolan, and Puerto Rican populations appear to have originated by vicariance, founder-events and within-island speciation, while multiple dispersal events (founder-events) between Cuba and Hispaniola during the Middle Miocene and the Late Miocene best explain diversity patterns in the genera *Antillattus* and *Truncattus*.

Keywords: Caribbean biogeography; molecular dating; ancestral range analysis; endemics; founder-event; intermediate dispersal model

Citation: Cala-Riquelme, F.; Wiencek, P.; Florez-Daza, E.; Binford, G.J.; Agnarsson, I. Island–to–Island Vicariance, Founder–Events and within–Area Speciation: The Biogeographic History of the *Antillattus* Clade (Salticidae: Euophryini). *Diversity* **2022**, *14*, 224. https://doi.org/10.3390/d14030224

Academic Editor: Martín J. Ramírez

Received: 20 November 2021
Accepted: 14 March 2022
Published: 18 March 2022

Publisher's Note: MDPI stays neutral with regard to jurisdictional claims in published maps and institutional affiliations.

1. Introduction

Since Darwin and Wallace, evolutionary biologists have been fascinated by the extraordinary diversity and richness of islands. Biogeography has been reinvigorated through the use of molecular methods to test divergence hypotheses [1–4] and matured through the successful reconciliation of theories that previously were treated as mutually exclusive: long-distance dispersal and vicariance. This progress has been aided by the growth of sophistication in testing long distance dispersal hypotheses—best supported when vicariance explanations are rejected by geological history (e.g., Matos–Maraví et al. [5])—and the development of models such as the intermediate dispersal hypothesis [6–13].

The Greater Antilles (Cuba, Hispaniola, Puerto Rico, and Jamaica) are one of the planet's recognized biodiversity hotspots [2,14]. The area is an excellent arena to test biogeographical hypotheses due to its complex geology (including, e.g., land bridge or Wallacean fragment islands, volcanic or Darwinian islands, and uplifted coral shelves), geography (including complex topography and diverse climates), and old age [2,11,15–27]. The uplift of the core Greater Antilles, arising from the earlier 'proto–Antilles ridge, began during the Middle Eocene (c. 48–37 Mya) and reached its maximum land area at the Eocene–Oligocene (c. 40–30 Mya) boundary [17,28–35]. Since that time, the Greater Antilles have remained above water with variation in island area and inter-island connections changing with sea level. For example, Hispaniola was physically connected to Puerto Rico and Cuba until the formation of the Mona Passage (late Oligocene to early Miocene, c. 30–23 Ma) and the Windward Passage (early-to-middle Miocene, c. 17–15 Ma), respectively [17,35–38].

The origin of the present-day terrestrial biota of the Greater Antilles has been hypothesized to extend back to the emergence of the proto-Antilles (c. 65 Mya), predicting the survival of relict lineages through periods of oceanic submergence of island fragments [28,32,39–41]. However, for most organisms, their origin more likely traces back to the permanent emergence of the Greater Antilles (c. 40 Mya) and could have involved both long-distance over-water dispersal events [42–44] (as occurred in *Solenodons* [45], Urocoptid snails [46], Calliphorid flies [47], and various spiders [26,48–52]) and vicariance. The oldest putative vicariance events are linked to the hypothesized existence of GAARlandia (GAAR = Greater Antilles Aves Ridge) a land bridge relatively briefly (c. 35 to 32 Mya) connecting the Greater Antilles and continental South America during the Eocene–Oligocene transition [32,33,44,53,54]. Though it remains under active debate [27], this hypothesis has received support in studies across a variety of taxa (e.g., freshwater fishes [55], lizards [16], bats [56], mammals [42,45], plants [57], and spiders [4,58]). However, a recent meta-analysis suggested that GAARlandia does not help explain the colonization of various land vertebrate lineages [27]. Regardless, both historical connections among islands leading to vicariant interchange of organisms, and long-distance dispersal are recognized as critically important components that must be considered together for a complete account of island biogeography [1,7–9,43,48,49,52,59].

In the last decades there has been a growing interest in studies on invertebrates [5,24, 52,60–67] including arachnids [4,18,22,23,26,48–50,58,68]. These studies have found mixed support for vicariance [22,69,70] and dispersal [4,16,18,45,48,50,51]; often a combination of the two [5].

The geographic distribution of spiders in the euophryine *Antillattus* clade of the family Salticidae make them an interesting model for testing hypotheses of Caribbean dispersal corridors [68]. Salticids are a diverse, globally distributed group of spiders (c. 6392 total species) [71] known as "jumping spiders" due to their semi-hydraulic locomotion system [72–74]. Within Salticidae, euophryines are a relatively young group (c. 33–30 Mya) [69]. Phylogenetic reconstruction shows that much like other salticid lineages [75–77], New and Old-World euophryines are grouped into separate clades, indicating that most euophryine diversification occurred intra-continentally [68]. In their landmark revisionary work on euophryines, Zhang and Maddison [68] highlighted the *Antillattus* clade as one of several salticid lineages that has diversified within the Caribbean. Members of the *Antillattus* clade (*Antillattus* Bryant [78], *Truncattus* Zhang and Maddison [79], and *Petemathis* Prószyński and Deeleman–Reinhold [68,80,81]) are small to medium-sized spiders of the Greater Antilles (Cuba, Hispaniola, and Puerto Rico) (Figures 1–3). During the morning, these spiders can be found in understory habitats and dense forests, and typically walk or jump between leaves, branches, and trunks. In the sunset and at night, they are found in their shelters, e.g., leaves, and under the bark).

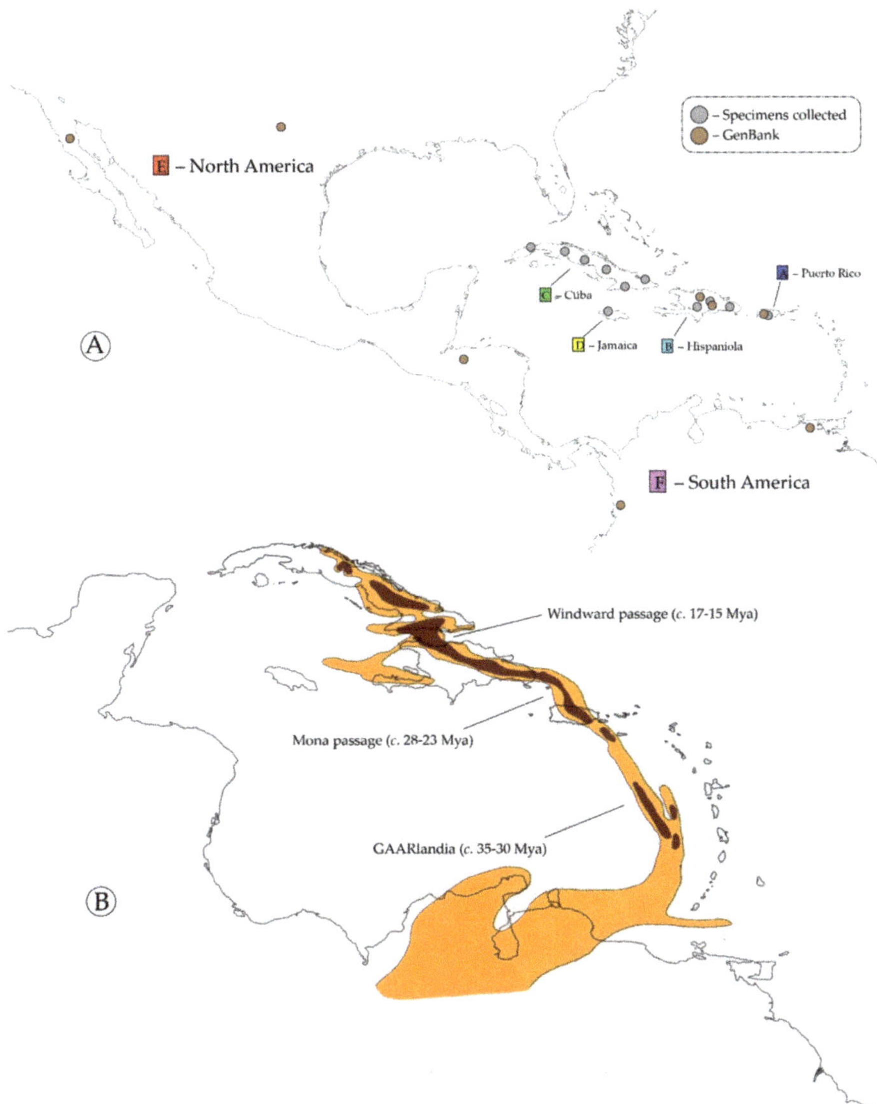

Figure 1. (**A**) Map of specimens collected for this work, including samples obtained from Genbank (Zhang and Maddison [69]). Area code used in the distribution ranges (A—Puerto Rico, B—Hispaniola, C—Cuba, D—Jamaica, E—North America, F—South America). (**B**) Schematic representations of the GAARlandia and Caribbean land areas available at certain time periods (Iturralde-Vinent [17], Iturralde-Vinent and MacPhee [32], MacPhee and Iturralde-Vinent [33]). Maps show simplified island positions in the respective time window used for the time-stratified analysis.

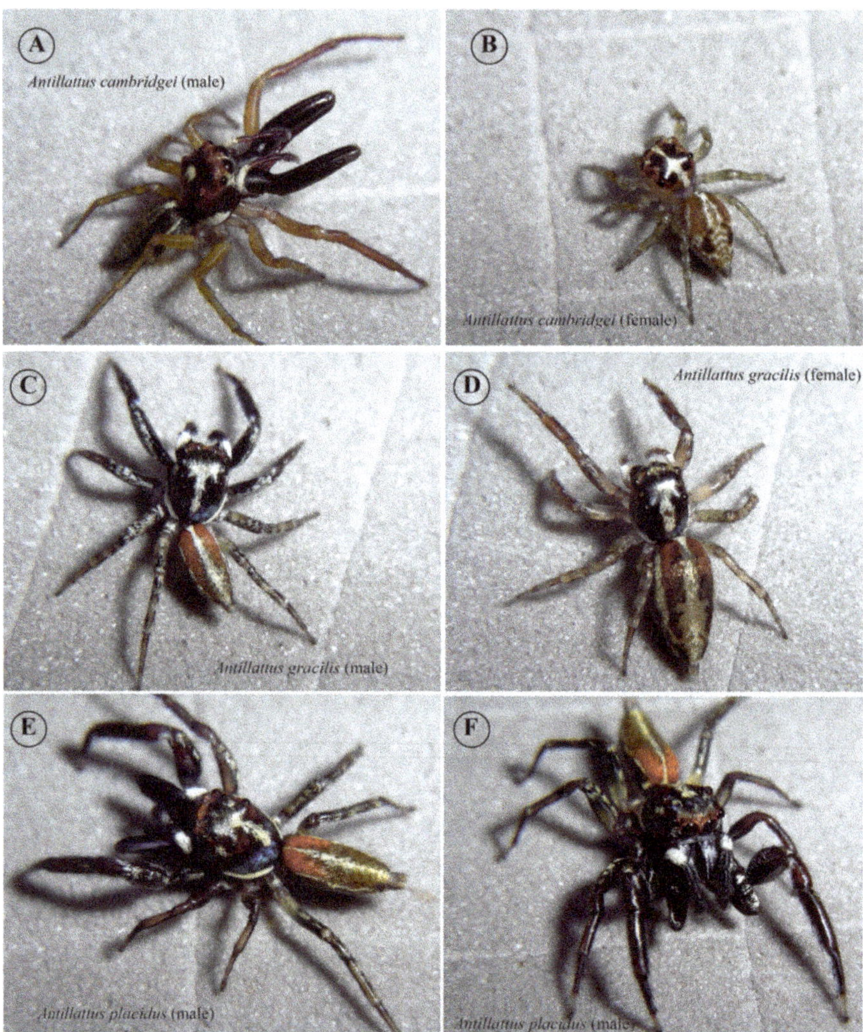

Figure 2. A–F *gracilis* group. (**A,B**) *Antillattus cambridgei*, male and female habitus. (**C,D**) *Antillattus gracilis*, male and female habitus. (**E,F**) *Antillattus placidus*, male and female habitus. Images by Wayne Maddison, released under a Creative Commons Attribution (CC–BY) 3.0 license.

The *Antillattus* clade is relatively late-diverging with an estimated origin in the Caribbean by dispersal in the Miocene [c. 22.34–19.74 Mya], a scenario that implies ancestors dispersed over the Greater Antilles via land connections prior to the opening of the Mona Passage and the Windward Passage [68]. Members of the *Antillattus* clade appear to have relatively low dispersal potential based on their biology and absence from Jamaica and the isolated volcanic islands of the Lesser Antilles—none of which formed a part of the hypothetical GAARlandia land bridge (Cuba, Hispaniola, Puerto Rico). We predict that the Mona Passage and Windward Passage may have been integral to the dispersal of the *Antillattus* clade among the Greater Antilles. Here, we evaluate the non-GAARlandia (overwater dispersal) and GAARlandia hypotheses to infer the timing and ancestral colonization route of Caribbean euophryines; analyze the relationship of the *Antillatus* clade to other Greater Antilles euophryines (*Popcornella*, *Corticattus*, and the *Agobardus* clade); and infer the details of diversification within the *Antillatus* clade. We use time-calibrated phylogenies to see if divergence times of taxa on Cuba, Hispaniola, and Puerto Rico correspond to estimated dates of the land connections (Mona Passage and Windward Passage), or if they are better explained by overwater dispersal. Finally, we apply biogeographical stochastic mapping (BSM) to estimate how the frequency of dispersal and vicariance events of the clade resulted in the present-day distribution and diversity.

Figure 3. **A–F** *darlingtoni* group. (**A,B**) *Antillattus applanatus*, male and female habitus. (**C,D**) *Antillattus darlingtoni*, male and female habitus. (**E,F**) *Antillattus maxillosus*, male and female habitus. Images by Wayne Maddison, released under a Creative Commons Attribution (CC–BY) 3.0 license.

2. Materials and Methods

Study Group and Taxon Sample

Antillattus clade intergeneric relationships and their outgroup structure are poorly known (see Zhang and Maddison [81]), while the broader phylogenetic placement of the *Antillattus* clade is better established (see Zhang and Maddison [68,81]). The *Antillattus* clade was instated as a clade separate from the insular Caribbean *Anasaitis-Corythalia* clade and closely related to the genera *Popcornella*, *Corticattus* and the *Agobardus* clade based on molecular and morphological studies (Zhang and Maddison [68]). These studies also resulted in the transfer of insular Caribbean species of *Pensacola* and *Cobanus*, and some species of *Agobardus* from Cuba, to the genus *Antillattus* (Zhang and Maddison [68]). Here, for phylogenetic inference, we included as outgroups the continental *Pensacola-Mexigonus* clade, and *Sidusa* clade, the Greater Antilles genera *Popcornella*, *Corticattus*, and the *Agobardus* clade (*Agobardus*, *Compsodecta* and *Bythocrotus*).

The *Antillattus* clade is currently composed of twenty-three species distributed as follows: ten species of *Antillattus* from Hispaniola and three species from Cuba, five species of *Truncattus* from Hispaniola, and five species of *Petemathis* from Puerto Rico. Here, we include a total of thirty-two taxa collected using beating and visual search methods in Cuba, Puerto Rico, and Hispaniola (Figures 1 and 4, Table 1). Material collected was fixed in the field in 95% ethanol. Caribbean voucher specimens will be deposited in the Smithsonian Institute, Washington DC. We collect and identify just over 60% of the known species for the *Antilattus* clade (nine *Antillattus*, three *Petemathis*, and three *Truncattus*), while the remaining sampled taxa could not be attributed to known species (Figures 2 and 3, Table 1).

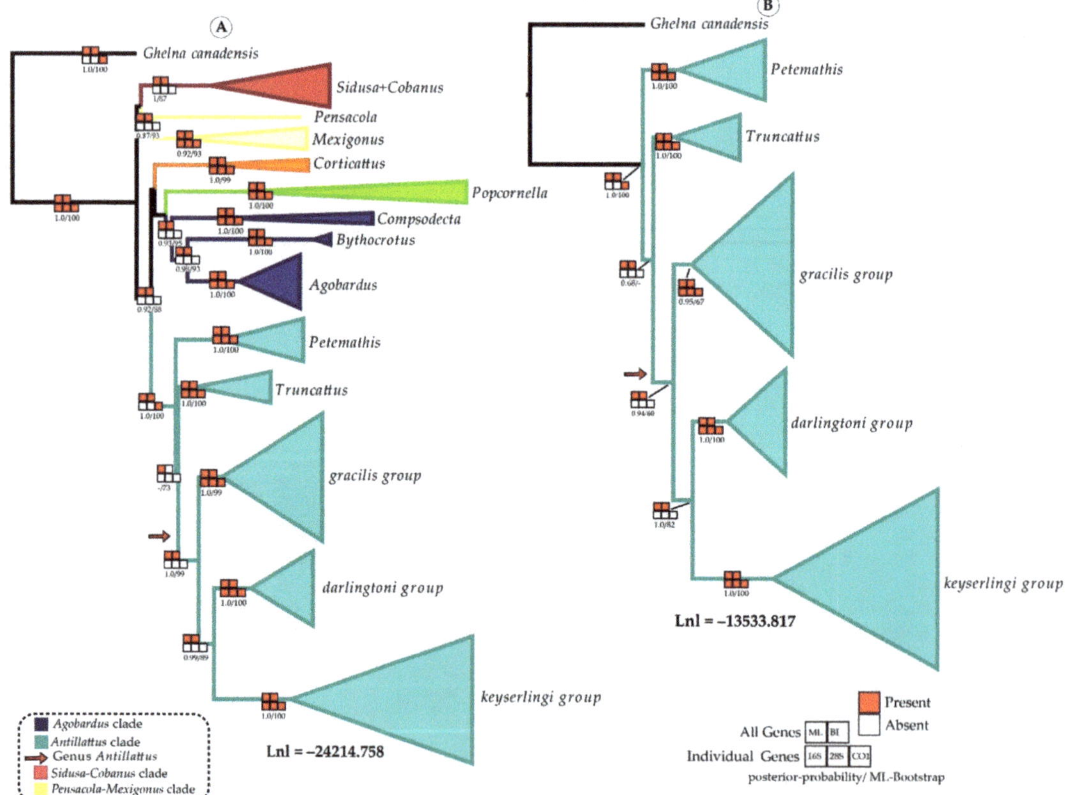

Figure 4. Summary of (**A**) ML (Lnl = −24,214.785) and BI (Harmonic-means −24,208.96) with outgroups and (**B**) ML (Lnl = −13,533.817) and BI (Harmonic–means = −13,597.32) without outgroups, based on analyses on the molecular datasets (28S, 16S-ND1 and CO1). Individual Gene refers to support for a clade in the ML tree of individual genes.

Table 1. Taxon sample with specific collection information and Genbank accession numbers of the previous published sequence. Checkmark (√) refers sequence obtained from this study.

Species	Voucher	Locality	CO1	16S-ND1	28S
Agobardus anormalis montanus	JXZ357	DOMINICAN REPUBLIC: Pedernales, (N18.128, W71.558)	KC615636	KC615802	KC615376
Agobardus bahoruco	JXZ324	DOMINICAN REPUBLIC: Pedernales, (N18.128, W71.558)		KC615844	KC615417
Agobardus cf. brevitarsus	JXZ311	DOMINICAN REPUBLIC: Pedernales, (N18.128, W71.558)	KC615637	KC615803	KC615637
Agobardus cordiformis	JXZ358	DOMINICAN REPUBLIC: Pedernales, (N17.965, W71.635)	KC615634	KC615800	KC615374
Agobardus gramineus	JXZ314	DOMINICAN REPUBLIC: Pedernales, (N17.965, W71.635)	KC615635	KC615801	KC615375
Agobardus oviedo	JXZ312	DOMINICAN REPUBLIC: Pedernales, (N17.802, W71.349)	KC615638	KC615804	KC615378
Antillattus [Cuba1]	CU787945 CU00107A	CUBA: Granma, Bartolomé Maso, (N20.009, W76.894)	√	√	√
Antillattus [Cuba2]	CU00025A CU00086A CU00090A CU00004A CU00016A	CUBA: Granma, Bartolomé Maso, (N20.013,W76.834)	√	√	
Antillattus [Cuba3]	CU787957 CU03506A	CUBA: Pinar del Rio, Viñales, (N22.657, W83.701)	√		
Antillattus [Cuba4]	CU00100A CU03361A CU03317A	CUBA: Guantánamo, Baracoa, (N20.331, W74.569)	√		
Antillattus [Cuba4]	CU03121A	CUBA: Guantánamo, Nibujón, (N20.052, W76.502)	√		
Antillattus cambridgei	JXZ321	DOMINICAN REPUBLIC: La Vega, (N19.033, W70.543)	KC615646	KC615818	KC615392
Antillattus cambridgei	DR784676 DR785410 DR785798 DR782454	DOMINICAN REPUBLIC: La Alta Gracia, (N19.067, W69.463)	√	√	
Antillattus cambridgei	DR785494 DR782541	DOMINICAN REPUBLIC: La Alta Gracia, (N19.893, W71.653)	√	√	
Antillattus cambridgei	DR782541 DR785783 DR785508	DOMINICAN REPUBLIC: La Alta Gracia, (N19.355, W070.111)	√	√	
Antillattus cambridgei	DR782598	DOMINICAN REPUBLIC: La Alta Gracia, (19.355N, W70.111)	√	√	√

Table 1. *Cont.*

Species	Voucher	Locality	CO1	16S-ND1	28S
Antillattus cambridgei	DR784852 DR785098 DR785696 DR785438 DR787296	DOMINICAN REPUBLIC: La Alta Gracia, (19.067, W69.463)	√	√	
Antillattus cambridgei	DR787296 DR787293 DR787254 DR787223	DOMINICAN REPUBLIC: La Vega, (N19.036, W70.543)	√	√	
Antillattus cambridgei	DR787328 DR787252 DR787285 DR787207 DR787327 DR787319 DR787324	DOMINICAN REPUBLIC: Santo Domingo, (19.051 N, W70.888)	√	√	
Antillattus cambridgei	DR787105	DOMINICAN REPUBLIC: San Juan, (N19.175, W71.049)	√	√	
Antillattus cf. applanatus	JXZ336	DOMINICAN REPUBLIC: Barahona, Cachote (N18.101, W71.194)	KC615699	KC615911	KC615699
Antillattus cubensis	CU003076 CU002975 CU003097 CU003360 CU002456 CU003486 CU02560A CU02975A CU03033A CU03076A CU03097A CU03360A	CUBA: Cienfuegos, Soledad, (N22.124, W80.325)	√		
Antillattus cubensis	CU03417A CU03488A	CUBA: Santiago de Cuba, San Luis, (N20.179, W75.783)	√		
Antillattus cubensis	CU3075A	CUBA: Santiago de Cuba, (N20.010, W76.037)	√		
Antillattus cubensis	CU02583A	CUBA: Guantánamo, Baracoa, (N20.331, W74.569)	√		
Antillattus cubensis	CU787598 CU783280 CU787621 CU787283 CU787277	CUBA: Granma, Bartolomé Maso, (N20.009, W76.894)	√		
Antillattus darlingtoni	JXZ341	DOMINICAN REPUBLIC: La Vega, Ébano Verde, (N19.033,W70.543)	KC615762	KC616005	KC615583
Antillattus darlingtoni	DR787120	DOMINICAN REPUBLIC: San Juan, Pico Duarte	√	√	

Table 1. *Cont.*

Species	Voucher	Locality	CO1	16S-ND1	28S
Antillattus darlingtoni	DR786937 DR784873	DOMINICAN REPUBLIC: Valle nuevo	√	√	
Antillattus darlingtoni	DR784828 DR784873	DOMINICAN REPUBLIC: La Vega, Ébano Verde, (N19.026, W19.0264)	√	√	
Antillattus gracilis	JXZ320	DOMINICAN REPUBLIC: La Vega, P.N.Armando Bermúdez, (N19.06, W70.86)		KC615817	KC615391
Antillattus gracilis	DR782845 DR787278	DOMINICAN REPUBLIC: Santo Domingo, Los Tablones (N19.051, W70.888)	√	√	
Antillattus keyserlingi	CU03135A	CUBA: Holguin, Frank Pais, (N20.529N, N75.768)	√		
Antillattus keyserlingi	CU02571A	CUBA: Santiago de Cuba, Gran Piedra, (N20.011, W75.623)	√		
Antillattus keyserlingi	CU787312	CUBA: Guantánamo, Baracoa, (20.331N, W74.569)	√		
Antillattus keyserlingi	CU00081A CU00088A CU02951A CU02985A CU03043A CU782822 CU783187 CU783232 CU783245 CU783281 CU783404 CU783425 CU787302 CU787433 CU787625	CUBA: Granma, Bartolomé Maso, (N20.052, W76.502)	√	√	√
Antillattus keyserlingi	CU02467A CU03538A CU03395A	CUBA: Holguin, Frank Pais,(N20.529, W75.768)	√	√	√
Antillattus keyserlingi	CU03036A CU03274A	CUBA: Granma, Bartolomé Maso, (N20.015–W76.839)	√		
Antillattus maxillosus	JXZ335	DOMINICAN REPUBLIC: La Vega, road Constanza to Ocoa, Valle Nuevo (N18.700, W70.606)	KC615708	KC615935	KC615510
Antillattus maxillosus	DR786952 DR786992 DR786981	DOMINICAN REPUBLIC: Valle nuevo, Villa Pajón (N18.82208, W070.6838)	√	√	
***Antillattus* [Cuba5]**	CU03373A CU03396A CU03539A CU03534A	CUBA: Pinar del Rio, Viñales, (N22.653, W83.699)	√		
Antillattus placidus	DR787249	DOMINICAN REPUBLIC: La Vega, Jarabacoa, (N19.036, W70.543)	√	√	

Table 1. *Cont.*

Species	Voucher	Locality	CO1	16S-ND1	28S
Antillattus placidus	DR782502 DR785683 DR785081	DOMINICAN REPUBLIC: La Alta Gracia, Yuma, (N19.355, W70.111)	√	√	
Antillattus scutiformis	JXZ326	DOMINICAN REPUBLIC: La Vega, road Constanza to Ocoa, Valle Nuevo (N18.848, W70.720)		KC615860	KC615433
Bythocrotus cf. crypticus	JXZ323	DOMINICAN REPUBLIC: El Seibo, Pedro Sanchez, (N18.86, W69.11)	KC615661	KC615839	KC615412
Bythocrotus crypticus	JXZ322	DOMINICAN REPUBLIC: Barahona, (N18.424, W71.112)	KC615660	KC615838	KC615411
Cobanus cambridgei	JXZ122	COSTA RICA: Prov. San José, (N9.65, W83.97)		KC615872	KC615445
Cobanus extensus	JXZ122	ECUADOR: Pichincha, near El Cisne, (N0.1493, W79.0317)		KC615872	KC615445
Cobanus mandibularis	JXZ245	PANAMA: Panamá: Gamboa, Pipeline Road, (N9.15840, W79.74252)		KC615876	KC615449
Cobanus unicolor	JXZ244	PANAMA: Chiriqui: Fortuna, Quebrada Samudio, (N8.73464, W82.24839)		KC615878	KC615451
Compsodecta festiva	JAM4122A	JAMAICA: Portland, Millbank, (N18.013, W76.379)	√		
Compsodecta haytiensis	JXZ325	DOMINICAN REPUBLIC: Barahona, Highway 44 south of Barahona (N18.138, W71.070)	KC615671	KC615859	KC615432
Compsodecta peckhami	JXZ327	DOMINICAN REPUBLIC: Pedernales, Rio Mulito (N18.155, W71.758)		KC615884	KC615457
Corticattus guajataca	JXZ305	PUERTO RICO: Isabela: Bosque de Guajataca (N18.421, W66.966)	KC615715	KC615945	KC615521
Corticattus latus	JXZ337	DOMINICAN REPUBLIC: Pedernales: Laguna de Oviedo (N17.802 W71.349)	KC615698	KC615908	KC615483
Mexigonus arizonenzis	JXZ163	USA: Arizona: Yavapai Co., Iron Springs (N34.58476, W112.57071)	KC615747	KC615988	KC615564
Mexigonus cf. minuta	d117	ECUADOR: Pichincha: Quito	√	√	√
Mexigonus morosus	JXZ362	USA: California: San Mate Co.,(N37.434, W122.311)		KC615990	KC615566

Table 1. *Cont.*

Species	Voucher	Locality	CO1	16S-ND1	28S
Pensacola signata	JXZ371	GUATEMALA: Depto. Petén, Reserva Natural Ixpanpajul		KC616006	KC615584
Petemathis portoricensis	PR782206	PUERTO RICO: Villalba: Toro negro, El Bolo Trail (N18.1777401, W66.488319)	√	√	
Petemathis portoricensis [Adjuntas]	JXZ306	PUERTO RICO: Adjuntas, HWY143 to Cerro Punta (N18.167, W66.576)	KC615716	KC615946	KC615522
Petemathis portoricensis [Maricao]	JXZ303	PUERTO RICO: Maricao, Bosque de Maricao (N18.150, W66.994)	KC615711	KC615940	KC615515
Petemathis tetuani	JXZ303	PUERTO RICO: Maricao, Bosque de Maricao (N18.150, W66.994)	KC615711	KC615940	KC615515
Petemathis tetuani	PR782277	PUERTO RICO: Villalba: Toro negro, El Bolo Trail, (N18.177, W66.488)	√	√	
Petemathis tetuani	PR392859	PUERTO RICO: Rio Grande, El Yunque, Mt. Britton, (N18.2957, W65.7906)	√	√	
Popcornella furcata	JXZ334	DOMINICAN REPUBLIC: La Vega, Reserva Científica Ébano Verde, (N19.04, W70.518)	KC615714	KC615944	KC615520
Popcornella spiniformis	JXZ339	DOMINICAN REPUBLIC: Barahona, Cachote (N18.098, W71.187)		KC615914	KC615489
Popcornella yunque	JXZ309	PUERTO RICO: Río Grande, El Yunque Nat. Forest, (N18.3174, W65.8314)		KC615937	KC615512
Sidusa [French guiana1]	JXZ128	FRENCH GUIANA: Commune Règina, les Nourages Field Station (N4.069, W52.669)	KC615770	KC616015	KC615593
Sidusa [French guiana2]	JXZ100	FRENCH GUIANA: Commune Règina, les Nourages Field Station, (N4.069, W52.669)	KC615679	KC615871	KC615444
Truncattus [Cuba1]	CU3492A	CUBA: Granma, Bartolomé Maso, National Park Pico Turquino (N 20.0526, W76.502)	√		
Truncattus [Cuba2]	CU787947 CU03405A	CUBA: Granma, Bartolomé Maso, National Park Pico Turquino (N20.0526, W76.5029)	√	√	√

Table 1. *Cont.*

Species	Voucher	Locality	CO1	16S-ND1	28S
Truncattus [Cuba3]	CU787949 CU00083A CU03065A	CUBA: Granma, Bartolomé Maso, National Park Pico Turquino (N20.0526, W76.5029)	√		
Truncattus [Cuba4]	CU00014A	CUBA: Granma, Bartolomé Maso, National Park Pico Turquino (N20.052, W76.502)	√		
Truncattus [Dominican Republic1]	DR787029	DOMINICAN REPUBLIC: Valle nuevo, Villa Pajón, (N18.82208, W070.6838)	√		
Truncattus cachotensis	JXZ338	DOMINICAN REPUBLIC: Barahona, Cachote, (N18.101, W71.194)	KC615701	KC615913	KC615488
Truncattus dominicanus	JXZ340	DOMINICAN REPUBLIC: La Vega, P.N.Armando Bermúdez,(N19.06, W70.86)	KC615703	KC615920	KC615495
Truncattus dominicanus	DR787325	DOMINICAN REPUBLIC: San Juan, Los tablones,(N19.0511, W70.888)	√	√	
Truncattus flavus	JXZ332	DOMINICAN REPUBLIC: La Vega, P.N.Armando Bermúdez, (N19.06, W70.86)	KC615707	KC615933	KC615508
Outgroups					
Agobardus anormalis montanus	JXZ357	DOMINICAN REPUBLIC: Pedernales, (N18.128, W71.558)	KC615636	KC615802	KC615376
Agobardus bahoruco	JXZ324	DOMINICAN REPUBLIC: Pedernales, (N18.128, W71.558)		KC615844	KC615417
Agobardus cf. brevitarsus	JXZ311	DOMINICAN REPUBLIC: Pedernales, (N18.128, W71.558)	KC615637	KC615803	KC615637
Agobardus cordiformis	JXZ358	DOMINICAN REPUBLIC: Pedernales, (N17.965, W71.635)	KC615634	KC615800	KC615374
Agobardus gramineus	JXZ314	DOMINICAN REPUBLIC: Pedernales, (N17.965, W71.635)	KC615635	KC615801	KC615375
Agobardus oviedo	JXZ312	DOMINICAN REPUBLIC: Pedernales, (N17.802, W71.349)	KC615638	KC615804	KC615378
Bythocrotus cf. crypticus	JXZ323	DOMINICAN REPUBLIC: El Seibo, Pedro Sanchez, (N18.86, W69.11)	KC615661	KC615839	KC615412
Bythocrotus crypticus	JXZ322	DOMINICAN REPUBLIC: Barahona, (N18.424, W71.112)	KC615660	KC615838	KC615411

Table 1. *Cont.*

Species	Voucher	Locality	CO1	16S-ND1	28S
Cobanus cambridgei	JXZ122	COSTA RICA: Prov. San José, (N9.65, W83.97)		KC615872	KC615445
Cobanus extensus	JXZ122	ECUADOR: Pichincha, near El Cisne, (N0.1493, W79.0317)		KC615872	KC615445
Cobanus mandibularis	JXZ245	PANAMA: Panamá: Gamboa, Pipeline Road, (N9.15840, W79.74252)		KC615876	KC615449
Cobanus unicolor	JXZ244	PANAMA: Chiriqui: Fortuna, Quebrada Samudio, (N8.73464, W82.24839)		KC615878	KC615451
Compsodecta festiva	JAM4122A	JAMAICA: Portland, Millbank, (N18.013, W76.379)	√		
Compsodecta haytiensis	JXZ325	DOMINICAN REPUBLIC: Barahona, Highway 44 south of Barahona (N18.138, W71.070)	KC615671	KC615859	KC615432
Compsodecta peckhami	JXZ327	DOMINICAN REPUBLIC: Pedernales, Rio Mulito (N18.155, W71.758)		KC615884	KC615457
Corticattus guajataca	JXZ305	PUERTO RICO: Isabela: Bosque de Guajataca (N18.421, W66.966)	KC615715	KC615945	KC615521
Corticattus latus	JXZ337	DOMINICAN REPUBLIC: Pedernales: Laguna de Oviedo (N17.802, W71.349)	KC615698	KC615908	KC615483
Mexigonus arizonenzis	JXZ163	USA: Arizona: Yavapai Co., Iron Springs (N34.58476, W112.57071)	KC615747	KC615988	KC615564
Mexigonus cf. minuta	d117	ECUADOR: Pichincha: Quito	KC615748	KC615989	KC615565
Mexigonus morosus	JXZ362	USA: California: San Mate Co.,(N37.434, W122.311)		KC615990	KC615566
Pensacola signata	JXZ371	GUATEMALA: Depto. Petén, Reserva Natural Ixpanpajul		KC616006	KC615584
Popcornella furcata	JXZ334	DOMINICAN REPUBLIC: La Vega, Reserva Científica Ébano Verde, (N19.04, W70.518)	KC615714	KC615944	KC615520
Popcornella spiniformis	JXZ339	DOMINICAN REPUBLIC: Barahona, Cachote (N18.098, W71.187)		KC615914	KC615489
Popcornella yunque	JXZ309	PUERTO RICO: Río Grande, El Yunque Nat. Forest, (N18.3174, W65.8314)		KC615937	KC615512

Table 1. *Cont.*

Species	Voucher	Locality	CO1	16S-ND1	28S
Sidusa [French guiana1]	JXZ128	FRENCH GUIANA: Commune Règina, les Nourages Field Station (N4.069, W52.669)	KC615770	KC616015	KC615593
Sidusa [French guiana2]	JXZ100	FRENCH GUIANA: Commune Règina, les Nourages Field Station, (N4.069, W52.669)	KC615679	KC615871	KC615444
Ghelna canadensis	d005	USA: North Carolina (N35.704, W82.373)	EF201651	JQ312080	KT462689

3. DNA Extraction, Amplification and Sequencing

DNA was isolated with a Qiagen DNeasy Tissue Kit (Qiagen, Valencia, CA, USA). We sequenced fragments of CO1, 16S-ND1, and 28S. We amplified CO1 using the LCO1490 (GGTCAACAAATCATAAAGATATTGG) [82] and C1–N–2776 (GGATAATCAGAATATCGTC-GAGG) [83] primers. The fragment of 16S-ND1 ribosomal RNA was amplified with the primers 16SA/12261 (CGCCTGTTTACCAAAAACAT) [82] and 16SB (CCGGTTTGAACTCA-GATC) [83]. The 28S ribosomal RNA fragment was amplified with the 28SO (TCGGAAG-GAACCAGCTACTA) and 28SC (GAAACTGCTCAAAGGTAAACGG) primers. For CO1, 16S-ND1, and 28S, the polymerase chain reactions (PCR) were performed with an initial denaturation at 94 °C for 2 min, followed by 40 cycles of denaturation at 94 °C for 25 s, annealing at 50 °C (first round)/44.5 °C (second round) for 25 s and extension at 65 °C for 2 min (first round)/1 min (second round); with a final extension at 72 °C for 10 min. We sequenced amplified fragments in both directions using Sanger sequencing at GENEWIZ's New Jersey facility. The forward and reverse reads were interpreted with Phred and Phrap [84,85] via Chromaseq v. 1.31 [86] in Mesquite v. 3.6 [87] using default parameters.

3.1. Phylogenetic Inference

We aligned sequences in MAFFT [88] using L–INS–I with a parameter 1PAM/k = 200, and a Gap opening penalty of 1.53. Gaps were treated as missing characters. The data resulting from the alignments were manually reviewed in Mesquite 3.6 (Maddison and Maddison [87]) with reference to the translation of amino acids using the "Color Nucleotide by Amino Acid" option. The dataset was partitioned by gene (and in the case of CO1 by codon), and the appropriate substitution model for each partition was selected with jModeltest 2.1.10 [89] using the Akaike information criterion [90] to select among the 24 models that can be implemented in MrBayes (Supplementary Table S1).

Maximum likelihood analyses were conducted in IQ–TREE v.2.0 [91]. ModelFinder [92], as implemented in IQ–TREE v.2.0 [91], was used to select the optimal partition scheme and substitution models for the molecular characters (iqtree–s dataMatrix.nex—runs 1000–m TESTMERGEONLY–spp setsBlock.nex–pre iqtreeAnalysis–nt AUTO). Finally, we used the CIPRES online portal [93,94] to run a Bayesian analysis with MrBayes v. 3.2.6 [95,96]. We ran the Markov chain Monte Carlo (MCMC) with four chains for 25,000,000 generations, sampling every 1000 generations, with a sampling frequency of 100 and a burn–in of 25%. The results were examined in Tracer v.1.7 [97] to verify proper mixing of chains, that stationarity had been reached, and to determine adequate burn-in. All resulting trees were interpreted in FIGTREE v.1.4.2 and edited in Adobe Illustrator CS6.

3.2. Time Calibration and Divergence Estimation

For the divergence time estimation analysis, the monophyly of *darlingtoni* group was constrained based on the results of the Bayesian and ML analyses. Node ages were estimated using a Bayesian, multi-gene approach in BEAST 1.10.4 [98] using a two-tier approach: (1) including outgroups, (2) excluding outgroups. Here, for the divergence estimation, we included as outgroups the South American representatives of *Pensacola-Mexigonus* clade (*Mexigonus* cf. *minuta*, *M. arizonensis*), and *Sidusa* clade and the Greater Antilles *Agobardus* clade (*Agobardus*, *Compsodecta* and *Bythocrotus*).

The dating analyses were run under a lognormal relaxed clock model [99] with a CO1 substitution rate parameter (ucld.mean) as a normal prior (mean = 0.0112 and s.d. = 0.001) [100] and an estimated substitution rate parameter for 28S and 16S-ND1. The lognormal relaxed clock model was selected between alternative clock models (non–clock, strict clock, relaxed clock) using a stepping-stone method [101] of Bayes Factors in MrBayes 3.2.7a [96,102]. The analysis ran for 20,000,000 generations with a birth–death process [103] under a GTR + G+I model, with default options for all other prior and operator settings. The birth–death model was used for the tree prior because it can simulate speciation and extinction rates over time; thus, at any point in time, every lineage can undergo speciation at rate λ or go extinct at rate μ [104].

We used a combination of calibrations with fossils and calibrations based on the results of Zhang and Maddison [68]. Our fossil calibration point is based on the Dominican amber genus *Pensacolatus* (type species *Pensacolatus coxalis* Wunderlich, 1988 [105]) (see Penney, [106]). Wunderlich [105] described *Pensacolatus* based on a Dominican amber fossil (20–15 Mya) and discusses similarity with the species described by Bryant [79] as *Pensacola* (Peckham and Peckham [107]). We confidently place *Pensacolatus coxalis* within the *Antillatus darlingtoni* group after thorough review of the original description of *P. coxalis* and comparison of morphological details with those compiled for taxa in this lineage in Zhang and Maddison [81]. Key characteristics in this assessment include one retromarginal tooth, post-epigastrium without a visible pre-spiracular bump, endite with an anterolateral cusp, palp with a proximal tegular lobe, and ventral tibial apophysis. Therefore, we use this fossil to calibrate the MRCA (Most Recent Common Ancestor) of the *darlingtoni* group (logNormal Prior [tmrca, mu = 0.01, sigma = 1.0, offset = 16]) (see [68,105]). Our second calibration is MRCA of *Antillattus* clade secondarily based on dating inferences within this linage from Zhang [108] [tmrca, normalPrior mean = 27.24 stdev = 5.0]. The convergence of parameters was examined in Tracer 1.7 [97] to determine burn–in and to check for stationarity. The maximum clade credibility tree was produced in TreeAnnotator v1.10.4, with 25% burn-in.

4. Biogeographical Estimation

For ancestral range estimation of the *Antillattus* clade, we used the tree of the divergence dating analysis resulting from the first tier approach (analysis with outgroups). We coded the Caribbean islands in their past shape, considering their historical composition of multiple paleo-islands [32]. The distribution ranges were divided into the following areas: A—Puerto Rico, B—Hispaniola, C—Cuba, D—Jamaica, E—North America, F—South America (Figure 1). We carried out the ancestral range estimation in the R package BioGeoBEARS v. 1.1.1 [109,110] to test different time periods and infer which are more likely with base of the model's configuration. This package tests three models in a maximum likelihood framework with various parameters that can be altered to test specific scenarios: a DEC model [110,111], a DIVALIKE model (likelihood version of the DIVA model [111,112]) and a BAYAREALIKE model (likelihood version of the BayArea model [113]). Moreover, each model is available in its original version and with an additional parameter +*j* (i.e., peripatric speciation) representing jump dispersal, or a founder event, which is speciation following long-distance dispersal [111].

To estimate the ancestral range distribution for *Antillattus* clade and outgroups, we conducted time-stratified analyses testing (1) non–GAARlandia (overwater dispersal), and (2) GAARlandia as the *Antillattus* clade ancestor colonization route using a set of 36 models that varied in the parameters [e—the rate of range contraction, d—the base rate of range expansion, and j—the weight of founder-event speciation at cladogenesis] and in the configuration of dispersal multiplier matrices used [109]. To estimate the ancestral range distribution among *Antillattus* clade without outgroups, we conducted time-stratified analyses testing (1) overwater dispersal, and (2) land connections prior to the opening of the Mona Passage and the Windward Passage using a set of 72 models that varied in the parameters and in the configuration of dispersal multiplier matrices used [109]. In both approaches, we tested three dispersal probability hypotheses: (a) the dispersal probability decreases with distance, (b) dispersal probability is independent of distance, and (c) the probability of overwater dispersal is essentially zero (Table 2) (see Crews and Esposito [52]). Dispersal probabilities were set as follows: they were set to 0.8 when two areas were adjacent, to 0.5 when two areas were weakly separated by a geographical barrier, to 0.2 when two areas were separated by water over a distance less than 200 km, to 0.05 for connection by island chain (e.g., Lesser Antilles) or intermediate island (e.g., Hispaniola between Cuba and Puerto Rico), to 0.001 for long-distance dispersal (areas separated by more than 200 km from sea), and to 0.0000001when dispersal was not possible by the lack area availability (we followed the BioGeoBEARS manual in setting extremely low rather than zero probabilities). Time periods were defined as follows to reflect the paleogeography of the area in each period [5,17,24]: (1) 23–15 Mya: Windward Passage, (2) 30–23 Mya: Mona Passage, (6) 32–35: GAARlandia hypothesis [32,33].

Table 2. Biogeographic specific scenarios analyzed in BioGeoBEARS for (a) *Antillattus* clade and outgroups and (b) *Antillattus* clade without outgroups. Each dispersal or vicariance scenario was tested using the six models available in BioGeoBEARS (DEC, DEC+J, DIVALIKE, DIVALIKE+J, BAYAREALIKE, BAYAREALIKE+J). Abbreviations: MO, Mona passage; WI, Windward passage.

	(1) Non-GAARlandia/(2) GAARlandia		
	A: Dispersal probability decreases as distance increases	B: Distance does not affect dispersal probability	C: Probability of overwater dispersal is very low
(A) GA1	A1a/A2a	A1b/A2b	A1c/A2c
	(1) Non–land connections/(2) Land connections		
	A: Dispersal probability decreases as distance increases	B: Distance does not affect dispersal probability	C: Probability of overwater dispersal is very low
(A) MO	A1a/A2a	A1b/A2b	A1c/A2c
(B) MO+WI	B1a/B2a	B1b/B2b	B1c/B2c

The +j parameter represents an approximation to model dispersal–dominated systems [109,114]; however, the validity of comparing models with and without +j parameter is controversial [115,116]. To conservatively address these issues [108,114–116], we use the best-fitting basic model and the best fitting model with +j parameter to discuss the ancestral range estimation, to estimate the number of lineages through time by area, and the number and type of biogeographical events [extinction, speciation (sympatric–subset speciation, within–area speciation, founder–event speciation), vicariance and dispersal events (anagenetic dispersal, range–expansion dispersal)]. Both the basic model with +j parameter were compared using likelihood values and the Akaike information criterion corrected for small sample sizes (AICc) [117]. Finally, to estimate the number of lineages through time, and the number and type of biogeographical events, we used the best model resulted in the analysis of *Antillattus* clade without outgroups. We ran biogeographical stochastic mapping (BSM) using the maximum clade credibility (MCC) tree [118,119]. Event frequencies were estimated by taking the mean and standard deviation of event counts from 100 BSMs.

5. Results

Phylogeny and Divergence Time

The combined molecular dataset consisted of 3071 sites (27,369 internal gaps), the best BI tree has a harmonic-means = −24208.96, and the best ML tree has an lnL = −24,214.758 (Figure 4). The *Antillattus* clade is supported as monophyletic (ML, bootstrap = 100%). The phylogeny suggests that the *Antillattus* clade is sister to other Caribbean (e.g., *Agobardus* clade) (ML, bootstrap = 88%) and continental clades (e.g., *Mexigonus-Pensacola* clade). The relationships among the three genera in the *Antillattus* clade are not well resolved. The genus *Petemathis* is resolved as sister to *Truncattus* + *Antillattus* with low support (bootstrap = 73%, pp =0). A second analysis without outgroups (Figure 4, lnL = −13,533.817, Harmonic–means = −13,597.32) support *Petemathis* as sister to *Truncattus* + *Antillattus* (ML, bootstrap = 100%, BI, pp = 1.0), while *Truncattus* is poorly resolved as sister to *Antillattus* (ML, bootstrap = 0%, BI, pp = 0.68). In both analyses, the genus *Antillattus* is monophyletic, however, the relationships within the genus are not well resolved. The representatives of the genus *Antillattus* were divided into three groups of species that we refer to as the *darlingtoni*, *keyserlingi*, and *gracilis* groups, with *gracilis* sister to the other two. The phylogeny recovered the genus *Petemathis*, the *darlingtoni* group, and the *keyserlingi* group as single-island endemic lineages. The *gracilis* group and *Truncattus* are found both on Hispaniola and Cuba.

In both BEAST analyses (including outgroups and excluding outgroups), the posterior probability values from our BEAST analyses are higher than those in the MrBayes analysis (Figures 4 and S1). For example, the genus *Truncattus* is recovered as sister group of the genus *Antillattus* with better support values (pp = 0.91). The chronogram of the *Antillattus* clade based on the birth–death process derived chronogram with a relaxed clock model (Figure 5), indicates that the MRCA of the *Antillattus* clade diverged during the Oligocene (c. 25 ± 3 Mya), and most of the subsequent divergences happened in the Miocene to present (c. present–21 Mya). The lineage leading to *Petemathis* diverged during the late Oligocene (c. 25 ±3 Mya). The divergence of the lineages leading to *Truncattus*, and the genus *Antillattus* were dated to the early Miocene (c. 21 ± 3 Mya and c. 19 ± 2 Mya respectively). Finally, the lineages leading to the *gracilis*, *keyserlingi*, and *darlingtoni* groups were dated to the early Miocene (c. 19 ± 2 Mya and c. 17 ± 2 Mya, respectively).

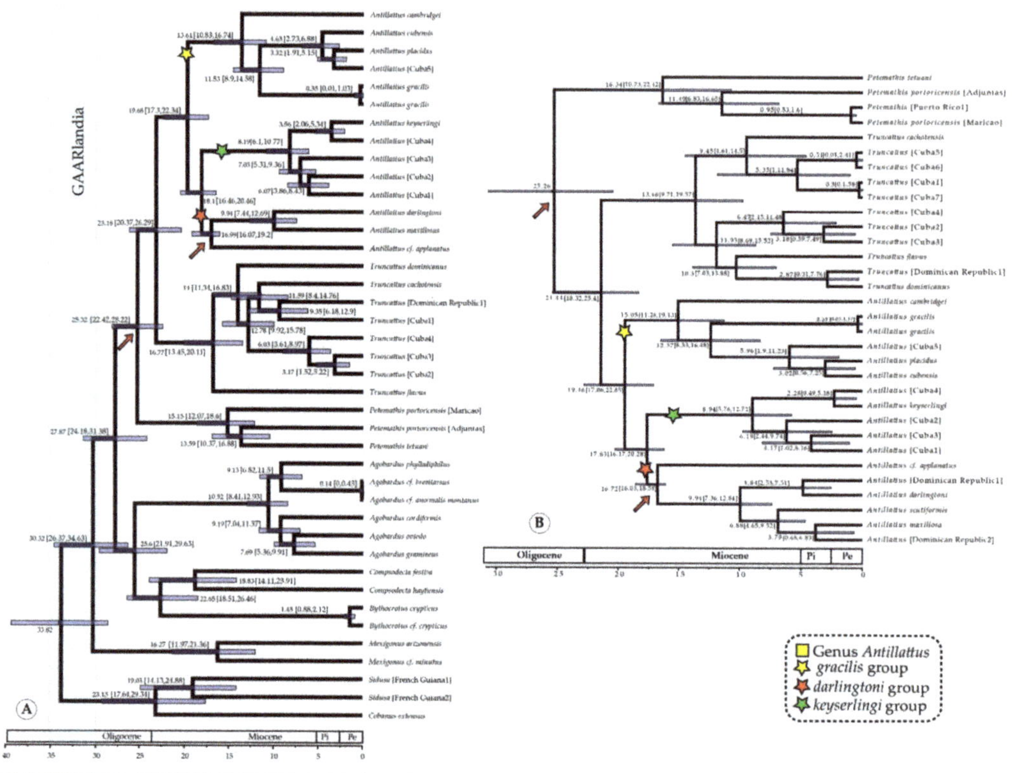

Figure 5. Beast divergence time estimations of all genes (CO1, 16S-ND1, 28S) using a Bayesian relaxed molecular clock (**A**) with outgroups and (**B**) without outgroups. The scale is in millions of years. Bars show 95% HPD [highest posterior density]. Stars indicate species groups within the genus *Antillattus* (blue star *gracilis* group, red star *darlingtonia* group, and green star *keyserlingi* group). Arrows indicate calibrated nodes.

6. Model Selection and Ancestral Range Estimation

The A2a DEC+J model (log likelihood: LnL = −33.87; parameter estimates: d = 0; e = 0; j = 0.22) and the A2a DEC model (log likelihood: LnL = −43.84; parameter estimates: d = 0.022; e = 0; j = 0) (Table 3, Supplementary Data S3) are consistent with the GAARlandia, and dispersal probability decreasing with distance. Both the basic and +j models resolve the most probable ancestral area for the extant species of the *Antillattus* clade is Northern South America and Hispaniola. The estimation of ancestral ranges among *Antillattus* clade, show that the favored model was the B2a DIVALIKE +j model (log likelihood: LnL = −17.6; parameter estimates: d = 0; e = 0; j = 0.29), while the best model within the basic models was the B2a DIVALIKE model (log likelihood: LnL = −25.01; parameter estimates: d = 0.048; e = 0; j = 0) (Table 3, Supplementary Data S3). Both models are consistent with the land connections prior to the Mona Passage and the Windward Passage hypothesis, and dispersal probability decreasing with the distance. Both the basic and +j models show again Hispaniola as a probable ancestral area (Figure 6).

Table 3. BioGeoBEARS' relative model probabilities for non-time-stratified analyses and time–stratified analyses corresponding to the most likelihood specific scenario A2a of the 12 specific scenarios tested for the non-GAARlandia and GAARlnadia hypotheses, and B2a of the 12 specific scenarios tested for the overwater dispersal and land connections prior to the Mona Passage and the Windward Passage hypotheses. The best performing model is marked with an asterisk for groups of analyses. LnL = log likelihood; n par = number of parameters in the analysis; d, e, j = parameters of the model (d = dispersal, e = extinction, j = founder event); AIC = Aikake information criterion; AICc = size–corrected AIC. * = Best-performing model for each groups of analyses.

	LnL	n par	d	e	j	AIC	AICc
Time–Constrained/GAARlandia (A2a)							
DEC	−43.84 *	2	0.021	<0.0001	0	91.69	92.01
DEC +J	−33.87 *	3	<0.0001	<0.0001	0.22	73.73	74.4
DIVALIKE	−47.05	2	0.035	<0.0001	0	98.1	98.43
DIVALIKE +J	−36	3	0.005	<0.0001	0.22	77.99	78.66
BAYAREALIKE	−60.83	2	0.021	0.031	0	125.7	126
BAYAREALIKE +J	−36.73	3	0.0038	<0.0001	0.21	79.45	80.12
Time–constrained/land connections prior to the Mona Passage and the Windward Passage (B2a)							
DEC	−25.1	2	0.033	<0.0001	0	54.19	54.61
DEC +J	−18.09	3	<0.0001	<0.0001	0.31	42.18	43.04
DIVALIKE	−25.01 *	2	0.048	<0.0001	0	54.01	54.43
DIVALIKE +J	−17.62 *	3	<0.0001	<0.0001	0.29	41.25	42.1
BAYAREALIKE	−34.22	2	<0.0001	0.041	0	72.44	72.86
BAYAREALIKE +J	−18.8	3	<0.0001	<0.0001	0.27	43.61	44.46

Estimation of Biogeographical Events

The DIVALIKE and DIVALIKE +j BioGeoBEARS stochastic map (BSM) based on 100 stochastic historical maps revealed that most probabilistic biogeographical events comprise within-area speciation (between 65 and 76% of probabilistic events in stochastic runs), founder-event (21%), range-expansion dispersal (15%), and vicariance (between 18 and 3%) (Table 4, Figure 6). The high number of within-area speciation probabilistic events in Hispaniola (between 43 and 48% of the total of the DIVALIKE and DIVALIKE +j within-area speciation probabilistic events), Cuba (between 40 and 44%), and Puerto Rico (between 12 and 13%) could be closely related to species richness. Most of the probabilistic estimated vicariance events among Cuba, Puerto Rico, and Hispaniola involved Hispaniola–Cuba (between 84 and 27% of the DIVALIKE and DIVALIKE +j vicariance probabilistic events), Hispaniola–Cuba–Puerto Rico (between 2 and 16%), Cuba–Puerto Rico (between 2 and 32%), and Hispaniola–Puerto Rico (between 1 and 26%).

Dispersal events are represented by range-expansions and founder-events (Table 4). Focusing on the range-expansion between Puerto Rico, Cuba, and Hispaniola, we found that the movement patterns varied enormously between areas and were highest among groups that have their ancestral range in Hispaniola (73% of the range-expansion probabilistic events and 45% of the founder-events probabilistic events). Range-expansion events only involved movements from Hispaniola–Cuba (73% of the range-expansion probabilistic events), and from Hispaniola–Puerto Rico (27%), while the range–expansion from Puerto Rico to Cuba and Hispaniola, and Cuba and Hispaniola to Puerto Rico were improbable (0% of simulations) (Supplementary Data S4). In contrast to the range–expansion events, founder-event speciation occurred in lineages that have their ancestral range in Cuba (51%), and the highest number of founder-event speciation involved movements from Cuba to Hispaniola (49% of the founder-event probabilistic events) and Hispaniola to

Cuba (45%), with other events playing little or no role (less 2%). Finally, the number of lineages estimated through time by zone showed the occurrence of a greater number of events [within-area speciation, dispersions, and vicariances] representing movement from Hispaniola (Figure 7).

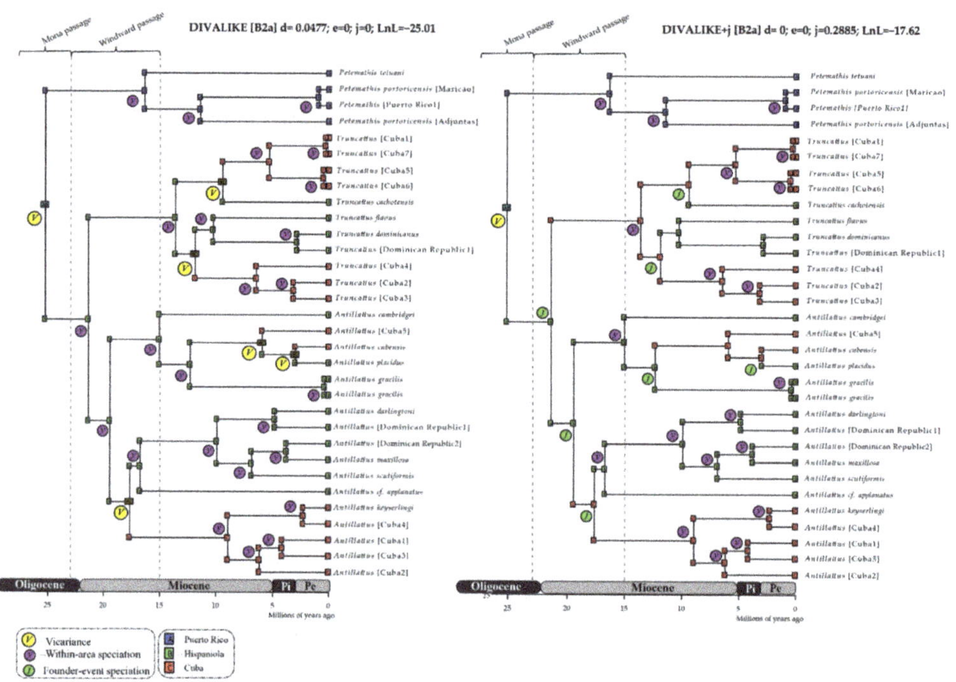

Figure 6. BioGeoBEARS phylograms corresponding to the time-stratified DIVALIKE and DIVA-LIKE+J B2a model (Table 1). The trees show the most probable geographic range pre- and post-split. The scale is in millions of years.

Table 4. Summary count of time-constrained Biogeographic Stochastic Mappings. DIVALIKE and DIVALIKE +j. Abbreviations: j, jump dispersal or founder-event speciation; a, range-switching dispersal; d, range-expansion dispersal; e, extinction; s, sympatric-subset speciation; v, vicariance; y, within-area speciation; Ÿd, allopatric dispersal; Ad, anagenetic dispersal; Ÿa: allopatric anagenetic; Ÿc: allopatric cladogenetic; sums, adds up all of the events across the stochastic maps.

DIVALIKE												
	j	a	d	e	s	v	y	Ÿd	Ad	Ÿa	Ÿc	Total events
means	0	0	5.5	0	0	6.61	24.39	5.5	5.5	5.5	31	36.5
stdevs	0	0	0.64	0	0	0.67	0.67	0.64	0.64	0.64	0	0.64
sums	0	0	550	0	0	661	2439	550	550	550	3100	3650

DIVALIKE+j												
	j	a	d	e	s	v	y	Ÿd	Ad	Ÿa	Ÿc	Total events
means	6.63	0	0	0	0	0.82	23.55	6.63	0	0	31	31
stdevs	1.04	0	0	0	0	0.64	0.94	1.04	0	0	0	0
sums	663	0	0	0	0	82	2355	663	0	0	3100	3100

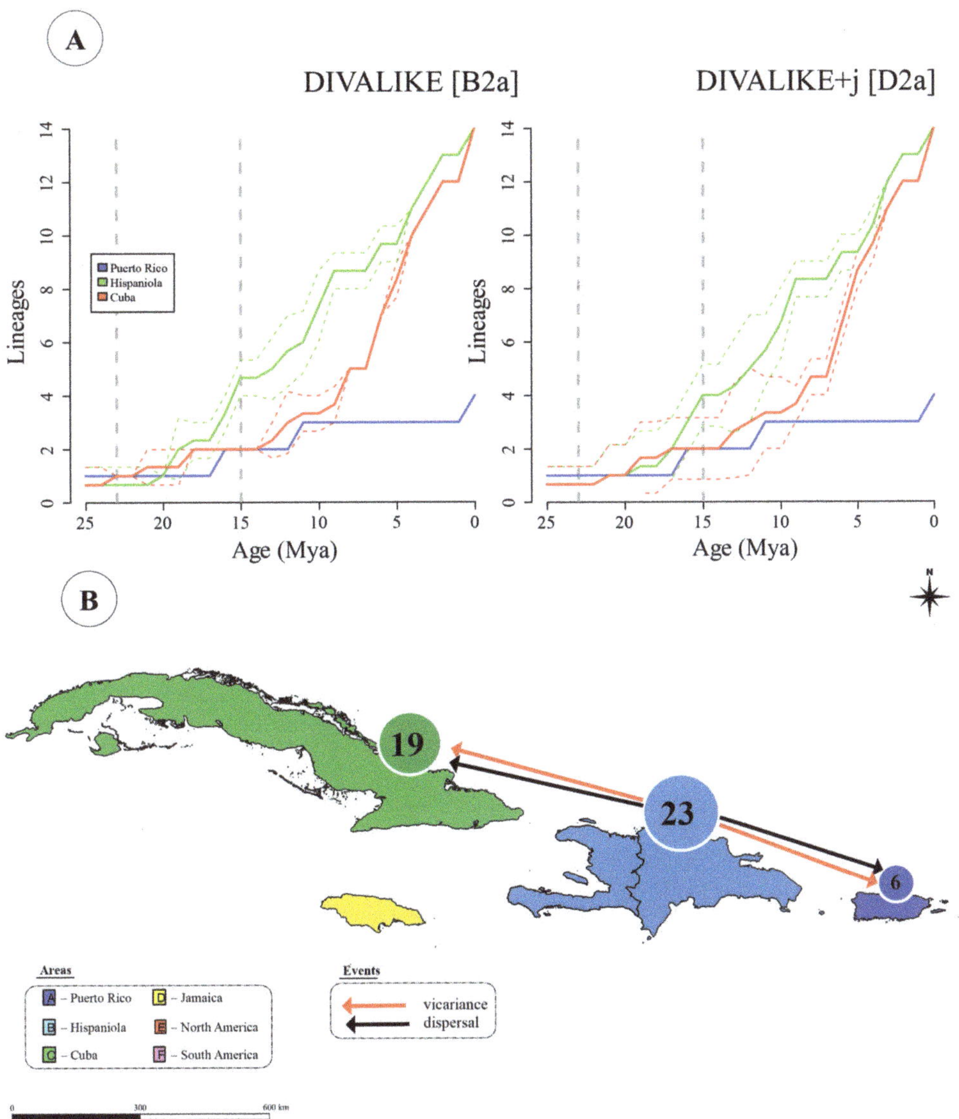

Figure 7. (**A**) Number of lineages occupying each area through time estimates under DIVALIKE and DIVALIKE +j and (**B**) the ancestral range. (**A**) Colored solid lines are the average of 100 biogeographic stochastic maps. Colored dashed lines are the 95% confidence interval. Gray dashed vertical line indicates the boundary between time slices in the time-stratified analysis. (**B**) Black arrows represent all dispersal events between areas. Red arrows represent a vicariance event between areas. Numbered circles indicate the inferred number of within-area speciation in each of the areas.

7. Discussion

7.1. Ancestral Range of the Antillattus Clade and GAARlandia

Our expanded Caribbean sampling within the *Antillattus* clade provides a more thorough analysis of diversification within the Caribbean and an opportunity to reassess its biogeographic origins. Our data suggest that the ancestor of the *Antillattus* clade colonized the Greater Antilles once from South America within a time frame consistent with GAARlandia; however, our continental outgroup taxon sampling limits confidence in inference of the source [32,33] (Figure 4). The estimated time of divergence of the *Antillattus* clade (25 ± 3 Mya) is consistent with dates inferred by Zhang and Maddison [68]. Neither the colonization of the Caribbean by *Antillatus* clade ancestors, nor the diversification of the group can in any way be linked to the colonization of the proto-Antillean volcanic arc in the Late Cretaceous (c. 65.5 Mya) [39,40].

7.2. Inter-Island Biogeographical History

The phylogenetic structure within the Greater Antilles reflects patterns consistent with historical island connectivity and breakup. Our estimation of the biogeographic history identified speciation within the Caribbean as the driving force of diversification (Table 4), consistent with the high levels of endemism in these spiders. For example, the species of the *darlingtoni* group are restricted to Hispaniola suggesting diversification exclusively within the island, while members of the genus *Truncattus* and the *gracilis* group are present in Cuba and Hispaniola, suggesting diversification both within and between these islands (Supplemental Data S4 and S5). The estimated divergence between Hispaniolan and Puerto Rican clades is c. 25.16 Mya consistent with the approximate timing of separation of these islands (c. 23–30 Mya). Similarly, Hispaniolan and Cuban clades split around 17–22 Mya coinciding with the geological separation of these islands. Hence, vicariance hypotheses can readily explain the distribution of major clades among islands. Similarly, the Puerto Rican genus *Petemathis* branched off from a Hispaniola lineage at c. 25 ± 3 Mya prior to the estimated timeframe of the Hispaniola and Puerto Rico split (20–30 Mya). While *Petemathis* only began to diversify later to around c.16.34 ± 6 Mya the split between Hispaniola and Puerto Rican lineages is easily explained by paleogeographical models and no long-distance dispersal is implied. Similarly, the *keyserlingi* and *darlingtoni* groups branched off from a Hispaniola lineage at c. 17.63 ± 2 Mya. The *keyserlingi* group only begins diversifying much later (c. 8.94 ± 3 Mya) and is restricted to Cuba. Of course, we cannot rule out earlier diversification of the group followed by extinction of early branches without much more detailed fossil record than is currently available. The *darlingtoni* group quickly diversified (c. 16.7 ± 1 Mya), presumably facilitating their colonization of Cuba before Hispaniola and Cuba split (c. 14–17 Mya).

On the other hand, the BSM analyses imply that dispersal between Hispaniola and Cuba continued happening after the geological separation of these islands suggesting that overwater dispersal also played an important role in shaping the current distribution and diversity of the linage (Figure 6, Supplementary Data S4). As in other groups of spiders, overwater dispersal is common in at least some lineages (e.g., Čandek et al. [50]; Crews and Esposito [52], Agnarsson et al. [46], Shapiro et al. [120], and can explain non-vicariant movement among Caribbean islands. Long-distance dispersal followed by range-expansion seems important in *Truncattus* (c. 13.66 ± 5 Mya) and the *gracilis* group (c. 15.5 ± 4 Mya). Similar studies show the occurrence of overwater dispersal/colonization events (e.g., founder-events and range-expansions) as the best explanation of among island movement after Hispaniola–Cuba split (butterflies *Calisto*: Matos–Maraví et al. [5]; aquatic beetles *Phaenonotum*: Deler–Hernández et al. [24]; weevils *Exophthalmus*: Zhang et al. [67]; mastiff bats *Molossus*: Loureiro et al. [121]).

7.3. From Hispaniola to Cuba and Puerto Rico

Our study indicates Hispaniola as a potential source for subsequent radiations throughout the Greater Antilles, with multiple exchanges between Hispaniola and Cuba (Figures 6 and 7, Supplementary Data S4 and S5). Other studies also support Hispaniola as a point of dispersal to other Antillean islands [122,123]. Fabre et al. [124] found evidence in Caribbean Capromyidae (hutias) supporting Hispaniola as a potential source of colonization to other Greater Antilles islands and the Bahamas. In their study, they suggest either (i) a vicariant event between eastern (Hispaniola) and western (Bahamas, Cuba, Jamaica) hutias or (ii) stepping-stone colonization from east to west. Čandek et al., [50] found that in *Cyrthognata* spiders dispersal from Hispaniola explains their colonization of the rest of the Caribbean Archipelago. The BioGeoBEARS ancestral range estimation of the GAARlandia DEC+j model for *Deinopis* (see, Chamberland et al., [4]) also supports the hypothesis that Hispaniola plays a pivotal role in Caribbean dispersal. McHugh et al. [48] and Shapiro et al. [120]) provide evidence that Caribbean *Micrathena* are not monophyletic and likely colonized the region multiple times, with evidence of interchanges between Cuba, Hispaniola and Puerto Rico. In the case of the origin of the *Antillattus* clade, the exact role of Hispaniola is less clear; however, the available evidence indicates that it may be the area of the Caribbean first colonized by the ancestor of the clade. Further studies of Caribbean biota will further clarify the role of Hispaniola in the overall biogeographical complexity of the Greater Antilles.

8. Conclusions

Our study sheds new light on the biogeography of the *Antillattus* clade and their Caribbean radiation. The phylogenetic and biogeographical evidence presented in this study fits the Caribbean palaeogeographical model of colonization and suggests a complex interplay of vicariance and overwater dispersal driven diversification in shaping the biota of this biodiversity hotspot. The ancestor of the *Antillattus* clade appears to have colonized the Greater Antilles (Puerto Rico, Hispaniola, and Cuba) during the timespan of GAARlandia and land connections prior to the Mona Passage and the Windward Passage. Our results suggest that the evolution of the *Antillattus* clade included both vicariant processes and long-distance dispersal with the majority of diversification attributed to within island speciation. Finally, among other insights, we have uncovered the importance of Hispaniola in the *Antillattus* clade colonization of the Caribbean, thereby providing further evidence that islands can function as key diversification hubs for archipelagos.

Supplementary Materials: The following are available online at https://www.mdpi.com/article/10.3390/d14030224/s1, https://drive.google.com/drive/folders/1HotuTFZ1UVIsRbIDCfOK5Xk4GmNr-jqg?usp=sharing (Accessed on 10 November 2021). Table S1: Substitution models selected by jModelTest and ModelFinder for each individual gene region and partition. Dataset including outgroups and excluding outgroups. Data S1: DNA analysis for dataset including outgroups and excluding outgroups. Data S2: Beast divergence time estimations (including outgroups and excluding outgroups) of all genes (CO1, 16S-ND1, 28S) using a Bayesian relaxed molecular clock. Data S3: BioGeoBEARS *Antillattus* clade (including outgroups and excluding outgroups). Data S4: B2a model (stochastic mapping (BSM)). Data S5: B2a model (number of lineages through time).

Author Contributions: Conceptualization, F.C.-R.; data curation, F.C.-R. and P.W.; formal analysis, F.C.-R. and P.W.; funding acquisition, G.J.B. and I.A.; investigation, F.C.-R.; methodology, F.C.-R. and E.F.-D.; project administration, G.J.B. and I.A.; resources, G.J.B. and I.A.; software, F.C.-R.; supervision, E.F.-D., G.J.B. and I.A.; validation, G.J.B. and I.A.; Visualization, F.C.-R., P.W., E.F.-D., G.J.B. and I.A.; writing—original draft, F.C.-R., P.W. and E.F.-D.; writing—review & editing, F.C.-R., G.J.B. and I.A. All authors have read and agreed to the published version of the manuscript.

Funding: Funding for this work came from NSF DEB-1314749 and DEB-1050253 to G. Binford and I. Agnarsson. Additional funds came from the Smithsonian Laboratories of Analytical Biology, a 2013 SI Barcode Network grant to J.A. Coddington and I. Agnarsson; and in part from National Geographic Society (WW-203R017) grant to I. Agnarsson.

Institutional Review Board Statement: All material was collected under appropriate collection permits and approved guidelines.

Informed Consent Statement: Not applicable.

Data Availability Statement: Code can be found at https://drive.google.com/drive/folders/1HotuTFZ1UVIsRbIDCfOK5Xk4GmNr-jqg?usp=sharing (accessed on 10 November 2021).

Acknowledgments: Many thanks to all the members of the CarBio team and our collaborators for their tireless effort in the field over the last years. Many thanks to Robert Anderson, Martín Fikáček, Andrew Smith, and Guanyang Zhang Rodrigo for their important help in the field. Special thanks to our collaborators in Cuba: Alexander Sanchez, Giraldo Alayón, Albert Deler–Hernandez, and rangers of "Pico Turquino National Park, "Maravillas de Viñales" National Park, "Sierra de Cubitas" Ecological Reserve, "Topes de Collantes" National Park, "Duaba–Yunque–Quibijan" Ecological Reserve, and "Soledad" Botanical Garden. In Hispaniola: Carlos Suriel, Solanlly Carrero Jiménez, and Gabriel de los Santos for their invaluable help with the organization and execution of logistically complex expeditions and ongoing collaboration. We are grateful to Laura May-Collado who did extraction and sequencing for many specimens. We especially acknowledge the comments of Sarah Crews and anonymous reviewers. We are grateful to the authorities and personnel of the Cuban Ministry of Science, Technology and Environment (CICA–CITMA), and the Empresa Nacional para la Protección de la Flora y la Fauna (ENPFF) for providing access to protected areas under their control.

Conflicts of Interest: The authors declare no conflict of interest.

References

1. Dávalos, L.M. Phylogeny and biogeography of Caribbean mammals. *Biol. J. Linn. Soc.* **2004**, *8*, 373–394. [CrossRef]
2. Ricklefs, R.; Bermingham, E. The West Indies as a laboratory of biogeography and evolution. *Philos. Trans. R. Soc. Lond. B Biol. Sci.* **2008**, *363*, 2393–2413. [CrossRef] [PubMed]
3. Agnarsson, I.; Kuntner, M. The generation of a biodiversity hotspot: Biogeography and phylogeography of the western Indian Ocean islands. In *Current Topics in Phylogenetics and Phylogeography of Terrestrial and Aquatic Systems*; Anamthawat-Jonsson, K., Ed.; Tech Publishers Press: Rijeka, Croatia, 2012; pp. 33–82.
4. Chamberland, L.; McHugh, A.; Kechejian, S.; Binford, G.J.; Bond, J.E.; Coddington, J.; Dolman, G.; Hamilton, C.A.; Harvey, M.S.; Kuntner, M.; et al. From Gondwana to GAAR landia: Evolutionary history and biogeography of ogre-faced spiders (Deinopis). *J. Biogeogr.* **2018**, *45*, 2442–2457. [CrossRef]
5. Matos-Maraví, P.; Águila, R.N.; Peña, C.; Miller, J.Y.; Sourakov, A.; Wahlberg, N. Causes of endemic radiation in the Caribbean: Evidence from the historical biogeography and diversification of the butterfly genus *Calisto* (Nymphalidae: Satyrinae: Satyrini). *BMC Evol. Biol.* **2014**, *14*, 199. [CrossRef]
6. Gillespie, R.G.; Roderick, G.K. Arthropods on islands: Colonization, speciation, and conservation. *Annu. Rev. Entomol.* **2002**, *47*, 595–632. [CrossRef]
7. De Queiroz, K. A unified species concept and its consequences for the future of taxonomy. *Proc. Calif. Acad. Sci.* **2005**, *56*, 196–215.
8. Cowie, R.H.; Holland, B.S. Dispersal is fundamental to biogeography and the evolution of biodiversity on oceanic islands. *J. Biogeogr.* **2006**, *33*, 193–198. [CrossRef]
9. Gillespie, R.G.; Baldwin, B.G.; Waters, J.M.; Fraser, C.I.; Nikula, R.; Roderick, G.K. Long-distance dispersal: A framework for hypothesis testing. *Trends Ecol. Evol.* **2012**, *27*, 47–55. [CrossRef]
10. Schönhofer, A.L.; McCormack, M.; Tsurusaki, N.; Martens, J.; Hedin, M. Molecular phylogeny of the harvestmen genus *Sabacon* (Arachnida: Opiliones: Dyspnoi) reveals multiple Eocene–Oligocene intercontinental dispersal events in the Holarctic. *Mol. Phylogenet. Evol.* **2013**, *66*, 303–315. [CrossRef]
11. Agnarsson, I.; Cheng, R.C.; Kuntner, M. A Multi-Clade Test Supports the Intermediate Dispersal Model of Biogeography. *PLoS ONE* **2014**, *91*, e86780. [CrossRef]
12. Toussaint, E.F.A.; Fikacek, M.; Short, A.E.Z. India-Madagascar vicariance explains cascade beetle bio-geography. *Biol. J. Linn. Soc.* **2016**, *118*, 982–991. [CrossRef]
13. Weaver, P.F.; Cruz, A.; Johnson, S.; Dupin, J.; Weaver, K.F. Colonizing the Caribbean: Biogeography and evolution of livebearing fishes of the genus *Limia* (Poeciliidae). *J. Biogeogr.* **2016**, *43*, 1808–1819. [CrossRef]
14. Mittermeier, R.A.; Robles-Gil, P.; Hoffman, M.; Pilgrim, J.; Brooks, T.; Mittermeier, C.G.; Lamoreux, J.; da Fonseca, G.A.B. *Hotspots Revisited: Earth's Biologically Richest and Most Endangered Terrestrial Ecoregions*; CEMEX, Agrupación Sierra Madre: Monterrey, Mexico, 2004.
15. Diamond, J.M.; Gilpin, M.E.; Mayr, E. Species-Distance Relation for Birds of Solomon Archipelago, and Paradox of Great Speciators. *Proc. Natl. Acad. Sci. USA* **1976**, *73*, 2160–2164. [CrossRef]
16. Glor, R.E.; Losos, J.B.; Larson, A. Out of Cuba: Overwater dispersal and speciation among lizards in the *Anolis carolinensis* subgroup. *Mol. Ecol.* **2005**, *14*, 2419–2432. [CrossRef]
17. Iturralde-Vinent, M.A. Meso-Cenozoic Caribbean paleogeography: Implications for the historical biogeography of the region. *Int. Geol. Rev.* **2006**, *48*, 791–827. [CrossRef]
18. Crews, S.C.; Gillespie, R.G. Molecular systematics of *Selenops* spiders (Araneae: Selenopidae) from North and Central America: Implications for Caribbean biogeography. *Biol. J. Linn. Soc.* **2010**, *101*, 288–322. [CrossRef]
19. Claramunt, S.; Derryberry, E.P.; Remsen, J.V., Jr.; Brumfield, R.T. High dispersal ability inhibits speciation in a continental radiation of passerine birds. *Proc. Royal Soc. B* **2012**, *279*, 1567–1574. [CrossRef]
20. Algar, A.C.; Mahler, D.L.; Glor, R.E.; Losos, J.B. Niche incumbency, dispersal limitation and climate shape geographical distributions in a species-rich island adaptive radiation. *Glob. Ecol. Biogeogr.* **2013**, *22*, 391–402. [CrossRef]
21. Helmus, M.R.; Mahler, D.L.; Losos, J.B. Island biogeography of the Anthropocene. *Nature* **2014**, *513*, 543–546. [CrossRef]

22. Dziki, A.; Binford, G.J.; Coddington, J.A.; Agnarsson, I. *Spintharus flavidus* in the Caribbean—A 30 million year biogeographical history and radiation of a 'widespread species'. *PeerJ* **2015**, *3*, e1422. [CrossRef]

23. Esposito, L.A.; Bloom, T.; Caicedo, L.; Alicia-Serrano, A.; Sanchez-Ruiz, J.; May-Collado, L.J.; Binford, G.; Agnarsson, I. Islands within islands: Diversification of tailless whip spiders (Amblypygi, Phrynus) in Caribbean caves. *Mol. Phylogenet. Evol.* **2015**, *93*, 107–117. [CrossRef] [PubMed]

24. Deler-Hernández, A.; Sýkora, V.; Seidel, M.; Cala-Riquelme, F.; Fikáček, M. Multiple origins of the *Phaenonotum* beetles in the Greater Antilles (Coleoptera: Hydrophilidae): Phylogeny, biogeography and systematics. *Zool. J. Linnean Soc.* **2017**, *183*, 97–120. [CrossRef]

25. Reynolds, R.G.; Strickland, T.R.; Kolbe, J.J.; Falk, B.G.; Perry, G.; Revell, L.J.; Losos, J.B. Archipelagic genetics in a widespread Caribbean anole. *J. Biogeogr.* **2017**, *44*, 2631–2647. [CrossRef]

26. Chamberland, L.; Salgado-Roa, F.C.; Basco, A.; Crastz-Flores, A.; Binford, G.J.; Agnarsson, I. Phylogeography of the widespread Caribbean spiny orb weaver *Gasteracantha cancriformis*. *PeerJ* **2020**, *8*, e8976. [CrossRef]

27. Ali, J.R.; Hedges, S.B. Colonizing the Caribbean: New geological data and an updated land-vertebrate colonization record challenge the GAARlandia land-bridge hypothesis. *J. Biogeogr.* **2021**, *48*, 2699–2707. [CrossRef]

28. Iturralde-Vinent, M.A. Aspectos geológicos de la biogeografía de Cuba. *Cienc. Tierra Espacio* **1982**, *5*, 85–100.

29. Mann, P.; Schubert, C.; Burke, K. Review of Caribbean neotectonics. In *The Caribbean Region: Geological Society of America*; Dengo, G., Case, J.E., Eds.; The Geology of North America: Boulder, CO, USA, 1990; pp. 307–338.

30. MacPhee, R.D.E.; Iturralde-Vinent, M.A. First Tertiary land mammal fossils from Greater Antilles: An Early Miocene sloth (Xenarhra, Megalonychidae) from Cuba. *Am. Mus. Novit.* **1994**, *3094*, 1–13.

31. MacPhee, R.D.E.; Iturralde-Vinent, M.A. Origin of the Greater Antilles land mammal fauna 1: New Tertiary land mammals from Cuba and Puerto Rico. *Am. Mus. Novit.* **1995**, *314*, 1–31.

32. Iturralde-Vinent, M.A.; MacPhee, R.D.E. Paleogeography of the Caribbean region: Implications for Cenozoic biogeography. *Bull. Am. Mus. Natl. Hist.* **1999**, *238*, 1–95.

33. MacPhee, R.D.E.; Iturralde-Vinent, M.A. A short history of Greater Antillean land mammals: Biogeography, paleogeography, radiation, and extinctions. *Tropics* **2000**, *10*, 145–154. [CrossRef]

34. White, J.L.; MacPhee, R.D.E. The sloths of the West Indies: A systematic and phylogenetic review. In *Biogeography of the West Indies: Patterns and Perspectives*; Woods, C.A., Sergile, F.E., Eds.; CRC Press: Boca Raton, FL, USA, 2001; pp. 201–235.

35. Pindell, J.L.; Barrett, S.F. Geological evolution of the Caribbean region: A plate tectonic perspective. In *The Caribbean Region: Geological Society of America*; Dengo, G., Case, J.E., Eds.; The Geology of North America: Boulder, CO, USA, 1990; pp. 405–432.

36. Van Gestel, J.P.; Mann, P.; Grindlay, N.R.; Dolan, J.F. Three-phase tectonic evolution of the northern margin of Puerto Rico as inferred from an integration of seismic reflection, well, and outcrop data. *Mar. Geol.* **1999**, *161*, 257–286. [CrossRef]

37. MacPhee, R.D.E.; Iturralde-Vinent, M.A.; Gaffney, E.S. Domo de Zaza, an early Miocene vertebrate locality in south-central Cuba, with notes on the tectonic evolution of Puerto Rico and the Mona Passage. *Am. Mus. Novit.* **2003**, *3394*, 1–42. [CrossRef]

38. Alonso, R.; Crawford, A.J.; Bermingham, E. Molecular phylogeny of an endemic radiation of Cuban toads (Bufonidae: Peltophryne) based on mitochondrial and nuclear genes. *J. Biogeogr.* **2012**, *39*, 434–451. [CrossRef]

39. Rosen, D.E. A vicariance model of Caribbean biogeography. *Syst. Zool.* **1975**, *24*, 431–464. [CrossRef]

40. Rosen, D.E. Geological hierarchies and biogeographical congruence in the Caribbean. *Ann. Mo. Bot. Gard.* **1985**, *72*, 636–659. [CrossRef]

41. Tada, R.; Iturralde-Vinent, M.; Matsui, T.; Tajika, E.; Oji, T.; Goto, K.; Nakano, Y.; Takayama, H.; Yamamoto, S.; Toyoda, K.; et al. K/T Boundary deposits in the Paleo-western Caribbean basin. *AAPG Mem.* **2003**, *26*, 23.

42. Hedges, S.B.; Hass, C.; Maxson, L. Caribbean biogeography: Molecular evidence for dispersal in West Indian terrestrial vertebrates. *Proc. Natl. Acad. Sci. USA* **1992**, *89*, 1909–1913. [CrossRef]

43. Hedges, S.B. Biogeography of the West Indies: An overview. In *Biogeography of the West Indies: Patterns and Perspectives*; Woods, C.A., Sergile, F.E., Eds.; CRC Press: Baton Rouge, LA, USA, 2001; pp. 15–33.

44. Hedges, S.B. Paleogeography of the Antilles and origin of West Indian terrestrial vertebrates. *Ann. Missouri Bot. Gard.* **2006**, *93*, 231–244. [CrossRef]

45. Sato, J.J.; Ohdachi, S.D.; Echenique-Diaz, L.M.; Borroto-Páez, R.; Begué-Quiala, G.; Delgado-Labañino, L.J.; Gámez-Díez, J.; Álbarez-Lemus, J.; Nguyen, T.S.; Yamaguchi, N.; et al. Molecular phylogenetic analysis of nuclear genes suggests a Cenozoic overwater dispersal origin for the Cuban solenodon. *Sci. Rep.* **2016**, *6*, 31173. [CrossRef]

46. Uit de Weerd, D.R.; Robinson, D.G.; Rosenberg, G. Evolutionary and biogeographical history of the land snail family Urocoptidae (Gastropoda: Pulmonata) across the Caribbean region. *J. Biogeogr.* **2016**, *43*, 763–777. [CrossRef]

47. Yusseff-Vanegas, S.; Agnarsson, I. Molecular phylogeny of the forensically important genus *Cochliomyia* (Diptera: Calliphoridae). *ZooKeys* **2016**, *609*, 107–120. [CrossRef] [PubMed]

48. Mchugh, A.; Yablonsky, C.; Binford, G.; Agnarsson, I. Molecular phylogenetics of Caribbean *Micrathena* (Araneae: Araneidae) suggests multiple colonization events and single island endemism. *Invertebr. Syst.* **2014**, *28*, 337–349. [CrossRef]

49. Agnarsson, I.; LeQuier, S.M.; Kuntner, M.; Cheng, R.C.; Coddington, J.A.; Binford, G. Phylogeography of a good Caribbean disperser: *Argiope argentata* (Araneae, Araneidae) and a new 'cryptic' species from Cuba. *ZooKeys* **2016**, *2016*, 25–44. [CrossRef] [PubMed]

50. Čandek, K.; Agnarsson, I.; Binford, G.J.; Kuntner, M. Biogeography of the Caribbean *Cyrtognatha* spiders. *Sci. Rep.* **2019**, *9*, 397. [CrossRef] [PubMed]
51. Čandek, K.; Binford, G.J.; Agnarsson, I.; Kuntner, M. Caribbean golden orbweaving spiders maintain gene flow with North America. *Zool. Scr.* **2020**, *49*, 210–221. [CrossRef]
52. Crews, S.C.; Esposito, L.A. Towards a synthesis of the Caribbean biogeography of terrestrial arthropods. *BMC Evol. Biol.* **2020**, *20*, 12. [CrossRef]
53. Ali, J.R. Colonizing the Caribbean: Is the GAARlandia land-bridge hypothesis gaining a foothold? *J. Biogeogr.* **2012**, *39*, 431–433. [CrossRef]
54. Philippon, M.; Cornée, J.-J.; Münch, P.; van Hinsbergen, D.J.J.; BouDagher-Fadel, M.; Gailler, L. Eocene intra-plate shortening responsible for the rise of a faunal pathway in the northeastern Caribbean realm. *PLoS ONE* **2020**, *15*, e0241000. [CrossRef]
55. Murphy, W.J.; Collier, G.E. Phylogenetic relationships within the aplocheiloid fish genus *Rivulus* (Cyprinodontiformes, Rivulidae): Implications for Caribbean and Central American biogeography. *Mol. Biol. Evol.* **1996**, *13*, 642–649. [CrossRef]
56. Dávalos, L.M. Short-faced bats (Phyllostomidae: Stenodermatina): A Caribbean radiation of strict frugivores. *J. Biogeogr.* **2007**, *34*, 364–375. [CrossRef]
57. Santiago-Valentin, E.; Olmstead, R.G. Phylogenetics of the Antillean Goetzeoideae (Solanaceae) and their relationships within the Solanaceae based on chloroplast and ITS DNA sequence data. *Am. Soc. Plant Taxon.* **2003**, *28*, 452–460.
58. Tong, Y.; Binford, G.; Rheims, A.R.; Kuntner, M.; Liu, J.; Agnarsson, I. Huntsmen of the Caribbean: Multiple tests of the GAARlandia hypothesis. *Mol. Phylogenet. Evol.* **2019**, *130*, 259–268. [CrossRef]
59. Censky, E.J.; Hodge, K.; Dudley, J. Over-water dispersal of lizards due to hurricanes. *Nature* **1998**, *395*, 556. [CrossRef]
60. Hall, J.P.W.; Robbins, R.K.; Harvey, D.J. Extinction and biogeography in the Caribbean: New evidence from a fossil riodinid butterfly in Dominican amber. *Proc. Soc. R. Soc. B-Biol. Sci.* **2004**, *271*, 797–801. [CrossRef]
61. Wahlberg, N. That awkward age for butterflies: Insights from the age of the butterfly subfamily Nymphalinae (Lepidoptera: Nymphalidae). *Syst. Biol.* **2006**, *55*, 703–714. [CrossRef]
62. Wahlberg, N.; Freitas, A.V. Colonization of and radiation in South America by butterflies in the subtribe Phyciodina (Lepidoptera: Nymphalidae). *Mol. Phylogenet. Evol.* **2007**, *44*, 1257–1272. [CrossRef]
63. Sourakov, A.; Zakharov, E.V. "Darwin's butterflies"? DNA barcoding and the radiation of the endemic Caribbean butterfly genus *Calisto* (Lepidoptera, Nymphalidae, Satyrinae). *Comp. Cytogenet.* **2011**, *5*, 191–210. [CrossRef]
64. Ceccarelli, F.S.; Zaldívar-Riverón, A. Broad polyphyly and historical biogeography of the Neotropical wasp genus *Notiospathius* (Braconidae: Doryctinae). *Mol. Phylogenet. Evol.* **2013**, *69*, 142–152. [CrossRef]
65. Lewis, D.S.; Sperling, F.A.H.; Nakahara, S.; Cotton, A.M.; Kawahara, A.Y.; Condamine, F.L. Role of Caribbean Islands in the diversification and biogeography of Neotropical *Heraclides* swallowtails. *Cladistics* **2015**, *31*, 291–314. [CrossRef]
66. Rodriguez, J.; Pitts, J.P.; von Dohlen, C.D. Historical biogeography of the widespread spider wasp tribe *Aporini* (Hymenoptera: Pompilidae). *J. Biogeogr.* **2015**, *42*, 495–506. [CrossRef]
67. Zhang, G.; Basharat, U.; Matzke, N.; Franz, N.M. Model selection in statistical historical biogeography of Neotropical insects—the *Exophthalmus* genus complex (Curculionidae: Entiminae). *Mol. Phylogenet. Evol.* **2017**, *109*, 226–239. [CrossRef]
68. Zhang, J.X.; Maddison, W.P. Molecular phylogeny, divergence times and biogeography of spiders of the subfamily Euophryinae (Araneae: Salticidae). *Mol. Phylogenet. Evol.* **2013**, *68*, 81–92. [CrossRef] [PubMed]
69. Chakrabarty, P. Systematics and historical biogeography of Greater Antillean Cichlidae. *Mol. Phylogenet. Evol.* **2006**, *39*, 619–627. [CrossRef] [PubMed]
70. Říčan, O.; Piálek, L.; Zardoya, R.; Doadrio, I.; Zrzavý, J. Biogeography of the Mesoamerican Cichlidae (Teleostei: Heroini): Colonization through the GAARlandia land bridge and early diversification. *J. Biogeogr.* **2013**, *40*, 579–593. [CrossRef]
71. World Spider Catalog. World Spider Catalog. Version 22.5. Natural History Museum Bern. Available online: http://wsc.nmbe.ch (accessed on 17 November 2021).
72. Spagna, J.C.; Peattie, A.M. Terrestrial locomotion in arachnids. *J. Insect Physiol.* **2012**, *58*, 599–606. [CrossRef]
73. Kropf, C. Hydraulic system of locomotion. In *Spider Ecophysiology*; Nentwig, W., Ed.; Springer: Berlin, Germany, 2013; pp. 43–56.
74. Brandt, E.E.; Roberts, K.T.; Williams, C.M.; Elias, D.O. Low temperatures impact species distributions of jumping spiders across a desert elevational cline. *J. Insect. Physiol.* **2020**, *122*, 104037. [CrossRef]
75. Bodner, M.R.; Maddison, W.P. The biogeography and age of salticid spider radiations (Araneae: Salticidae). *Mol. Phylogenet. Evol.* **2012**, *65*, 213–240. [CrossRef]
76. Maddison, W.P.; Hedin, M.C. Jumping spider phylogeny (Araneae: Salticidae). *Invertebr. Syst.* **2003**, *17*, 529–549. [CrossRef]
77. Maddison, W.P.; Bodner, M.R.; Needham, K.M. Salticid spider phylogeny revisited, with the discovery of a large Australasian clade (Araneae: Salticidae). *Zootaxa* **2008**, *1893*, 46–64. [CrossRef]
78. Bryant, E.B. The salticid spiders of Hispaniola. *Bull. Mus. Comp. Zool.* **1943**, *92*, 445–529.
79. Zhang, J.X.; Maddison, W.P. New euophryine jumping spiders from the Dominican Republic and Puerto Rico (Araneae: Salticidae: Euophryinae). *Zootaxa.* **2012**, *3476*, 1–54. [CrossRef]
80. Prószyński, J.; Deeleman-Reinhold, C.L. Description of some Salticidae (Aranei) from the Malay Archipelago. II. Salticidae of Java and Sumatra, with comments on related species. *Arthropoda Sel.* **2012**, *21*, 29–60. [CrossRef]
81. Zhang, J.X.; Maddison, W.P. Genera of euophryine jumping spiders (Araneae: Salticidae), with a combined molecular-morphological phylogeny. *Zootaxa* **2015**, *3938*, 1–147. [CrossRef]

82. Folmer, O.; Black, M.; Hoeh, W.; Lutz, R.; Vrijenhoek, R. DNA primers for amplification of mitochondrial cytochrome oxidase subunit I from diverse metazoan invertebrates. *Mol. Mar. Bio. Biotechnol.* **1994**, *3*, 294–299.

83. Hedin, M.C.; Maddison, W.P. A combined molecular approach to phylogeny of the jumping spider subfamily Dendryphantinae (Araneae: Salticidae). *Mol. Phylogenet. Evol.* **2001**, *18*, 386–403. [CrossRef]

84. Green, P. Phrap, Version 0.990329. 1999. Available online: http://www.phrap.org (accessed on 2 June 2021).

85. Green, P.; Ewing, B. Phred, Version 0.020425 c. 2002. Available online: http://www.phrap.org (accessed on 2 June 2021).

86. Maddison, D.R.; Maddison, W.P. 2021. Chromaseq: A Mesquite Package for Analyzing Sequence Chromatograms. Version 1.31. 2018. Available online: http://chromaseq.mesquiteproject.org (accessed on 2 June 2021).

87. Maddison, W.P.; Maddison, D.R. Mesquite: A Modular System for Evolutionary Analysis. Version 3.6. 2018. Available online: http://www.mesquiteproject.org (accessed on 2 June 2021).

88. Katoh, K.; Standley, D.M. MAFFT multiple sequence alignment software version 7: Improvements in performance and usability. *Mol. Biol. Evol.* **2013**, *30*, 772–780. [CrossRef]

89. Darriba, D.; Taboada, G.L.; Doallo, R.; Posada, D. jModelTest 2: More models, new heuristics and parallel computing. *Nat. Methods* **2012**, *9*, 772. [CrossRef]

90. Posada, D.; Buckley, T.R. Model selection and model averaging in phylogenetics: Advantages of Akaike Information Criterion and Bayesian Approaches over Likelihood Ratio Tests. *Syst. Biol.* **2004**, *53*, 793–808. [CrossRef]

91. Nguyen, L.T.; Schmidt, H.A.; von Haeseler, A.; Minh, B.Q. IQ-TREE: A fast and effective stochastic algorithm for estimating maximum-likelihood phylogenies. *Mol. Biol. Evol.* **2015**, *32*, 268–274. [CrossRef]

92. Kalyaanamoorthy, S.; Minh, B.Q.; Wong, T.F.K.; von Haeseler, A.; Jermiin, L.S. ModelFinder: Fast model selection for accurate phylogenetic estimates. *Nat. Methods* **2015**, *14*, 587–589. [CrossRef]

93. Altekar, G.; Dwarkadas, S.; Huelsenbeck, J.P.; Ronquist, F. Parallel Metropolis coupled Markov chain Monte Carlo for Bayesian phylogenetic inference. *Bioinformatics* **2004**, *20*, 407–415. [CrossRef] [PubMed]

94. Miller, M.; Pfeiffer, W.; Schwartz, T. Creating the CIPRES Science Gateway for inference of large phylogenetic tres. In Proceedings of the 2010 Gateway Computing Environments Workshop (GCE 2010), New Orleans, LA, USA, 14 November 2010; pp. 1–8. [CrossRef]

95. Huelsenbeck, J.P.; Ronquist, F. MRBAYES: Bayesian inference of phylogenetic trees. *Bioinformatics* **2001**, *17*, 754–755. [CrossRef] [PubMed]

96. Ronquist, F.; Huelsenbeck, J.P. MrBayes 3: Bayesian phylogenetic inference under mixed models. *Bioinformatics* **2003**, *19*, 1572–1574. [CrossRef]

97. Rambaut, A.; Suchard, M.A.; Xie, D.; Drummond, A.J. Tracer v1.6. 2014. Available online: http://beast.bio.ed.ac.uk/Tracer (accessed on 2 June 2021).

98. Drummond, A.J.; Suchard, M.A.; Xie, D.; Rambaut, A. Bayesian phylogenetics with BEAUti and the BEAST 1.7. *Mol. Biol. Evol.* **2012**, *29*, 1969–1973. [CrossRef] [PubMed]

99. Battistuzzi, F.U.; Billing-Ross, P.; Paliwal, A.; Kumar, S. Fast and slow implementations of relaxed-clock methods show similar patterns of accuracy in estimating divergence times. *Mol. Biol. Evol.* **2011**, *28*, 2439–2442. [CrossRef] [PubMed]

100. Bidegaray-Batista, L.; Arnedo, M.A. Gone with the plate: The opening of the Western Mediterranean basin drove the diversification of ground-dweller spiders. *MBC Evol. Biol.* **2011**, *11*, 317. [CrossRef]

101. Xie, W.; Lewis, P.O.; Fan, Y.; Kuo, L.; Chen, M.H. Improving marginal likelihood estimation for Bayesian phylogenetic model selection. *Syst. Biol.* **2011**, *60*, 150–160. [CrossRef]

102. Ronquist, F.; Teslenko, M.; van der Mark, P.; Ayres, D.L.; Darling, A.; Höhna, S.; Larget, B.; Liu, L.; Suchard, M.A.; Huelsenbeck, J.P. MrBayes 3.2: Efficient Bayesian phylogenetic inference and model choice across a large model space. *Syst. Biol.* **2012**, *61*, 539–542. [CrossRef]

103. Gernhard, T. The conditioned reconstructed process. *J. Theor. Biol.* **2008**, *253*, 769–778. [CrossRef]

104. Stadler, T. On incomplete sampling under birth–death models and connections to the sampling-based coalescent. *J. Theor. Biol.* **2009**, *261*, 58–66. [CrossRef]

105. Wunderlich, J. *Die Fossilen Spinnen im Dominikanischen Bernstein*; Published by the author Straubenhardt: Baden-Württemberg, Germany, 1988.

106. Penney, D. *Dominican Amber Spiders: A Comparative Palaeontological-Neontological Approach to Identification, Faunistics, Ecology and Biogeography*; Siri Scientific Press: Rochdale, UK, 2008.

107. Peckham, G.W.; Peckham, E.G. On some new genera and species of Attidae from the eastern part of Guatemala. *Proc. Nat. Hist. Soc. Wisconsin.* **1885**, *2*, 62–86.

108. Zhang, J. Phylogeny and Systematics of the Jumping Spider Subfamily Euophryinae (Araneae: Salticidae), with Consideration of Biogeography and Genitalic Evolution. Ph.D. Thesis, University of British Columbia, Vancouver, BC, Canada, 2012. Available online: https://open.library.ubc.ca/collections/ubctheses/24/items/1.0072804 (accessed on 17 November 2021).

109. Matzke, N.J. Model selection in historical biogeography reveals that founder-event speciation is a crucial process in Island Clades. *Syst. Biol.* **2014**, *63*, 951–970. [CrossRef]

110. Matzke, N.J. BioGeoBEARS: BioGeography with Bayesian (and Likelihood) Evolutionary Analysis with R Scripts. Version 1.1.1. 2018. Available online: https://doi.org/10.5281/zenodo.1478250 (accessed on 4 September 2021).

111. Ree, R.H.; Smith, S.A. Maximum likelihood inference of geographic range evolution by dispersal, local extinction, and cladogenesis. *Syst. Biol.* **2008**, *57*, 4–14. [CrossRef]
112. Ronquist, F. Dispersal-vicariance analysis: A new approach to the quantification of historical biogeography. *Syst. Biol.* **1997**, *46*, 195–203. [CrossRef]
113. Landis, M.J.; Matzke, N.J.; Moore, B.R.; Huelsenbeck, J.P. Bayesian analysis of biogeography when the number of áreas is large. *Syst. Biol.* **2013**, *62*, 789–804. [CrossRef]
114. Klaus, K.V.; Matzke, N.J. Statistical Comparison of Trait-Dependent Biogeographical Models Indicates That Podocarpaceae Dispersal Is Influenced by Both Seed Cone Traits and Geographical Distance. *Syst. Biol.* **2020**, *69*, 61–75. [CrossRef]
115. Ree, R.H.; Sanmartín, I. Conceptual and statistical problems with the DEC+J model of founder-event speciation and its comparison with DEC via model selection. *J. Biogeogr.* **2018**, *45*, 741–749. [CrossRef]
116. Matzke, N.J. Statistical comparison of DEC and DEC+J is identical to comparison of two ClaSSE submodels, and is therefore valid. *OSF Preprint* **2021**, 1–40. [CrossRef]
117. Matzke, N.J. Probabilistic historical biogeography: New models for founder-event speciation, imperfect detection, and fossils allow improved accuracy and model-testing. *Front. Biogeogr.* **2013**, *5*, 242–248. [CrossRef]
118. Matzke, N.J. Stochastic Mapping under Biogeographical Models. PhyloWiki BioGeoBEARS Website. 2016. Available online: http://phylo.wikidot.com/biogeobears#stochastic_mapping (accessed on 27 September 2021).
119. Magalhaes, I.L.F.; Santos, A.J.; Ramírez, M.J. Incorporating topological and age uncertainty into event-based biogeography supports paleo-islands in Galapagos and ancient connections among Neotropical dry forests. *Diversity* **2021**, *13*, 418. [CrossRef]
120. Shapiro, L.; Binford, G.J.; Agnarsson, I. Single-Island Endemism despite Repeated Dispersal in Caribbean *Micrathena* (Araneae: Araneidae): An Updated Phylogeographic Analysis. *Diversity* **2022**, *14*, 128. [CrossRef]
121. Loureiro, L.O.; Engstrom, M.D.; Lim, B.K. Optimization of Genotype by Sequencing data for phylogenetic purposes. *MethodsX* **2020**, *7*, 100892. [CrossRef] [PubMed]
122. Hedges, S.B.; Woods, C.A. Caribbean hot spot. *Nature* **1993**, *364*, 375. [CrossRef]
123. Woods, C.; Borroto, P.R.; Kilpatrick, C. Insular patterns and radiation of West Indian rodents. In *Biogeography of the West Indies: Patterns and Perspectives*; Woods, C.A., Sergile, F.E., Eds.; CRC Press: Boca Raton, FL, USA, 2001; pp. 333–351.
124. Fabre, P.H.; Vilstrup, J.T.; Raghavan, M.; Der Sarkissian, C.; Willerslev, E.; Douzery, E.J.P.; Orlando, L. Rodents of the Caribbean: Origin and diversification of hutias unravelled by next-generation museomics. *Biol. Lett.* **2014**, *10*, 20140266. [CrossRef] [PubMed]

Article

New Distributional Records of *Phidippus* (Araneae: Salticidae) for Baja California and Mexico: An Integrative Approach

Luis C. Hernández Salgado [1], Dariana R. Guerrero Fuentes [2], Luz A. Garduño Villaseñor [1], Lita Castañeda Betancur [1], Eulogio López Reyes [1] and Fadia Sara Ceccarelli [3,*]

[1] Departamento de Biología de la Conservación, Centro de Investigación Científica y Educación Superior de Ensenada, Ensenada 22860, BC, Mexico; luiscarlos@cicese.edu.mx (L.C.H.S.); luzgv@cicese.edu.mx (L.A.G.V.); lita3023@hotmail.com (L.C.B.); elopez@cicese.mx (E.L.R.)

[2] Colección Nacional de Arácnidos, Instituto de Biología, Universidad Nacional Autonóma México, Mexico City 04510, DF, Mexico; darguerrero03@gmail.com

[3] CONACYT-Departamento de Biología de la Conservación, Centro de Investigación Científica y Educación Superior de Ensenada, Ensenada 22860, BC, Mexico

* Correspondence: ceccarelli@cicese.mx

Abstract: Because of its heterogeneity in ecoregions and its varied topography, the Mexican peninsula of Baja California (BCP) is an area of high diversity for many taxa, including spiders. However, a paucity of studies means that the diversity of BCP's spiders is generally poorly known. The North American jumping spider genus *Phidippus* comprises over 60 species, of which approximately 45% are found in Mexico. Among those, 6 have been recorded to date from the BCP but adding up the species recorded in nearby states, up to 20 more can be expected. As part of a larger study on the evolution and biogeography of the North American genus *Phidippus*, the aim here was to explore the diversity of the genus in the BCP using an integrative taxonomic approach and to present new distributional records. Until now, at least ten species have been collected from the BCP, one of which is a new record for Mexico, three new records for the BCP, and at least one undescribed species.

Keywords: distribution; diversity; Salticidae

Citation: Hernández Salgado, L.C.; Guerrero Fuentes, D.R.; Garduño Villaseñor, L.A.; Castañeda Betancur, L.; López Reyes, E.; Ceccarelli, F.S. New Distributional Records of *Phidippus* (Araneae: Salticidae) for Baja California and Mexico: An Integrative Approach. *Diversity* **2022**, *14*, 159. https://doi.org/10.3390/d14030159

Academic Editor: Matjaž Kuntner

Received: 28 September 2021
Accepted: 16 November 2021
Published: 24 February 2022

Publisher's Note: MDPI stays neutral with regard to jurisdictional claims in published maps and institutional affiliations.

1. Introduction

Biogeographical and biodiversity studies rely primarily on knowing the complete distribution of focal taxa as well as the total number of species present in an area. To date, numerous taxonomic groups and vast areas are poorly known owing to a paucity of studies. One such area is the Mexican Baja California Peninsula (BCP), which is the world's second longest peninsula, situated between the latitudes 23° N and 32° N, which means that its climates range from temperate to subtropical. The peninsula's location and orography allow for a variety of ecoregions to exist within the region, such as Mediterranean coastal scrub, mountain coniferous forest, and deserts [1]. While the BCP's relative isolation confers a high number of endemics, especially in plant species [2], the peninsula lies parallel to the mainland, and therefore dispersals and species interchanges are possible from the east as well as from the north. So, considering the geographic and phytogeographic similarities with southern California and Arizona in the United States, as well as Sonora and Sinaloa in Mexico, shared distributional patterns between taxa are expected [3]. Nevertheless, most studies of the region's fauna have focused on vertebrate taxa, while ecologically important groups, such as spiders, have received proportionally much less attention. For the BCP, and particularly its southernmost region, there are approximately 411 described spider species [4].

The spider family Salticidae has the largest number of genera (646) and species (>6000) within the Araneae, comprising about 13% of the order's species [5,6]. A well-known and charismatic salticid genus, *Phidippus* has 76 described species and includes some of the largest jumping spiders in the world, some reaching up to 20 mm in length [7,8]. This genus is distributed from North to Central America, including the Caribbean.

In 2004, Edwards carried out a complete revision of the genus, including a phylogenetic hypothesis based on morphological data. Based on this revision, Mexico contains 36 species of *Phidippus*, while the BCP has the following 6 recorded species [7,9,10]: *P. boei* Edwards, 2004, *Phidippus californicus* Peckham & Peckham, 1901, *P. carneus* Peckham & Peckham, 1896, *P. johnsoni* (Peckham & Peckham, 1883), *P. nikites* Chamberlin & Ivie, 1935 and *P. phoenix* Edwards, 2004. However, considering geographic distances and similarities in habitats between California, Arizona, Sonora, and Sinaloa, of the 26 *Phidippus* species found in the region, several could be present in the BCP. Despite their size and charismatic color patterns, new species of *Phidippus* are still being found and described, the latest as recently as 2020 [9]. Thus, there also exists a possibility that in the BCP there may be new *Phidippus* species that are not formally described.

In addition to traditional morphological taxonomic methods, in the past decade and a half, a commonly used data source for species identification in animals, including spiders, has been the DNA "barcoding" region [11–15]. Despite its advantages, such as its universality and a discernable threshold between inter- and intraspecific nucleotide diversity for most taxonomic groups [16,17], the "barcode" region has also been the source of controversy, owing to limitations in correctly delimiting species arising from mitochondrial mechanisms such as incomplete lineage sorting and introgression [18–20] and dependence on the analytic method [21]. Therefore, integrative approaches are preferred over standalone methods [22]. In this study, a combination of morphological examinations, especially adult male and female genitalia [23] and DNA "barcoding" for species identification, is used to explore the diversity of *Phidippus* in the peninsula.

2. Materials and Methods

2.1. Fieldwork

Fieldwork was carried out from 2017 to 2021 in a wide range of habitats from shrubland, palm oasis, and pine forests to highly modified rural and urban sites throughout the Baja California Peninsula. The sampling was carried out manually, and *Phidippus* specimens collected were preserved in 96% ethanol at −20 °C. The specimens were placed in the Museum of Arthropods of Baja California (MABC), located at the Ensenada Center for Scientific Research and Higher Education (CICESE) in Baja California, Mexico.

2.2. Morphological Examinations

To identify the collected individuals to species level, the taxonomic work of Edwards [7] was used as a reference to identify adult male and female specimens. Body terminology is standard for spiders; genitalia terminology follows Maddison [24]. The following abbreviations are used in the text: ALE—anterior lateral eyes, AER—anterior eyes row, PLE—posterior lateral eyes, PME—posterior median eyes.

The male and female adult genitalia were dissected and examined under a stereoscope and immersed in 96% alcohol to determine the species. The epigynes were previously cleared following the protocol proposed by Guerrero-Fuentes and Francke [25] but omitting the steps involving hydrochloric and glacial acetic acid. Digital photos of selected jumping spiders were taken using a LUMIX DFC490 camera mounted on a Nikon Z16 APO-A stereo microscope.

2.3. DNA "Barcode" Analysis

For the species collected in the BCP, the prosoma and legs were used for DNA extraction, using the DNeasy Blood & Tissue Kit by Qiagen. Polymerase Chain Reaction (PCR) was carried out following the protocol proposed by the Canadian Centre for DNA Barcoding (CCDB) [22]. The primers used for COI amplification were LCO-1490: 5′-GGTCAACAAATCATAAAGATATTGG-3′; HCO-2198: 5′-TAAACTTCAGGGTGACCAAA AAATCA-3′, C1-N-2191 (Nancy): 5′-CCCGGTAAAATTAAAATATAAACTTC-3′ and C-

1-J-1751 (Ron): 5′-GGAGCTCCTGACATAGCATTCCC-3′. PCR products for 121 newly collected specimens from the BCP were sent to Macrogen, Inc., Korea, for sequencing.

The sequences were edited and assembled with Geneious Prime 2021.2.2 and Sequencher v 4.1.4. The 121 newly generated "barcode" sequences, all the same length, were uploaded to the Bold Systems v4 database [26] and a Barcode Gap Analysis (BGA) was carried out using the Kimura 2 Parameter substitution model with MUSCLE alignment and pairwise gap deletion to corroborate species identities, particularly in groups where there were no adult specimens. Additionally, 23 reference sequences from previously identified individuals from other localities (Guerrero-Fuentes, in prep), were used in this study to corroborate morphological identifications. The reference sequences were selected for species known to occur in the BCP or nearby states from the U.S.A. (California and Arizona) and Mexico (Sonora and Sinaloa). A list of the species used for reference can be found in Table 1. In addition to the species listed in Table 1, reference sequences belonging to *P. bidentatus* and *P. cruentus* were included because these two species are widespread in Mexico, and their complete distribution is likely to be unknown. Since most of the reference sequences were obtained with a different primer set, they are missing the first ca. 240 nucleotides of the DNA "barcoding" region. So, rather than using BOLD tools, where missing data are detrimental to calculating genetic distances, a Bayesian phylogenetic tree was reconstructed, since Bayesian phylogenetic relationships can be accurately inferred despite missing data in the matrix [27]. The newly obtained sequences, the reference sequences, plus a sequence belonging to the jumping spider species *Habronattus borealis* as an outgroup taxon were aligned using the MAFFT v 7 server [28]. Nucleotide substitution models and codon partitioning schemes were selected using Partition Finder v. 1.1.1 [29] under an AICc model selection, which resulted in the following suggested models and partitions: TrN + G for COI codon position 1, HKY + I for codon position 2, and TVM + G for codon position 3. Bayesian phylogenetic inference was then applied in MrBayes v. 3.2.6 [30], running 4 parallel Markov chains for 50 million generations, with a tree sampled every 5000th generation. A consensus tree was then built after discarding the first 25% as burn-in, and the tree was evaluated by looking for supported nodes (with posterior probabilities greater than 0.95), particularly with regards to species-level clades which included reference sequences.

In cases where there were inconsistencies between the Bayesian tree clades, BGA, and morphological identifications, decisions were made based on the reliability of morphological characters and the DNA "barcoding" sequence fragment. The newly obtained sequences were deposited in the NCBI's GenBank database (accession numbers can be found in the Supplementary Materials Table S1).

Table 1. List of *Phidippus* species currently recorded for the U.S. states of Arizona and California and the Mexican states of Sonora and Sinaloa, with a note for the species for which a reference sequence of COI was used.

Extended List of Species *	Species Included as Reference Sequence?
Phidippus adumbratus Gertsch, 1934	YES
Phidippus apacheanus Chamberlin & Gertsch, 1929	YES
Phidippus ardens Peckham & Peckham, 1901	NO
Phidippus asotus Chamberlin & Ivie, 1933	YES
Phidippus audax (Hentz, 1845)	YES
Phidippus aureus Edwards, 2004	YES
Phidippus boei Edwards, 2004	YES
Phidippus californicus Peckham & Peckham, 1901	YES
Phidippus carneus Peckham & Peckham, 1896	YES
Phidippus clarus Keyserling, 1885	YES
Phidippus comatus Peckham & Peckham, 1901	YES
Phidippus concinnus Gertsch, 1934	YES
Phidippus felinus Edwards, 2004	NO
Phidippus johnsoni (Peckham & Peckham, 1883)	YES

Table 1. *Cont.*

Extended List of Species *	Species Included as Reference Sequence?
Phidippus kastoni Edwards, 2004	YES
Phidippus nikites Chamberlin & Ivie, 1935	YES
Phidippus octopunctatus (Peckham & Peckham, 1883)	YES
Phidippus olympus Edwards, 2004	NO
Phidippus phoenix Edwards, 2004	YES
Phidippus pius Scheffer, 1906	YES
Phidippus tigris Edwards, 2004	NO
Phidippus toro Edwards, 1978	NO
Phidippus tux Pinter, 1970	NO
Phidippus tyrannus Edwards, 2004	NO
Phidippus tyrelli Peckham & Peckham, 1901	NO
Phidippus pacosauritus Edwards, 2020	NO

* Species currently recorded from the BCP are shown in **bold**.

3. Results

Specimens belonging to the genus *Phidippus* were found in 16 different localities of the Baja California Peninsula (BCP; Supplementary Information Table S1). For this study, a total of 121 newly collected individuals belonging to the genus *Phidippus* were used for DNA "barcoding", of which 75 were adults and were thus examined morphologically and assigned to 9 described and 1 undescribed species.

3.1. Taxonomy

Phidippus adumbratus Gertsch, 1934. New record for Mexico
Supplementary Materials Figure S1
Examined material. MEXICO, Baja California: 3 females (Ph083, Ph088, Ph0147), Sierra Blanca (32.0749° N, −116.4528° W, 600 m), Municipio de Ensenada, 27.I.2020, B. Meza leg.; 1 male (Ph076), Mesa Escondido (29.8027° N, −114.7355° W, 872 m), San Antonio de Las Minas, Municipio de Ensenada, 29.IX.2020, D. Ward Jr leg.
Distribution. California, USA. Baja California, Mexico.
Diagnosis.
Male. Carapace dorsal view, ocular quadrangle covered with gray iridescent scales, median and posterior bands with orange to reddish scales; in frontal view, cheek band strongly marked with white scales and extended to PLE, eyes area with white scales too. Chelicerae are iridescent and striped with a vertical fringe of white and brown setae. Leg fringes are strongly dense and alternating black and white. Abdomen dorsally covered with red scales, basal white band present. Palp with white dorsal stripe from femur to cymbium; embolus is a long, very thin, and recurved spike; palea wider than long, ectal and retrolateral margins smooth.
Female. Carapace covered with sparse white scales, in dorsal view with median ocular band with reddish scales; ocular quadrangular with sparse and tan scales, in frontal view, cheek band and eyes area with white scales. Abdomen covered red with spots, basal, and lateral bands white. Epigynum with long length flaps and straight posteriorly, septum rudimentary to absent, without sagittal ridge, middle slightly depressed, copulatory ducts with one pair of supernumerary bends.
Phidippus boei Edwards, 2004
Supplementary Materials Figure S2
Examined material. MEXICO, Baja California: 1 female (Ph013), Santa Catarina (29.5981° N, −114.2245° W), Municipio de San Quintín, 05.VI.2019.
Distribution. Southern California, USA; Baja California and Baja California Sur, Mexico.
Diagnosis.
Female. Carapace black; in frontal view, cheek band weakly marked with gray scales. Abdomen is dorsally totally black covered with red scales except on median black stripe, without spots. Epigynum with large length flaps and straight posteriorly, septum absent to distinct and without sagittal ridge, middle slightly depressed, copulatory ducts without supernumerary bends.

Phidippus californicus Peckham & Peckham, 1901
Supplementary Materials Figure S3
Examined material. MEXICO, Baja California: 2 male, 1 female (Ph073, Ph099, Ph103), Rancho Mil (32.1205° N, −115.2611° W), Ejido El Mayor, Municipio de Mexicali, 23.III.2020, E. López and H. P. Murillo leg.; 1 female (Ph106), 23.IV.2020, same locality, E. López leg.

Distribution. Arizona, California, New Mexico, Oregon, Texas, and Utah, USA; Baja California, Baja California Sur, Chihuahua, Sinaloa, and Sonora, Mexico.

Diagnosis.

Male. Carapace black with white broad submarginal band from PME to thoracic slope or absent; in frontal view, cheek band weakly marked with gray scales. Chelicerae are iridescent and glabrous. Leg fringes are dense and alternating black and white. Abdomen is dorsally covered with red scales except the medial black stripe, basal white band present, white spots or absent. Palp with white dorsal stripe from femur to tibia, cymbium with black setae; embolus is a long, thin, and slightly recurved spike; palea as long as wide, ectal and retrolateral margins smooth.

Female. Carapace in dorsal view with median ocular band present or absent; ocular quadrangular with sparse white scales, in frontal view, cheek band broad and white. Abdomen covered red; spots, basal, and lateral bands white. Epigynum with medium length flaps and straight posteriorly, without septum and sagittal ridge, middle slightly depressed, copulatory ducts with one pair of supernumerary bends.

Phidippus comatus Peckham & Peckham, 1901. New record for Baja California
Supplementary Materials Figure S4
Examined Material. MEXICO, Baja California: 1 male (MABC-B001), Road to Pino Suárez (32.4099° N, −116.2761° W), Japá, Municipio de Tecate, 02.V.2018, E. López leg.

Distribution. Saskatchewan, Canada; Arizona, California, New Mexico, Nevada, Oregon, Texas, Utah, Washington, and Wyoming, USA; Chihuahua, Coahuila, Durango, Guanajuato, and Hidalgo, Mexico.

Diagnosis.

Male. Carapace, the median ocular tufts replaced by a dense, horizontal setal crests; ocular quadrangle covered with tan scales, median ocular band white, submarginal band broad from ALE to thoracic slope, cheek band white. Chelicera iridescent and vertically striped with white setae. Leg fringes are dense and alternating black and white. Femur I, ventrally with dark metallic blue distal bulge with gray tuft. Abdomen is dorsally covered with tan scales, basal band white, and white spots are present. Palp with white dorsal stripe from femur to the basal edge of the cymbium; embolus is a long, and recurved spike; palea wider than long, ectal and retrolateral margins smooth.

Phidippus johnsoni (Peckham & Peckham, 1888)
Supplementary Materials Figure S5
Examined material. MEXICO, Baja California: 1 female (Ph023), El Mogor (32.0339° N, −116.6038° W, 376 m), Valle de Guadalupe, Municipio de Ensenada, 07.V.2018, R. Santos, K. Munguía and E. López leg.; 1 female (Ph015), same locality, 11.I.2019, F. S. Ceccarelli, E. López, K. Munguía and H. P. Murillo leg.; 1 male (Ph022), same locality, 23.IV.2019, E. López leg.; 1 female (Ph014), same locality, 17.VII.2019, J. Quintana leg.; 1 male, 1 female (Ph011, Ph058), same locality, 13.XI.2019, K. Munguía, F.S. Ceccarelli and E. López leg.; 1 male, 3 females (Ph025, Ph091, Ph109, Ph115), same locality, 17.XII.2019, E. López and A. López leg.; 2 males (Ph100, Ph116), same locality, 15.I.2020, K. Munguía and E. López leg.; 1 male, 1 female, juvenile (Ph059, Ph093, Ph037), same locality, 23–24.I.2020, V. Aguilera and E. López leg.; 1 male (Ph098), same locality, 29.I.2020, E. López leg.; 1 female (Ph085), same locality, 14.II.2020; 1 female (Ph016), Xanic (32.0952° N, −116.5862° W), Valle de Guadalupe, 01.XI.2019, E. López leg.; 1 male (Ph067), Sexto Ayuntamiento (31.8812° N, −116.6447° W), Municipio de Ensenada, 22.02.2020, A. Alfaro leg.; 1 male (Ph064), El Sauzal (31.8679° N, −116.6690° W 32 m), Municipio de Ensenada, 20.02.2020, 1 male (Ph028), 27.02.2020, 1 male (Ph062), 01.03.2020, 1 male (Ph153), 10.III.2020, same locality, L. A. Garduño leg.; 1 male (Ph066), UABC (31.8635° N, −116.6664° W), Ciudad de Ensenada,

Municipio de Ensenada, 11.III.2020, C. Baiza leg.; 2 females (Ph057, Ph092), Colonia Popular 89, (31.8958° N, −116.5622° W), Ciudad de Ensenada, Municipio de Ensenada, 10.IV.2019, H. P. Murillo leg.; 2 males (Ph035, Ph043), Fraccionamiento Carlos Pacheco I, Ciudad de Ensenada (31.8837° N, −116.6141° W), Municipio de Ensenada, 06.III.2020,E. López leg.; 1 male (Ph075), Ciudad de Ensenada (31.8637° N, −116.6476° W), Municipio de Ensenada, 16.III.2020, A. Aquino leg.; 1 male (Ph074), Playa San Miguel (31.9014° N, −116.731° W), Municipio de Ensenada, 22-III−2020, 1 male (Ph078), same locality, 05.IV.2020, F. S. Ceccarelli leg.; 1 female (Ph068) Ojos Negros (31.8180° N, −116.3865° W), Municipio de Ensenada, 15.II.2020; 1 male (Ph095) Maneadero (31.7189° N, −116.60° W), Municipio de Ensenada, 17-III-2019, D. Parra.

Distribution. Abundant throughout southwest Canada, western USA, and northwest Mexico.

Diagnosis

Male. Carapace is totally black in dorsal and lateral views, cheek band poorly marked with gray and iridescent scales in frontal view. Chelicerae are iridescent and glabrous. Leg fringes are poorly dense with black and white setae. Abdomen are covered with red scales on lateral edges or totally red; in some specimens, spots are barely visible. Palp without dorsal stripe; embolus is a short and thin spike; palea is distinctly longer than wide, ectal margin is squared and retrolateral margin is notched.

Female. Habitus like the male. Abdomen may present white spots; red scales are only on lateral edges. Epigynum with short length flaps and posteriorly divergent, septum and sagittal ridge present, middle slightly depressed, copulatory ducts with supernumerary bends.

Phidippus nikites Chamberlin & Ivie, 1935

Supplementary Materials Figure S6

Examined material. MEXICO, *Baja California*: 1 male (Ph021), El Mogor (32.0339° N, −116.6038° W, 376 m), Valle de Guadalupe, Municipio de Ensenada, 16.X.2019, E. López; 1 female (Ph080), 15.06.2020, same locality. *Baja California Sur*: 1 male (Ph135), Vizcaíno (27.2734° N, −113.5338), Municipio de Mulegé, 20.IX.2020, H. P. Murillo leg.

Distribution. California, Idaho, Nevada, and Oregon, USA; Baja California, Mexico.

Diagnosis.

Male. Carapace dorsum is totally covered with red scales; in the frontal view, cheek band is weakly marked with gray scales. Chelicerae are iridescent and glabrous. Leg fringes are poorly dense, alternating black and white, with a few reddish to orange scales. Abdomen dorsally covered with red scales. Palp without dorsal stripe, cymbium with black setae; embolus is a short and recurved blade; palea longer than wide, ectal margin extended distally and retrolateral margin notched.

Female. General color pattern is like the male. Carapace, cheek band broad and red. Abdomen is covered red; spots are not visible. Epigynum with short length flaps and convergent posteriorly, without septum and sagittal ridge, middle depressed, copulatory ducts with one to two pairs of supernumerary bends.

Phidippus octopunctatus (Peckham & Peckham, 1883). New record for Baja California.

Supplementary Materials Figure S7

Examined material. MEXICO, *Baja California*: 1 female, (Ph150), El Mogor (32.0339° N, −116.6038° W, 376 m), 20.VIII.2020, E. López leg.; 1 male (Ph087), El Sauzal (31.8679° N, −116.6690° W, 32 m), Municipio de Ensenada, 10.IX.2020, L. A. Garduño leg.

Distribution. Widespread from western to central USA; northern to central Mexico.

Diagnosis.

Male. Carapace dorsum is totally covered with gray scales; in the frontal view, cheek band is weakly marked with gray scales. Chelicerae are black, dull, and glabrous. Leg fringes are poorly dense, alternating black and white. Abdomen dorsally covered with gray scales. Palp with white or gray dorsal stripe, cymbium almost black, with some white setae; embolus is a long and slightly recurved spike; palea wider than long, ectal and retrolateral margins smooth

Female. General color pattern is like the male. Carapace, cheek band broad, with gray and white scales. Abdomen covered with gray scales, without spots or bands. Epigynum without flaps, anterior depressed, copulatory ducts with one pair of supernumerary bends.

Phidippus phoenix Edwards, 2004

Supplementary Materials Figure S8

Examined material. MEXICO, *Baja California*: 1 male (Ph070), Ampliación La Moderna (31.8693° N, −116.6431° W), Ciudad de Ensenada, Municipio de Ensenada, 16.III.2020; 1 male (Ph077), 01.III.2020; 1 male, 1 female (Ph071, Ph072), 23.III.2020; 1 male (Ph079), 24.IV.2020, same locality, F. S. Ceccarelli leg.; 1 female (Ph096), Punta Colonet (31.0764° N, −116.2761° W), Municipio de Ensenada, 05.VIII.2019, B. Meza leg.; 6 males, 8 females (Ph031, Ph032, Ph034,Ph045, Ph046, Ph047, Ph048, Ph050, Ph052, Ph053, Ph054, Ph055, Ph056, Ph060), same locality, 25.I.2020, L. A. Garduño, E. López, and H. P. Murillo leg.; 1 female (Ph102), Santa Catarina (29.598151° N, −114.224484° W), Municipio de San Quintín, 06.V.2019, K. Munguía leg.; 1 female (Ph010), 05.VI.2019, same locality, K. Munguía, E. López, and H. P. Murillo, leg.; 1 male (Ph090), CICESE (31.8657° N, −116.6625° W), El Sauzal, Ciudad de Ensenada, Municipio de Ensenada, 15.II. 2020, L. Sankey leg.; 1 male (Ph081), 01.III.2020, same locality, E. López leg.; 3 males (Ph030, Ph039, Ph069), 10–17.IV.2020, same locality, L. A. Garduño leg.

Distribution. Southern Arizona and southern California, USA; Baja California and Baja California Sur, Mexico.

Diagnosis.

Male. Carapace in dorsal view with white median ocular band; in frontal view, cheek band very broad with white scales. Chelicerae are iridescent and fringed, with white scales and setae. Leg fringes are dense and white. Abdomen dorsally covered with red scales. Palp with white dorsal stripe from femur to cymbium; embolus is a long and recurved spike; palea is wider than long, ectal and retrolateral margins smooth.

Female. Carapace in dorsal view with white median ocular band; in frontal view, cheek and submarginal bands fused, and colored white. Abdomen with lateral edges red or white; spots and lateral bands white. Epigynum with medium length flaps and posteriorly divergent, septum rudimentary and without sagittal ridge, middle slightly depressed, copulatory ducts with one pair of supernumerary bends.

Phidippus tux Pinter, 1970. New record for Baja California.

Supplementary Materials Figure S9

Examined material. MEXICO, *Baja California Sur*: 1 female (Ph158) Oasis Carambuche (26.1293° N, −112.0167° W, 300 m), La Purísima, Municipio de Comondú, IX.2020, H. P. Murillo leg.

Distribution. Arizona, USA; Nayarit, Jalisco and Sonora, Mexico.

Diagnosis.

Female. Carapace dorsum is totally covered with yellow scales; in the frontal view, cheek band is strongly marked with yellow to white scales, the area of eyes covered with brown to tan scales. Abdomen totally covered with yellow scales, or partially covered with a posterior abdominal area dark and U-shaped, spots and bands are white. Epigynum with medium length and wide flaps, divergent posteriorly; with septum and sagittal ridge, middle depressed, copulatory ducts without supernumerary bends.

The expanded distributions of the new records, *P. adumbratus*, *P. comatus*, *P. octopunctatus* and *P. tux*, can be found in Figure 1.

3.2. DNA "Barcoding"

Based on the BGA in the Bold Systems v4 database, the 121 individuals belong to 10 species (see Supplementary Materials Table S2). Individuals tentatively assigned to undescribed species (*Phidippus* spp. 1 and 3) were found to have a distance below 2% to their nearest neighbor, which grouped them with *P. boei*. The remaining species (9 identified and 1 unidentified) were consistently delimited as separate from each other, based on the BGA. The nucleotide alignment upon which the COI phylogenetic tree was based consisted of 145 taxa and 1204 sites, which included the "barcode" region as well as ca. 500 additional nucleotides because the reference sequences were amplified using different primer sets (Guerrero-Fuentes et al., in prep.). The Bayesian consensus tree recovered 12 lineages for the *Phidippus* species from the BCP (Figure 2). The sequences from the individuals that formed clades with the identified reference species, and which were also identified morphologically, belonged to the following six species: *P. adumbratus*, *P. boei*, *P. californicus*, *P. nikites*, *P. octopunctatus* and *P. phoenix*. A further two species, which were identified morphologically as *P. comatus* and *P. johnsoni*, did not form a clade in the tree with their respective reference species. Rather, *P. comatus* from the BCP fell into an unresolved group (albeit with the *P. comatus* reference species) and the *P. johnsoni* were in a clade with the *P. concinnus* reference species, sister to the *P. johnsoni* reference species. Reference sequences were not available for *P. tux*, so its identification was based on female genital morphology and other morphological traits. A further three well supported clades within the tree did not have any associated reference sequences. For two of these clades, morphological identification was not possible owing to a lack of adult specimens. For a third clade (*Phidippus* sp. 2), preliminary morphological examinations pointed to the species belonging to the *insignarius* group following Edwards' [7] revision for the genus.

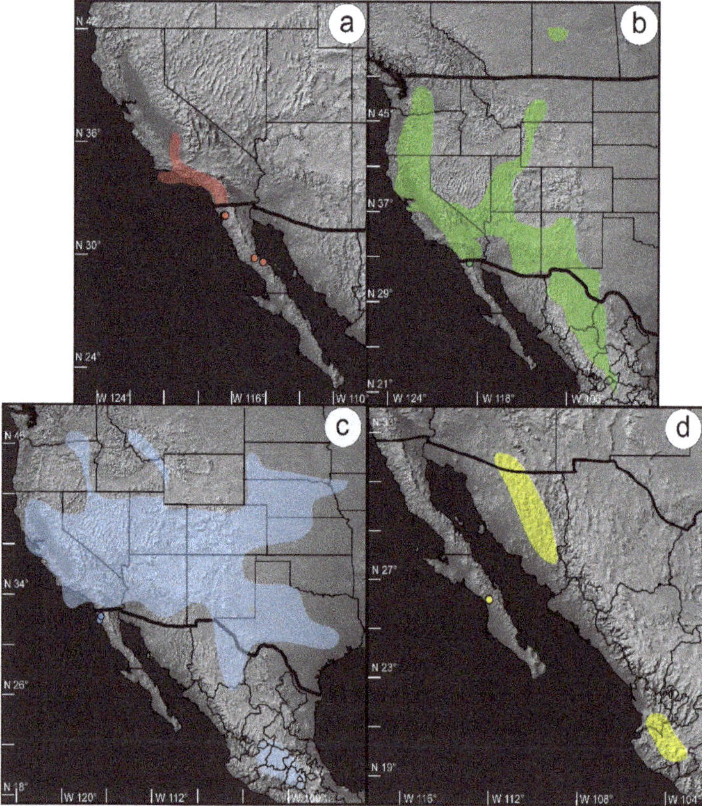

Figure 1. Distributional maps of the four *Phidippus* species with new distributional records in this study, namely, *P. adumbratus* (**a**), *P. comatus* (**b**), *P. octopunctatus* (**c**) and *P. tux* (**d**). Colored polygons represent the previously known distributions, based on Edwards (2004) and colored circles with black outlines represent the localities of the samples from this study. Country borders are shown with thick black lines, and thin black lines represent state borders for Mexico and the U.S.A. and Canadian provinces.

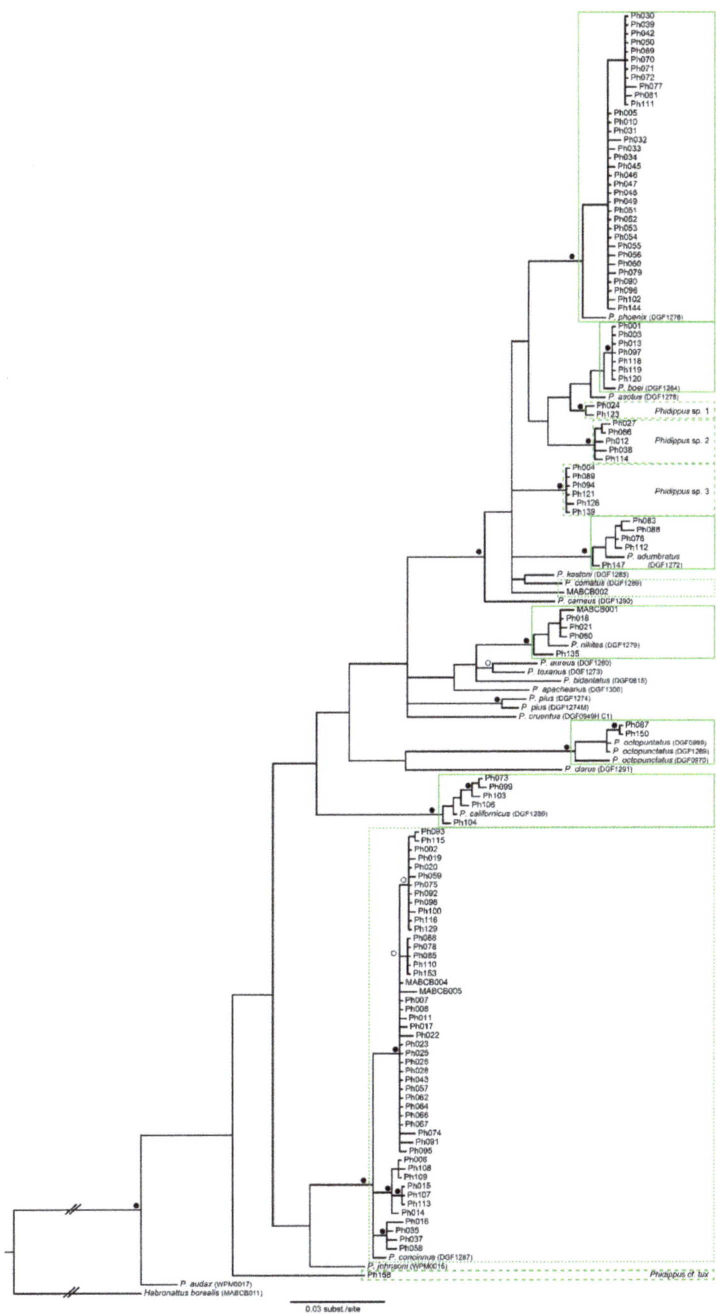

Figure 2. Bayesian phylogenetic tree of COI sequences of *Phidippus* individuals from this study, marked with DNA codes (cross-reference with Supplementary Materials Table S1) and reference sequences with codes in brackets following the species names. Green boxes with solid lines outline the species identified by morphology, while dotted lines represent species with uncertainties based on DNA sequences and/or unidentified specimens. Solid black circles at nodes represent posterior probability values >0.95 and white circles posterior probabilities between 0.9 and 0.95.

4. Discussion

In this study, no comments will be made with regards to the phylogenetic relationships between the species since many nodes are not supported, as COI is generally not ideal for resolving deeper relationships, added to the fact that the results here show a gene tree and not a species tree. It has been reported that the complex evolutionary dynamics of COI could contribute to misleading node resolution in jumping spiders [31–34]. In any case, in this study the tree was built for species grouping, which worked for most species, including the grouping of juveniles (which cannot be reliably identified using morphological features) with adults. The Barcode Gap Analyses helped to further discriminate between species. DNA barcoding has been shown to be a generally reliable method for discriminating species [12–14,35] and as part of a combined (integrative taxonomic) approach; the molecular and morphological data complemented each other for assigning individuals to *Phidippus* species.

Inconsistencies between morphological and COI tree-based identifications were found for two species and may be attributed to certain limitations commonly found in COI, such as introgressive hybridization and incomplete lineage sorting [18]. In the case of *P. comatus*, increased sampling throughout the species' distributional range, as well as more specimens belonging to sister taxa, may help resolve the nodes. The morphological examination of the male specimen from the BCP left no doubt about its identity, and the fact that it was collected near its known distributional range and in a similar environment and altitude to those reported by Edwards [7] ruled out the possibility of an accidental record, even though only a single individual was collected.

The incongruence between the COI of the BCP's *P. johnsoni* grouping (with *P. concinnus*) and morphological identification pointed to a slightly more complex situation. Genital examination clearly confirmed the distinct identities of *P. johnsoni* and *P. concinnus*. *P. johnsoni* is a widespread species, found from western Canada, throughout western USA, to as far south as northwestern Mexico. Close morphological examinations of *P. johnsoni* pointed to sympatric morphotypes with regard to abdomen color patterns, but a lack of population genetic studies on this species made it difficult to reach conclusions about the structuring of these populations and whether *P. johnsoni* may in fact be a species complex with conserved genital morphology. Thus, *P. concinnus* may have diverged from a most recent common ancestor of a *P. johnsoni* population. This would explain why the BCP *P. johnsoni* sequences cluster with *P. concinnus* instead of with its reference *P. johnsoni* sequence, which came from an individual collected in Grant County, Washington state, USA.

Since no reference sequences were available for *P. tux*, its identity was based solely on morphological examination of one single specimen collected during this study. A single individual might lead to doubts about whether it is an accidental record, perhaps a case of accidental faunal translocation. Further sampling in and around the locality where this *P. tux* individual was found, as well as from its complete distributional range, will be necessary for an in-depth study of how it got to the BCP. If indeed there is an established population of *P. tux* in the BCP, and since the locality lies in the southern part of the peninsula, it is likely that the population is genetically closer to the *P. tux* populations from Mexico's west coast states and may either be a relictual population as a result of the peninsula's separation from the mainland during the last 10 million years [3,36], or it may be an established population following westward dispersal from the mainland to the peninsula.

The other confirmed species for which new records from Mexico and the BCP are presented are *P. octopunctatus*, which has a widespread distribution, and the new records are near the limits of its known distribution, and *P. adumbratus*, which until now was recorded from the California floristic province, an area of high endemism [37,38] located along the coast of the North American Pacific. In this study, *P. adumbratus* is newly recorded for Mexico, as well as from a new ecoregion, namely, Baja California's Central Desert ecoregion, as defined by Gonzalez-Abraham et al. [1]. Perhaps this distribution could be explained by Hill and Edwards' [39] hypothesis on the dispersal routes of *Phidippus* species since the Last Glacial Maximum (LGM; ~20 Ka), whereby *P. adumbratus* may have migrated from the southern part of the BCP northwards, reaching California as the climate changed and warmed.

Three clades in this study's COI Bayesian phylogeny did not include any of the reference species, perhaps because they are species for which reference sequences were not available, or perhaps they are, to date, undescribed species. For two of the unidentified morphospecies, adult specimens were not available; however, the BGA placed them with *P. boei*. A third species probably belongs to the *insignarius* group following Edwards' [7] classification. However, to accurately determine whether these individuals belong to undescribed species, further sampling and a full taxonomic work will be necessary.

In the Baja California Peninsula, many unexplored places are difficult to access for sampling, and many *Phidippus* species, despite their relatively large size, are difficult to find in the field owing to habits such as hiding at the base of dense cactus spines, which complicates collecting. Although this study contributed to knowledge of the diversity of spiders in the BCP and the distributional range and richness of *Phidippus*, which increased from six to nine species, increased sampling efforts are required to uncover the BCP's true richness and diversity of *Phidippus* spiders. After the present contribution, the number of known spider species for the BCP increased from 396 to 400, and the diversity of Salticidae in the BCP increased from 37 to 41 species, based on the most recent data published on the diversity of the peninsula's spiders [4,10]. Furthermore, several *Phidippus* species are more widely distributed than previously thought. This new information has direct implications for both ecological and historical biogeographic studies. For ecological biogeography, such as Species Distribution Modeling, a higher number of known distributional datapoints allow for models with greater accuracy and precision [40]. As for historical biogeographical implications, the fact that the northern part of the BCP was found to harbor a great diversity of *Phidippus* suggests that it could be an ancestral area for at least some taxa. Additionally, several taxa may have dispersed to the BCP, as proposed by Hill and Edwards [39]; however, this hypothesis only considers a fraction of the diversity present in the BCP. Given the fact that the species found in the BCP belong to different species groups as defined by Edwards [7], species richness as well as phylogenetic diversity [41] for *Phidippus* is likely to be high for this area. A more in-depth molecular phylogenetic study of *Phidippus* and related genera will shed more light on the historical biogeographic and macroevolutionary processes of these spiders.

Supplementary Materials: The following are available online at https://www.mdpi.com/article/10.3390/d14030159/s1. Table S1: Individual codes, species identification, collection information and GenBank accession numbers for the COI sequences of the *Phidippus* species collected in the Baja California Peninsula for this study; Table S2: Barcode Gap Analysis result. Figure S1: Photographs of *Phidippus adumbratus* Gertsch, 1934; Figure S2: Photographs of *Phidippus boei* Edwards, 2004; Figure S3: Photographs of *Phidippus californicus* Peckham & Peckham, 1901; Figure S4: Photographs of *Phidippus comatus* Peckham & Peckham, 1901; Figure S5: Photographs of *Phidippus johnsoni* (Peckham & Peckham, 1888); Figure S6: Photographs of *Phidippus nikites* Chamberlin & Ivie, 1935; Figure S7: Photographs of *Phidippus octopunctatus* (Peckham & Peckham, 1883); Figure S8: Photographs of *Phidippus phoenix* Edwards, 2004; Figure S9: Photographs of *Phidippus tux* Pinter, 1970.

Author Contributions: Conceptualization, L.C.H.S., L.A.G.V. and F.S.C.; methodology, L.C.H.S., L.A.G.V., D.R.G.F., E.L.R., F.S.C. and L.C.B.; formal analysis, L.C.H.S. and F.S.C.; investigation, L.C.H.S., L.A.G.V., D.R.G.F., F.S.C.; resources, F.S.C.; data curation, L.C.H.S., L.A.G.V., D.R.G.F., F.S.C.; writing—original draft preparation, L.C.H.S., L.A.G.V., D.R.G.F., F.S.C.; writing—review and editing, L.C.H.S., L.A.G.V., D.R.G.F., F.S.C.; visualization, L.C.H.S. and F.S.C.; supervision, F.S.C.; project administration, F.S.C. and E.L.R.; funding acquisition, F.S.C. All authors have read and agreed to the published version of the manuscript.

Funding: This work was funded by Fondo Sectorial de Investigación para la Educación SEP-CONACYT Investigación Científica Básica 2017–2018 grant A1-S-15134 to F.S.C.; Jiji Foundation grant of the International Community Foundation awarded to F.S.C. (award number 20180127) and Becas Nacionales CONACYT to Luis Carlos Hernández.

Institutional Review Board Statement: Not applicable.

Data Availability Statement: DNA sequences derived from this study are deposited in the NCBI's GenBank database (accession numbers can be found in the Supplementary Materials Table S1).

Acknowledgments: We thank Alberto López Aleman, Ana Alfaro, Andrés Martínez Aquino, Bill and Mary Clark, Bryan Meza, Christian Baiza, D. Parra, Damian Elías, David M Ward Jr, Diego de Pedro, Edna Arvizu, Elvia Plascencia, Guilhermi Azevedo, Heriberto Pérez Murillo, Jorge Quintana, Khutzy Munguía Ortega, Le Roy Sankey, Mario Salazar Ceseña, Marshal Hedin, and Sergio Gonzalez Piñuelas, for helping us collect *Phidippus*. We also thank G.B. Edwards for his comments on the identification of species, in particular, for confirming the identity of *Phidippus tux*, and Aurora Cabrera and Ana Salgado for their contribution to the English language revision of this manuscript. We also thank Matjaž Kuntner and three anonymous referees for their constructive comments on earlier versions of the manuscript. We would also like to acknowledge the financial support granted by the Posgrado en Ciencias de la Vida, CICESE, to L.C.H.S. for a research stay.

Conflicts of Interest: The authors declare no conflict of interest.

References

1. González-Abraham, C.E.; Garcillán, P.P.; Ezcurra, E.; Trabajo de Ecorregiones, G.d.T.d.E. Ecorregiones de La Península de Baja California: Una Síntesis. *Bot. Sci.* **2010**, *87*, 69. [CrossRef]
2. Rebman, J.P.; Gibson, J.; Rich, K. Annotated Checklist Of The Vascular Plants Of Baja California, Mexico. *Proc. San Diego Soc. Nat. Hist.* **2016**, *45*, 1–24.
3. Riddle, B.R.; Hafner, D.J.; Alexander, L.F.; Jaeger, J.R. Cryptic Vicariance in the Historical Assembly of a Baja California Peninsular Desert Biota. *Proc. Natl. Acad. Sci. USA* **2000**, *97*, 14438–14443. [CrossRef] [PubMed]
4. Jiménez, M.L.; Palacios-cardiel, C.; Maya-morales, J.; Edwin Berrian, J.; Ibarra Núñez Jiménez, G. Nuevos Registros De Arañas (Arachnida: Araneae) Para La Región Del Cabo, Península De Baja California, México New Records Of Spiders (Arachnida: Araneae) For Cape Region, Baja California Peninsula, Mexico. *Acta Zool. Mex.* **2018**, *34*, 1–13. [CrossRef]
5. Peng, X.J.; Tso, I.M.; Li, S.Q. Five New and Four Newly Recorded Species of Jumping Spiders from Taiwan (Araneae: Salticidae). *Zool. Stud.* **2002**, *41*, 1–11.
6. World Spider Catalog. Available online: https://wsc.nmbe.ch/ (accessed on 31 August 2020).
7. Edwards, G.B. Revision of the Jumping Spiders of the Genus Phidippus (Araneae: Salticidae). *Occas. Pap. Florida State Collect. Arthropods* **2004**, *11*, 1–158.
8. Maddison, W.P. A Phylogenetic Classification of Jumping Spiders (Araneae: Salticidae). *J. Arachnol.* **2015**, *43*, 231. [CrossRef]
9. Edwards, G.B. Description of Phidippus Pacosauritus Sp. Nov. (Salticidae: Salticinae: Dendryphantini: Dendryphantina) with a Reanalysis of Related Species in the Mystaceus Group. *Peckhamia* **2020**, *221.1*, 1–18.
10. Richman, D.B.; Cutler, B.; Hill, D.E. Salticidae of North America, Including Mexico Salticidae of North America, Including Mexico. *Peckhamia* **2012**, *95.3*, 1–88.
11. Barrett, R.D.H.; Hebert, P.D.N. Identifying Spiders through DNA Barcodes. *Can. J. Zool.* **2005**, *83*, 481–491. [CrossRef]
12. Hebert, P.D.N.; Stoeckle, M.Y.; Zemlak, T.S.; Francis, C.M. Identification of Birds through DNA Barcodes. *PLoS Biol.* **2004**, *2*, 1657–1663. [CrossRef]
13. Blagoev, G.A.; deWaard, J.R.; Ratnasingham, S.; deWaard, S.L.; Lu, L.; Robertson, J.; Telfer, A.C.; Hebert, P.D.N. Untangling Taxonomy: A DNA Barcode Reference Library for Canadian Spiders. *Mol. Ecol. Resour.* **2016**, *16*, 325–341. [CrossRef] [PubMed]
14. Naseem, S.; Tahir, H.M. Use of Mitochondrial COI Gene for the Identification of Family Salticidae and Lycosidae of Spiders. *Mitochondrial DNA Part A DNA Mapp. Seq. Anal.* **2016**, *29*, 96–101. [CrossRef] [PubMed]
15. Astrin, J.J.; Höfer, H.; Spelda, J.; Holstein, J.; Bayer, S.; Hendrich, L.; Huber, B.A.; Kielhorn, K.H.; Krammer, H.J.; Lemke, M.; et al. Towards a DNA Barcode Reference Database for Spiders and Harvestmen of Germany. *PLoS ONE* **2016**, *11*, e0162624. [CrossRef] [PubMed]
16. Čandek, K.; Kuntner, M. DNA Barcoding Gap: Reliable Species Identification over Morphological and Geographical Scales. *Mol. Ecol. Resour.* **2014**, *15*, 268–277. [CrossRef] [PubMed]
17. Coddington, J.A.; Agnarsson, I.; Cheng, R.C.; Čandek, K.; Driskell, A.; Frick, H.; Gregorič, M.; Kostanjšek, R.; Kropf, C.; Kweskin, M.; et al. DNA Barcode Data Accurately Assign Higher Spider Taxa. *PeerJ* **2016**, *2016*, e2201. [CrossRef] [PubMed]
18. Funk, D.J.; Omland, K.E. Species-Level Paraphyly and Polyphyly: Frequency, Causes, and Consequences, with Insights from Animal Mitochondrial DNA. *Annu. Rev. Ecol. Evol. Syst.* **2003**, *34*, 397–423. [CrossRef]
19. Vences, M.; Thomas, M.; Bonett, R.M.; Vieites, D.R. Deciphering Amphibian Diversity through DNA Barcoding: Chances and Challenges. *Philos. Trans. R. Soc. B Biol. Sci.* **2005**, *360*, 1859–1868. [CrossRef] [PubMed]
20. Ceccarelli, F.S.; Sharkey, M.J.; Zaldívar-Riverón, A. Species Identification in the Taxonomically Neglected, Highly Diverse, Neotropical Parasitoid Wasp Genus Notiospathius (Braconidae: Doryctinae) Based on an Integrative Molecular and Morphological Approach. *Mol. Phylogenet. Evol.* **2012**, *62*, 485–495. [CrossRef]

21. Meier, R.; Blaimer, B.B.; Buenaventura, E.; Hartop, E.; von Rintelen, T.; Srivathsan, A.; Yeo, D. A Re-Analysis of the Data in Sharkey et Al.'s (2021) Minimalist Revision Reveals That BINs Do Not Deserve Names, but BOLD Systems Needs a Stronger Commitment to Open Science. *Cladistics* **2021**, 1–12. [CrossRef] [PubMed]
22. Schlick-Steiner, B.C.; Steiner, F.M.; Seifert, B.; Stauffer, C.; Christian, E.; Crozier, R.H. Integrative Taxonomy: A Multisource Approach to Exploring Biodiversity. *Annu. Rev. Entomol.* **2010**, *55*, 421–438. [CrossRef]
23. Huber, B.A. The significance of copulatory structures in spider systematics. In *Biosemiotik-Praktische Anwendung und Konsequenzen Fur Die Einzelwissenschaften*; Schult, J., Ed.; VWB Verlag: Berlin, Germany, 2004; pp. 89–100.
24. Maddison, W.P. Pelegrina Franganillo and Other Jumping Spiders Formerly Placed in the Genus Metaphidippus (Araneae: Salticidae). *Bull. Museum Comp. Zool.* **1996**, *154*, 215–368.
25. Guerrero-Fuentes, D.R.; Francke, O.F. Taxonomic Revision of Anicius Chamberlin, 1925 (Araneae: Salticidae), with Five New Species of Jumping Spiders from Mexico. *Zootaxa* **2019**, *4638*, 485–506. [CrossRef] [PubMed]
26. Ratnasingham, S.; Hebert, P.D.N. BOLD: The Barcode of Life Data System: Barcoding. *Mol. Ecol. Notes* **2007**, *7*, 355–364. [CrossRef] [PubMed]
27. Wiens, J.J.; Moen, D.S. Missing Data and the Accuracy of Bayesian Phylogenetics. *J. Syst. Evol.* **2008**, *46*, 307–314. [CrossRef]
28. Katoh, K.; Standley, D.M. MAFFT Multiple Sequence Alignment Software Version 7: Improvements in Performance and Usability. *Mol. Biol. Evol.* **2013**, *30*, 772–780. [CrossRef]
29. Lanfear, R.; Calcott, B.; Ho, S.Y.W.; Guindon, S. PartitionFinder: Combined Selection of Partitioning Schemes and Substitution Models for Phylogenetic Analyses. *Mol. Biol. Evol.* **2012**, *29*, 1695–1701. [CrossRef] [PubMed]
30. Ronquist, F.; Teslenko, M.; Van Der Mark, P.; Ayres, D.L.; Darling, A.; Höhna, S.; Larget, B.; Liu, L.; Suchard, M.A.; Huelsenbeck, J.P. Mrbayes 3.2: Efficient Bayesian Phylogenetic Inference and Model Choice across a Large Model Space. *Syst. Biol.* **2012**, *61*, 539–542. [CrossRef] [PubMed]
31. Bodner, M.R.; Maddison, W.P. The Biogeography and Age of Salticid Spider Radiations (Araneae: Salticidae). *Mol. Phylogenet. Evol.* **2012**, *65*, 213–240. [CrossRef]
32. Hedin, M.; Maddison, W.P. A Combined Molecular Approach to Phylogeny of the Jumping Spider Subfamily Dendryphantinae (Araneae: Salticidae). *Mol. Phylogenet. Evol.* **2001**, *18*, 386–403. [CrossRef] [PubMed]
33. Maddison, W.P.; Li, D.; Bodner, M.; Zhang, J.; Xu, X.; Liu, Q.; Liu, F. The Deep Phylogeny of Jumping Spiders (Araneae, Salticidae). *ZooKeys* **2014**, *87*, 57–87. [CrossRef]
34. Maddison, W.P.; Maddison, D.R.; Derkarabetian, S.; Hedin, M. Sitticine Jumping Spiders: Phylogeny, Classification, and Chromosomes (Araneae, Salticidae, Sitticini). *ZooKeys* **2020**, *2020*, 1–54. [CrossRef] [PubMed]
35. Adeniran, A.A.; Hernández-Triana, L.M.; Ortega-Morales, A.I.; Garza-Hernández, J.A.; Cruz-Ramos, J.; J. de la Chan-Chable, R.J.; Vázquez-Marroquín, R.; Huerta-Jiménez, H.; Nikolova, N.I.; Fooks, A.R.; et al. Identification of Mosquitoes (Diptera: Culicidae) from Mexico State, Mexico Using Morphology and COI DNA Barcoding. *Acta Trop.* **2021**, *213*, 105730. [CrossRef] [PubMed]
36. Mulcahy, D.G.; Macey, J.R. Vicariance and Dispersal Form a Ring Distribution in Nightsnakes around the Gulf of California. *Mol. Phylogenet. Evol.* **2009**, *53*, 537–546. [CrossRef]
37. Burge, D.O.; Thorne, J.H.; Harrison, S.P.; O'Brien, B.C.; Rebman, J.P.; Shevock, J.R.; Alverson, E.R.; Hardison, L.K.; Rodríguez, J.D.; Junak, S.A.; et al. Plant Diversity and Endemism in the California Floristic Province. *Madroño* **2016**, *63*, 3–206. [CrossRef]
38. Vanderplank, S.E.; Rebman, J.P.; Ezcurra, E. Where to Conserve? Plant Biodiversity and Endemism in Mediterranean Mexico. *Biodivers. Conserv.* **2018**, *27*, 109–122. [CrossRef]
39. Hill, D.E.; Edwards, G.B. Origins of the North American Jumping Spiders (Araneae: Salticidae). *Peckhamia* **2013**, *107*, 1–67.
40. Stockwell, D.R.B.; Peterson, A.T. Effects of Sample Size on Accuracy of Species Distribution Models. *Ecol. Modell.* **2002**, *148*, 1–13. [CrossRef]
41. Faith, D.P. Conservation Evaluation and Phylogenetic Diversity. *Biol. Conserv.* **1992**, *61*, 1–10. [CrossRef]

Article

Beta Diversity along an Elevational Gradient at the Pico da Neblina (Brazil): Is Spider (Arachnida-Araneae) Community Composition Congruent with the Guayana Region Elevational Zonation?

André A. Nogueira [1,*], Antonio D. Brescovit [2], Gilmar Perbiche-Neves [1,3] and Eduardo M. Venticinque [4]

[1] Programa de Pós-Graduação em Ciências Biológicas (Zoologia), Instituto de Biociências, Campus de Botucatu, Universidade Estadual Paulista–UNESP, Botucatu 18618-689, SP, Brazil; gpneves@ufscar.br
[2] Laboratório de Artrópodes, Instituto Butantan, Av. Vital Brazil 1500, São Paulo 05503-900, SP, Brazil; Antonio.brescovit@butantan.gov.br
[3] Laboratório de Plâncton, Departamento de Hidrobiologia, Centro de Ciências Biológicas e da Saúde CCBS, Universidade Federal de São Carlos, São Carlos 13565-905, SP, Brazil
[4] Centro de Biociências, Departamento de Ecologia, Campus Universitário—Lagoa Nova, Universidade Federal do Rio Grande do Norte—UFRN, Natal 59072-970, RN, Brazil; eduardo.venticinque@ufrn.br
* Correspondence: andrearanhas@gmail.com

Citation: Nogueira, A.A.; Brescovit, A.D.; Perbiche-Neves, G.; Venticinque, E.M. Beta Diversity along an Elevational Gradient at the Pico da Neblina (Brazil): Is Spider (Arachnida-Araneae) Community Composition Congruent with the Guayana Region Elevational Zonation? *Diversity* 2021, 13, 620. https://doi.org/10.3390/d13120620

Academic Editor: Matjaž Kuntner

Received: 15 October 2021
Accepted: 16 November 2021
Published: 26 November 2021

Publisher's Note: MDPI stays neutral with regard to jurisdictional claims in published maps and institutional affiliations.

Abstract: Beta diversity is usually high along elevational gradients. We studied a spider community at the Pico da Neblina (Brazil), an Amazonian mountain which is one of the southern components of the Guayana region. We sampled six elevations and investigated if beta diversity patterns correspond to the elevational division proposed for the region, between lowlands (up to 500 m), uplands (500 m to 1500 m), and highlands (>1500 m). Patterns of dominance increased with elevation along the gradient, especially at the two highest elevations, indicating that changes in composition may be accompanied by changes in species abundance distribution. Beta diversity recorded was very high, but the pattern observed was not in accordance with the elevationaldivision proposed for the region. While the highlands indeed harbored different fauna, the three lowest elevationshad similar species compositions, indicating that the lowlands spider community extends into the uplands zone. Other measures of compositional change, such as similarity indices and species indicator analysis, also support this pattern. Our results, in addition to a revision of the literature, confirm the high diversity and endemism rates of montane spider communities, and we stress the importance of protecting those environments, especially considering the climate crisis.

Keywords: Arachnida; Araneae; biodiversity; community ecology; elevation; Pantepui; species turnover

1. Introduction

Montane biotas have always been a main source of interest for biologists. The drastic environmental changes observed in relatively short distances make mountain systems ideal for studies on diversity patterns [1–4] and for use as natural laboratories [5–7]. Montane biota also draws attention due to its usually large species richness and endemism levels [8,9]. Finally, due to the strong environmental gradients, montane biotas are also characterized by large beta diversity values, with several compositional changes in their communities along the gradients.

There is evidence of a gradual compositional change along elevational gradients [1,10–12] instead of important and localized discontinuities at a specific point. This could be attributed to the fact that some important environmental factors, such as temperature, decline continuously with elevation [13,14]. However, there are some empirical examples of more abrupt changes within elevational gradients [15,16]; evidence of well-defined elevational zonation of biotic communities.

227

In this study, we quantify the beta diversity component of a spider meta-community at an elevational gradient in Amazonia, the Pico da Neblina, the highest Brazilian mountain, as well as highest place in South America outside the Andes [17]. This mountain is part of the Guayana region, a very remote place in northern South America famous for its table-top mountains. A physiographical division based on elevation and temperature proposed for the region distinguished three main units, lowlands (up to 500 m, >24 °C annual average), uplands (from 500 to 1500 m, 18–24 °C), and highlands (>1500 m, 8–18 °C) [18], and the elevational distribution of the vegetation seems to support this division [19,20].

Spiders are good models for diversity studies [21,22]. One of the most species-rich orders (circa 49,700 species [23]), spiders are conspicuous elements in most ecosystems and can be sampled in large numbers in relatively short time and using low-cost sampling methods, such as nocturnal hand collecting, pitfall traps, and beating trays [24,25].

Our main interest is to verify the patterns of beta diversity of the spider community along the elevational gradient, and to assess if they are in accordance with the elevational division proposed for the Guayana region. We expect that changes in the composition ofspider communities are connected to those observed for vegetation, since the structure of the vegetation is considered an important environmental factor for spider species [26,27]. We also measure and discuss patterns of dominance (i.e., the relative importance of the most abundant species) and community structures along the gradient, as there is evidence for larger dominance at higher elevations [5,11,28]. Finally, we use species indicator analysis (SIA) to identify the degree of associations of the species with different elevations or elevational zones, which may represent an additional test of the fit of our data to the elevational division proposed for the region.

It is worth noting that knowledge on spider mountain fauna has been increasing in the last decades, with a few studies on the diversity of spider communities along elevational gradients [29–31], as well as several species described from montane ecosystems [32–34]. The results reported in these studies provide examples of the large diversity characteristic of mountain ecosystems, and also reports important compositional changes along the gradients. This increase of knowledge on montane spider fauna will allow us to evaluate our results in a broader context, and to look for more general patterns of spider beta diversity in elevational gradients, as well as to make some comments on the conservation of these mountain communities.

2. Material and Methods

2.1. Study Area

The study was conducted at the Pico da Neblina (00°48′07″ N e 66°00′40″ W). It belongs to the Pico da Neblina National Park (municipality of São Gabriel da Cachoeira, state of Amazonas, Brazil), one of the largest conservations units in Brazil (2,260,344.15 ha), and also to the Yanomami Indigenous Land, with which the Park overlaps. The Pico da Neblina lies in a mountain region which represents the watershed between the Amazon and Orenoco basins, as well as the boundary between Brazil and Venezuela [35].

As mentioned above, the Pico da Neblina is part of one of the southern and more isolated mountain components of the Guayana region (Figure 1), and is characterized by extensive high elevation plateaus (2000 to 2400 masl), although it does not present the typical tepui, table-top shape [18]. The annual average rainfall at lowlands is 3000 mm/year, without a dry season, and the average humidity is about 85–90% [33]. Rainfall increases with elevation until around 1800 m, being gradually replaced by a constant mist, and the average humidity reaches almost 100% [35].

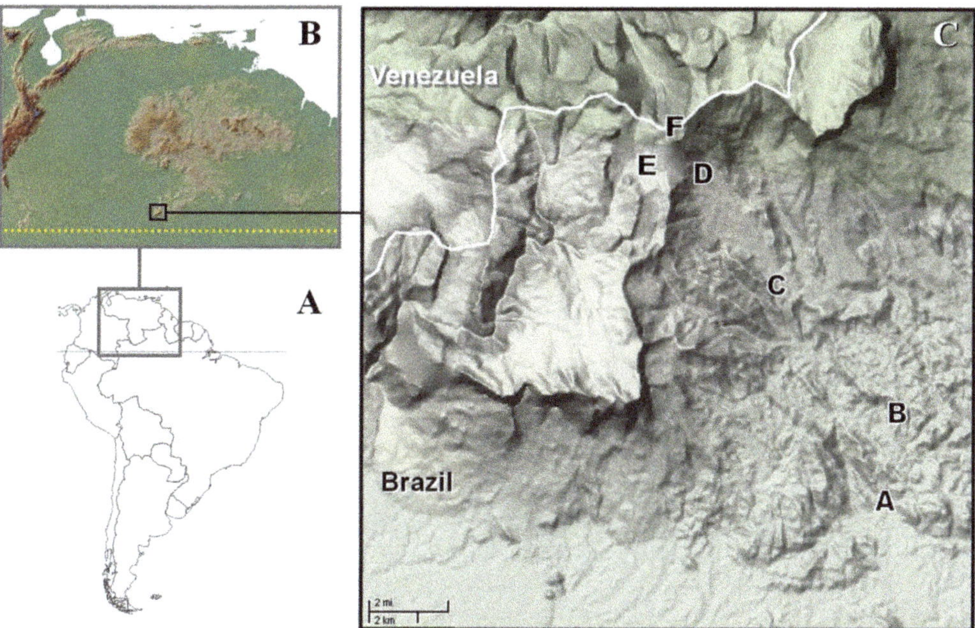

Figure 1. Study area. (**A**) South America; (**B**) northern South America (rectangle of map A, enlarged). The mountain range at the left of the map represents the northern part of the Andes, and the mountainous region in the center of the map is the Guayana Shield, showing the study area in its southern part, and dotted yellow line represents the equator. (**C**) Closer view of the study area (rectangle of map B enlarged), the Pico da Neblina. Letters represent the elevations sampled: A—100 m, B—400 m, C—860 m, D—1550 m, E—2000 m, F—2400 m.

Vegetation in the lowlands is composed ofa tall, evergreen forest, and the uplands are covered by montane forests, which havedecreasing biomass and tree size, especially when declivity is accentuated, leading to shallower soils [36]. In the highlands, forests are replaced by more open types of vegetation like high elevation scrublands and broad leave meadows, which grow on organic peat soils and on rocky substrates. At the Neblina, forests formation occurs almost up to 2000 m, and above that elevation their high-altitude formations stand out for their diversity and endemism [19]. Species from the families Bromeliacea, Rapateacea and Theacea are among the most characteristics elements of this flora. The flora of the region, especially from high altitudes is renowned for its high diversity and endemism [19], and the Guayanas highlands are considered as a discontinuous biogeographical province, also called Pantepui. Detailed information on the geology and vegetation of the region can be found in Berry and collaborators [37] and Berry and Riina [19].

2.2. Sampling Methods

Spiders were collected with two traditional methods in spider inventories [24]; beating tray and manual active search (i.e., hand collecting). In the first method, the understory vegetation, such as shrubs and small trees, is sampled through the beating of leaves, branches, vines, and other parts of the vegetation with a stick, while holding a 1 m^2 tray under it. The spiders falling in the tray are collected, and the sampling unit consisted of 20 of those beating events, in different plants, along a 30 m long transect. In the second method, spiders from the forest floor and from the understory are directly collected with the help of tweezers and/or plastic vials. The sampling unit represents one hour of search along an approximate area of 300 m^2 (30 × 10 m). The first method was employed during the day, from 8:00 to 11:00 h, and the second at night from 19:30 to 23:00 h. All spiders collected with both methods were immediately fixed in 70% ethanol.

Sampling was carried out by three collectors at six elevations, 100, 400, 860, 1550, 2000 and 2400 m. At each elevation we investigated three sites, about 100 m apart from each other. We obtained a total of 54 samples by elevation (27 of each method) resulting in a final count of 324 samples (162 of each method) for the Pico da Neblina. The sampling expedition occurred from 22 September 2007 to 13 October 2007, the local "dry season". We only identified adult spiders. Specimens were sorted into morphospecies usually by the first author and then identified until the lowest taxonomic level by specialists. Voucher specimens weredeposited at the collection of the Instituto Nacional de Pesquisas da Amazônia (INPA), at Manaus (AM), and duplicates weredeposited at the Instituto Butantan (IBSP), São Paulo (SP) and at the Museu Paraense Emílio Goeldi (MPEG), Belém (PA).

2.3. Analyses

For each elevation (considering the pooled data from the three sites by elevation), we calculated and present the following parameters: species richness, abundance, diversity, dominance, and proportion of singletons. In this study, we will refer to the number of species captured in our samples as species richness, and the number of specimens captured in our samples as abundance, although we are aware that the results of our sampling represent only an estimation of the real richness and abundance of the community.

As a diversity measure we used the exponential of the Shannon–Wiener Index, also known as the "numbers equivalent" or "effective number of species" of a given community. It represents the number of equally likely elements needed to produce the value of the diversity index. The use of number equivalents (D) over raw diversity indexes has been recommended [38] as this transformation allows a more intuitive interpretation. Unlike raw diversity indexes, which are nonlinear, number equivalents possess doubling propriety, i.e., if two completely distinct communities of equal size with a diversity D = X are combined, their diversity will be D = 2X [38]. Finally, it is convenient to stress the importance of using a measure of diversity that takes into account species abundance. Changes in relative abundances can be as perceptible as changes in species composition and their study allow a more accurate picture of the community than just species richness, for which dominant and rare species, often represented by just one individual, are given the same weight [38,39]. To measure the dominance, we used the Berger–Parker index [40], which is based on the proportional abundance of the single most dominant species. Dominance patterns by elevation were also assessed by the visual inspections of rank abundance plots. The proportion of singletons refers to the number of species represented by just one individual in a given elevation, regardless of the species total abundance. We also present the proportional distribution of absolute singletons, i.e., species represented by just one individual considering the total inventory, by elevation.

We calculated the beta diversity for three levels: among the three sampling sites within the same elevations; between different elevations, pooling the communities at each elevation and generating a distance matrix; and for the total inventory, including all elevations. We used a beta diversity based on the number equivalents (D), where beta D = gamma D/alpha D [39]. This procedure allows obtaining independent alpha and beta components, a logical principle often violated by traditional diversity indices [39]. The alpha is calculated as the sum of the weighted Shannon–Wiener index of each community (sites or elevations), and the weight represents the proportional abundance of each community in relation to the pooled abundance of all communities being compared.Gamma diversity is obtained by simply calculating the Shannon–Wiener index for the pooled community in question. After we obtained the alpha and gamma diversity, we converted them to its equivalent numbers (D gamma and D alpha) to calculated the beta diversity, which is expressed in number of communities, ranging from 1 (when all communities compared are identical) to N, which is the total number of communities being compared, when they are all completely different [39]. In our case, the maximum possible beta diversity, N, is either two, for pairwise comparisons between different elevations; three, for comparisons within elevations; or six, considering the six elevations of the whole gradient.

We used the Bray–Curtis index of similarity to generate a distance matrix for the 18 sampling sites and for the 6elevations sampled. We also constructed a matrix based on the proportion of species shared (in relation to the total richness for the pair of sites or elevations) between the 18 sampling sites and the 6elevations. We used the Bray–Curtis matrix (18 sampling sites) to perform a NMDS [41] and checked the stress, a measure of the fit between the final solution of the analysis and the original distance matrix of the community. This ordination technique has already been positively evaluated [42,43] even for dealing with species of rich and under sampled communities [44].

To assess the relation between beta diversity, similarity, spatial distance, and elevation, we generated distances matrices of those parameters for the six elevations and performed partial Mantel tests, based on 10,000 permutations for each test ($\alpha = 0.05$). Partial Mantel tests, through the Pearson correlation coefficient, assess the relation between two distances matrices, while controlling for the effect of a third matrix [45]. We related the two similarity indices between each other and to the two distance measures (spatial and elevation), successively controlling for space and elevation, and also without a control factor. We followed an approach suggested by Legendre [46] and performed the permutation on the residuals of a null model assuming the absence of effect of the third factor in a partial regression.

To verify if the changes in composition werein accordance with the division proposed for the Guayana region, we performed an analysis of similarity (ANOSIM), a non-parametric permutation procedure to test for significant differences in composition among differently grouped sampling units [47].We compared the fauna of three elevational groups; Lowlands (100 m and 400 m), Uplands (860 m and 1550 m), and Highlands (2000 m and 2400 m). To measure the similarity, we used the Bray–Curtis index. Significance levels were adjusted by a Bonferroni correction.

Finally, we used species indicator analysis (SIA) [48] to verify the association of the 157 most abundant species (represented by at least five individuals) with the different elevations. SIA calculates an indicator value (IV) based on the frequency and abundance with which a species occurs at the sites of a given category, and then tests if the IV differs significantly from random based on a Monte Carlo permutation (n = 1000). The higher the frequency and exclusivity of distribution in a given category, the higher the IVwill be of a species, which ranges from 0 (absence of a category) to 100 (present in all sites of a category). SIA hasthe advantages that they treat each species independently, and allows the comparison of the adequacy of the data to different typologies of the categories being compared, through the sum of species indicator values [48].

We analyzed the distribution of species under three different partitions of the gradient, which ranged from a coarser to a more refined elevational typology. This represents another approach to check the adjustment of our community with the elevational division proposed for the region. The first partition split the gradient in two categories, Lower Half (100 m, 400 m and 860 m) and Upper Half (1550 m, 2000 m and 2400 m). The second, based on the division proposed for the Guayana region, considered three categories: Lowlands (100 m and 400 m), Uplands (860 m and 1550 m) and Highlands (2000 m and 2400 m). In the last partition, we considered each elevation as a category. The presence of the species was verified at every sampling site at each elevation.

We used the software EstimateS [49] to obtain the number of shared species, and the software PAST [50] to calculate the Bray–Curtis similarity, Shannon–Wiener, and Berger–Parker Indexes, and also to perform the ANOSIM. The NMDS was performed with the R program [51], and the Partial Mantel with the software PASSAGE [52]. We ran the ISA with the software PC-ORD [53].

3. Results

We obtained 3140 adult spiders, which were assigned to 529 morphospecies from 39 families. A complete list of the species collected at the Pico da Neblina is presented in Nogueira et al. [54].

3.1. Diversity and Dominance

Richness, diversity, and proportion of absolute singletons decreased with increasing elevation (Table 1), while the abundance and proportion of singletons by elevation showed a more variable pattern (Table 1). Dominance increased with elevation, although not monotonically. Notably, dominance sharply increased in the two highest elevations, especially the last one, at 2400 m, where the single most dominant species accounted for more than 50% of the total abundance (Table 1 and Figure 2).

Table 1. Diversity measures of the spider community from the Pico da Neblina (AM, Brazil).Elevations sampled, S (%S)—richness and proportional richness, N(%N)—abundance and proportional abundance, D—exponential Shannon–Wiener or numbers equivalent, BP dom—Berger–Parker dominance index, % singl/S alt—proportion of singletons in relation to the richness of each elevation, % singl/S tot—proportion of singletons in relation to total number of singletons.

Elevation (m.a.s.l.)	S(%S)	N(%N)	D	BP dom	% singl/S alt	% singl/S tot
100	224 (42.4)	688 (21.9)	142.74	0.03	48.21	32.49
400	194 (36.7)	590 (18.8)	98.59	0.09	51.03	24.87
860	171 (32.4)	713 (22.7)	82.02	0.09	43.86	16.75
1550	115 (21.8)	597 (19)	61.68	0.07	41.74	14.21
2000	69 (13.1)	295 (9.4)	26.31	0.21	50.72	10.66
2400	24 (4.5)	257 (8.2)	6.10	0.53	37.5	1.02

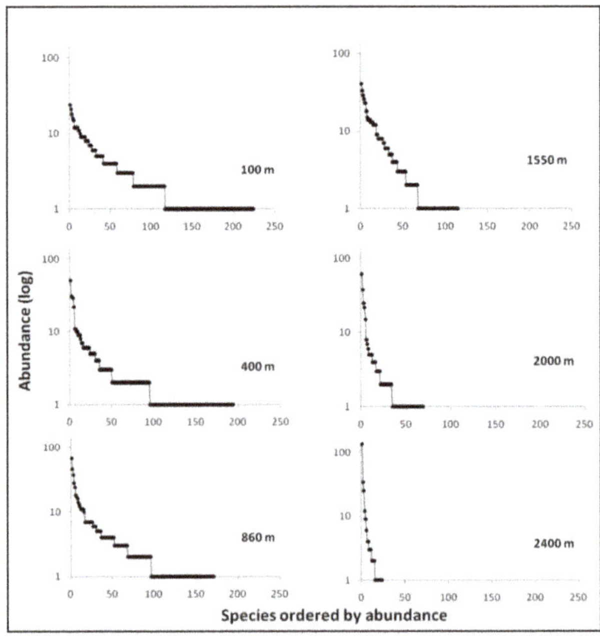

Figure 2. Rank abundance plot of the spider community for each elevation. Species are ordered by decreasing abundance.

Our results also showa positive association between species from the genus *Chrysometa* and elevation. The dominant species from the three highest elevations sampled belong to this genus, and two of them, *Chrysometa petrasierwaldae* Nogueira et al. 2011 (137 ind.) and *C. nubigena* Nogueira et al. 2011 (96 ind.), were the most abundant of the whole inventory.

3.2. Beta Diversity

There were important compositional changes between elevations along the gradient, which may occur abruptly at certain places. Beta diversity between different elevations varied from 1.45 to 1.90 (Table 2), with an average value of 1.73. These large values indicate an impressive complementarity of communities between different elevations, considering that the maximum beta diversity possible when comparing two sites is 2.

Table 2. Matrix of beta diversity expressed in number equivalents (D beta) of the spider community sampled at six elevations at the Pico da Neblina (AM, Brazil). Comparisons between adjacent elevations are shaded in gray.

	100	400	860	1550	2000
400	1.50				
860	1.60	1.45			
1550	1.90	1.89	1.71		
2000	1.80	1.85	1.80	1.65	
2400	1.79	1.84	1.76	1.79	1.62

The Mantel partial test showed that the beta diversity and Bray–Curtis similarity indexes were highly related, even when controlled for space or elevation (Table 3 and Figure 3). Both indexes were related to elevation in a significant way, as similarity decreased and beta diversity increased with increasing elevational difference between the elevations sampled, although the relation was stronger with similarity. The relation with differences in spatial distances, on the other hand, was not significant for either index (Table 3).

Table 3. Results of Mantel and partial Mantel tests performed for the spider community from the Pico da Neblina (AM, Brazil). Comparison—matrix being compared; control—matrix controlled or partial factor; R—Pearson correlation coefficient; P—significance level of the result. BCS—Bray–Curtis Similarity index.

Comparsion	Control (Partial)	R	P
β diversity × BCS	space	−0.896	0.0051
β diversity × BCS	elevation	−0.824	0.0018
β diversity × BCS		−0.921	0.002
β diversity × Elevation	space	0.554	0.0507
β diversity × Elevation		0.727	0.0102
β diversity × Space		0.576	0.0532
BCS × Elevation	space	−0.829	0.0028
BCS × Elevation		−0.797	0.0014
BCS × Space		−0.494	0.0964

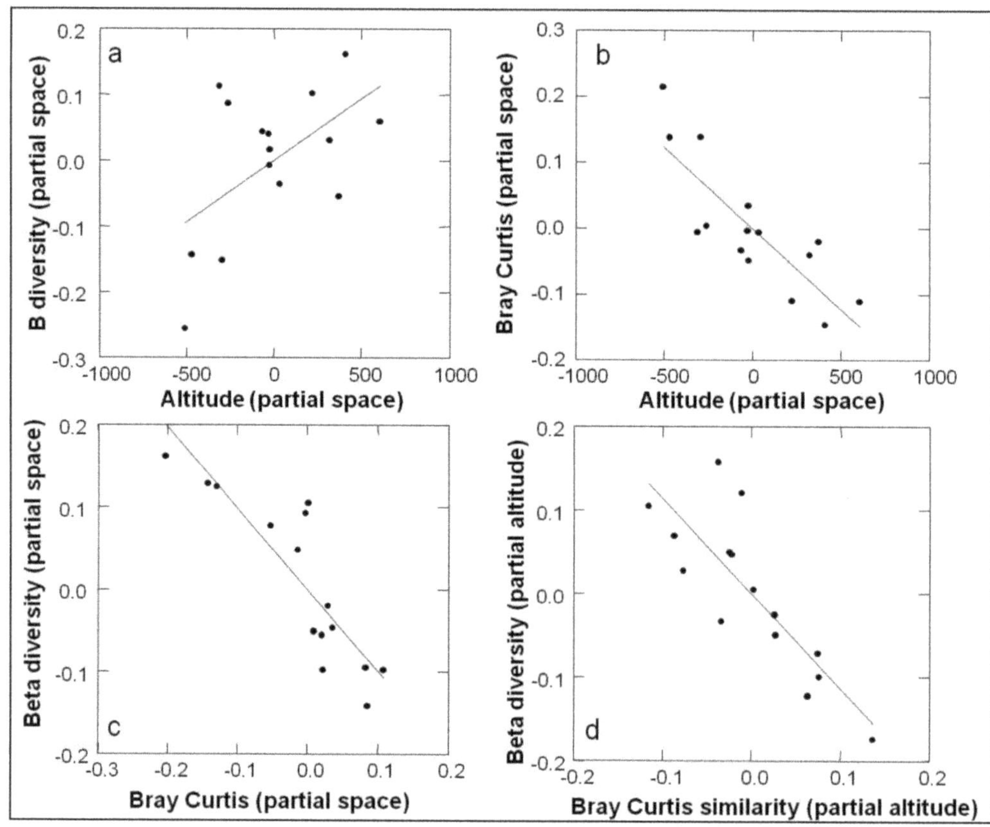

Figure 3. Results of the partial Mantel test. (**a**) Relation between elevational difference and beta diversity, controlled by spatial distance.(**b**) Relation between elevation and Bray–Curtis similarity index, controlled by spatial distance.(**c**) Relation between Bray–Curtis similarity index and beta diversity, controlled by spatial distance.(**d**) Relation between Bray–Curtis similarity index and beta diversity, controlled by elevational difference.

The NMDS (Stress 5.97) indicates that changes in composition are not gradual along the gradient (Figure 4). Along the first axis, which accounted for 58% of total variation, it is possible to observe two main groups of sites, one formed by the three lowest elevations, and the other by the two highest elevations sampled. The fourth elevation, at 1550, is fairly isolated from both groups and occupies an intermediate position.

The ANOSIM revealed significant differences in the composition of spider communities at the Pico da Neblina (R = 0.8362, $p < 0.001$). The comparison performed between the three groups of elevations—lowlands, uplands, and highlands—also indicated significant differences between all of them ($p < 0.01$ for all comparisons). The magnitude of the relation was very high when comparing the highlands with the lower groups (R = 1 with lowlands an R = 0.94 with uplands), but considerably smaller for the lowlands in relation to the uplands (R = 0.55).

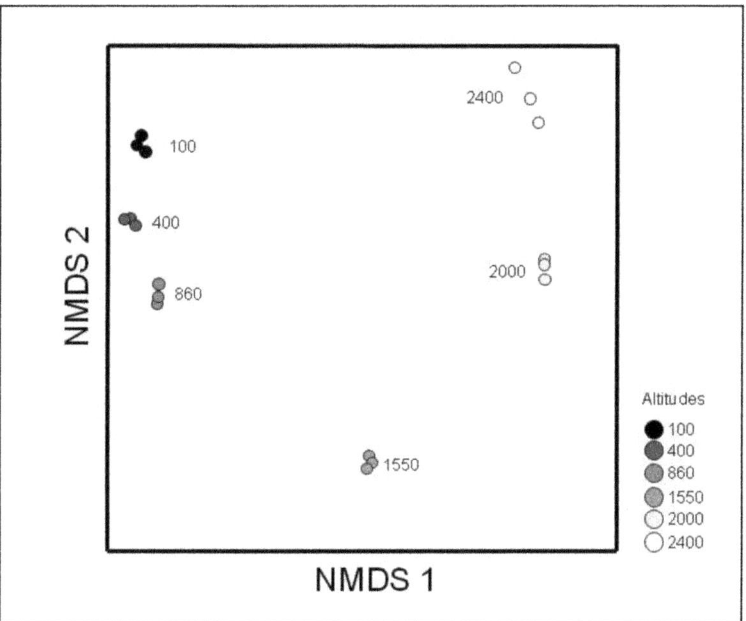

Figure 4. Graphic representation of the first two axes of a NMDS performed for all species of spiders at the 18 sites sampled, in 6different elevations. First axis explained 58% of total variation and second axis 20% Stress = 5.97.

3.3. Species Indicator Analysis

Of the 157 species represented by at least five individuals, 100 were assigned as indicators for at least one of the three partitions proposed for the data. Table 4 summarizes the number of indicator species for each category in each partition.

Table 4. Number of indicator species of spiders, designated by the species indicator analysis, for the three partitions of the gradient, at the Pico da Neblina (AM, Brazil): Lower and Upper half; Guayana region and Elevations. IS—number of species assigned as significant ($p < 0.05$) indicators; S—total richness of the category;% IS—proportional number of indicator species in relation to richness of categories; aver IV—average indicator value; IV 100—species with indicator value of 100;Total IV—sum of indicator values for each partition.

Partition	Category	IS	S	% IS	Aver IV	IV 100	Total IV
Lower and	LH	42	110	38.2	68	3	
upper half	UH	6	62	9.7	64.9	0	
	Total	48	157	30.6	68.2	3	3272.6
Guayana region	Lowlands	22	96	22.9	72.1	1	
	Uplands	15	103	14.6	78.8	1	
	Higlands	3	24	12.5	92.9	1	
	Total	40	157	25.5	75.9	3	3034.5
Elevations	100	21	72	29.2	89.2	12	
	400	9	72	12.5	82.3	4	
	860	12	80	15.0	80.5	3	
	1550	19	52	36.5	90.9	9	
	2000	6	20	30.0	81.7	2	
	2400	4	12	33.3	96.6	3	
	Total	71	157	45.2	87.1	33	6183.1

The families Anyphaenidae and Tetragnathidae showed the most consistent association with higher elevation environments, for the three partitions, while for some other families only the partition by elevation revealed some indicator species (Table 5). Lowest and medium elevations were characterized by indicator species representing a larger number of families, amongst which Araneidae and Ctenidae stand out (Table 5).

Table 5. Results of species indicator analysis by families. Number of indicator species of spiders for the three partitions of the gradient at the Pico da Neblina (AM, Brazil), by families. Categories: Partition 1, LH—lower half, UH—upper half; Partition 2, L—lowlands, U—uplands, H—highlands. IS—indicator species; S—richness of the families considering only species included in the analysis (at least five individuals).

| | Partition 1 | | | Partition 2 | | | | Partition 3 | | | | | | |
| Family | Category | | | Category | | | | Category | | | | | | |
	IS (S)	LH	UH	IS (S)	L	U	H	IS (S)	100	400	860	1550	2000	2400
Anyphaenidae	1 (4)		1	1 (4)			1	3 (4)				1		2
Araneidae	13 (35)	12	1	7 (35)	5	1	1	14 (35)	4	4	3	2	1	
Corinnidae	2 (4)	2		1 (4)		1		1 (4)			1			
Ctenidae	4 (6)	4		4 (6)	2	2		4 (6)		2	1	1		
Deinopidae	1 (1)	1						1 (1)			1			
Hahniidae								1 (1)					1	
Linyphiidae	1 (4)		1	1 (4)		1		2 (4)			1	1		
Lycosidae	1 (1)	1		1 (1)	1									
Mimetidae								3 (6)				2		1
Oonopidae								1 (1)					1	
Pholcidae	4 (8)	4		4 (8)	2	2		3 (8)		1	2			
Pisauridae	1 (2)	1		1 (2)	1			2 (2)	2					
Salticidae	3 (15)	3		2 (15)	1	1		5 (15)	2	1		1	1	
Scytodidae				1 (2)		1		1 (2)				1		
Senoculidae				1 (1)		1		1 (1)				1		
Sparassidae	3 (4)	3		3 (4)	3			2 (4)	1			1		
Tetragnathidae	3 (13)	2	1	3 (13)		2	1	6 (13)	2			2	1	1
Theridiidae	3 (32)	2	1	5 (32)	2	3		15 (32)	6	1	3	5		
Theridiosomatidae	3 (5)	2	1	1 (5)	1			3 (5)	2				1	
Thomisidae	1 (2)	1		1 (2)	1									
Uloboridae	4 (9)	4		3 (9)	3			3 (9)	2			1		

4. Discussion

Our results revealed very important changes in composition along the elevational gradient. However, the general beta diversity pattern was not entirely adjusted to the elevational division proposed for the study region. While the highlands zone (above 1500 m) did present a unique set of spider species, the lowland (up to 500 m) spider fauna expanded up until the uplands (from 500 to 1500 m), producing thus a different pattern than that observed for the vegetation, where a clear distinction between the lowland and upland communities can be perceived. Our results also showed important differences in the diversity and dominance levels of the communities, as discussed above.

4.1. Diversity and Dominance

The drastic increase in dominance above 1550 m shows that elevation acts not only on the number and the identity of species of a community, but also on the distribution of species abundance, which indicates a different sort of influence of the environmental, mainly temperature gradient, on the biotic community. The fewer and mostly different species from higher elevations also partition the total abundance in a much more uneven way.

An increasing dominance level at higher elevations has already been observed for spiders [28,29] and various other groups [11,55,56]. This pattern could represent an example of the positive relation between evenness of community species abundance and productivity [56], assuming a negative relation between altitude and productivity [4,14]. Evenness can be also positively related with habitat structure [57], which also applies to our study, as forests are structurally more complex than more open types of vegetation, such as those from the highest elevations sampled.

An alternative explanation could be based on richness decrease, whichwould lead to a decrease in interspecific competition, which would allow the remaining species to increase its abundance. This process, known as density-dependence, suggests that species density is dependent on competition [58], and has already been used to explain similar dominance patterns for ants in a tropical elevational gradient [59].

4.2. Beta Diversity

The changes in composition of the spider communities along the gradient have been large, but it was possible to see that those changes were more intense at some specific elevations. This indicates the occurrence of an elevational zonation for the spider community of the Pico da Neblina. Nonetheless, the patterns of compositional changes observed along the gradient do not seem to fit the elevational division proposed for the Guayana region, in lowlands, uplands and highlands, despite the significant differences found between these categories by the ANOSIM.

The main differences between the elevational division tested and the changes in composition observed in our data are related to the uplands categories. The third elevation sampled (860 m) is more similar to the lower sites, from the lowland category, than with the other elevation (1550 m) in the upland category, as displayed in the NMDS. This indicates that the upland category is heterogeneous and represents an inadequate elevational division for the spider fauna. Instead, results show that mainly lowland fauna extend up until the third elevation, at 860 m, and the next elevation at 1550 m represents a compositional rupture, although it is still covered by forests.

The fauna from 1550 m arenot very similar with that from higher elevations either, even with its upper neighbor, at 2000 m. This is not a surprise, given the drastic differences in climatic factors and especially in the structure of the vegetation, going from forested habitats to open physiognomies at the highland sites. The structure of the vegetation is indeed considered as one of the most important environmental factors for spider communities [26,27,60], and may have a large influence on the composition of the communities [61–63]. The distinction of the highland spider community may also reflect the distinction of the flora of these elevations, reputed by its endemism [18,19] and peculiar formations. The dominance pattern of the highlands fauna is another character that distinguished themfrom the lower sites, while the fauna ofthe fourth elevation, at 1550, represent anotherintermediate condition.

The coincidence of significant ruptures in composition and main vegetation types has already been reported for other studies performed at elevational gradients [64–67]. In other cases, however, changes were more gradual [12] or were not directly associated to predominant patterns of the vegetation, as observed for ground-dwelling spiders [29].

We can conclude that the elevational division proposed for the Guayana region does not fit well with our data, because communities from lowlands extend higher than expected. The spiders from the highland sites effectively represent a distinct compositional group, and changes from forested sites to open vegetation coincide with the largest ruptures observed across our gradient.

The distribution of indicator species among families furnishes the identity of the main components of the fauna from each elevation. The lowlands are dominated by species from several families, the main contributors being Ctenidae, Pholcidae, Sparassidae, Uloboridae, and especially Araneidae. All of these families and several others contributed to theindicator species for the three partitions, which indicates that even within families the range of elevational distribution presents a great variation. It is worth mentioning that the proportion of indicator species by partition is very unbalanced for some families, such as Salticidae and Theridiidae, whose species designed as indicators are concentrated in the more refined partition by elevation. It may indicate that species from these families are characterized by short elevational ranges and more specific habitat requirements.

Species designated as indicators of the higher elevations, considering the categories from the three partitions (Upper half, Highlands, 2000 m and 2400 m), belonged to only 10 families. All of those species were web builders or hunters occupying vegetation, which signals the absence of ground dwelling spider, like the Ctenidae, as significant components of these environments. Even families usually associated to the ground or leaf litter, as Hahniidae and Oonopidae [68], were represented by species occurring on the vegetation at 2000 m (pers. comm.). The family Anyphaenidae, with four species and two species of the genus *Chrysometa* (Tetragnathidae) were the most characteristic elements of that fauna. A similar result was observed in the Udzungawa mountains (Tanzania), where they recorded the spider species distribution along anelevational gradient. The authors also found that higher elevations were dominated by vegetation dwellers and orb-weavers, while ground-dwelling species were more abundant at lower elevations [31].

The dominance of the tetragnathid orb-weaver genus *Chrysometa* at higher elevations, mentioned above and discussed in Nogueira et al. [69], contrasted with the distribution of species of Araneidae, the most species-rich orb-weaver family. The elevational replacement among these groups have already been noticed at the Colombian Andes ([28] and references), and constitutes evidence that flying insects, the main target of this kind of web [70], are still an available resource at higher elevations. This suggests that *Chrysometa* species can tolerate climatic conditions that represent a constraint to most Neotropical Araneidae species distribution. *Araneus bogotensis* (Keyserling, 1864) (Araneidae) constitute a notable exception, as it was selected as an indicator of high elevation sites for the three partitions.

4.3. Spider in Mountains

The knowledge on montane spider faunas has greatly increased in the new millennia, especiallyoverthe last decade. Montane faunas are reputed for theirdiversity and endemism [6,71,72], and indeed, new species of spiders are being described from mountains of several places worldwide, mostly from under sampled places such as Asia [32–34] and Africa ([73], new genus from the East Arc Mountains, Tanzania). It is worth noting that most of those new species described are from upper parts of the mountains, rather than from the lower elevations.

We can present a similar result in our study. We described eight new species of the orb-weaver genus *Chrysometa* (Tetragnathidae) [69], and only one of these species were found in the lower half of the gradients. The remaining seven species were from the upper half, distributed from 1200 to 2400 m, the highest elevation sampled. The recurring finding of new species is evidence of the high endemism levels of mountain fauna, especially at higher elevations [74].

Another evidence of this pattern was provided by Chaladze et al. [75], who developed a Spider Diversity Model for the Caucasus Ecoregion. He showed that, although endemism is related to overall species richness, there are hotspots of endemic species (= higher proportion) in highmountain areas. A checklist of a Macedonia mountain spider fauna [76] also reported that most of the endemic species for this region are constituted of mountain elements. In our case, it is impossible to make similar evaluations of the regional spider fauna, since the amazon spider fauna is still very poorly known, and thus it is difficult to assess the real distribution of most species. However, high endemism levels have been observed for the herpetofauna [77], one of the few animal groups studied in more detail at the Pico da Neblina.

Other recent studies performed more complete and standardize inventories along elevational gradients. These studies show a great variation in their results, as some were performed in tropical mountains [30,31] while other were conducted in temperate regions [29,78,79]. Another source of variation are the sampling methods, since some studies used only pitfall traps, and thus could only assess the soil and litter fauna, while others employed other methods that enabled then to assess a larger portion of the community, as web-builders and vegetation dwellers.

However, some common patterns concerning beta-diversity emerged from these studies. For example, most of the compositional differences along the gradient are due to turnover, rather than nestedness. So, high elevation communities, even when presenting low richness, are not formed from a subset of more tolerant species from lower parts of the gradient, bur are constituted by a different, usually unique, group of species [30].

While high turnover rates usually produce regular patterns of a distance–decay process (i.e., dissimilarity in communities is strongly connected with physical distance), some studies also stress that the distribution of spiders is also tightly connected with microhabitat conditions (mainly type of vegetation) [31,79], which diminishes the intensity of the relation between similarity and distance. This indicates that most species do not seem to face dispersal limitation, and thus beta diversity patterns would be more influenced by local processes, such as resource availability and microclimates, as well as historical factors [30].

4.4. Montane Spider Conservation

The conservation of spider mountain fauna should be considered a priority, based on theirhigh diversity and endemism levels; characteristics confirmed by our results and by the literature. The fauna from high elevations is particularly sensitive, as they usually present a restrict distribution [69,80] and are habitat specialists, two characteristics associate with extinction proneness [81–83]. A small range size increases the chance of population reduction and eventual extinction, and the specialization to a particular type of microhabitat tend to make a species be more sensitive to environmental changes [84].

A conservationist focus should also be stimulated due to the fact that the ongoing climate crisis will probably enhance negative effects on these communities. Climate change may be considered one of the worst human-induced environmental changes, due to its irreversibility [85]. The warming climate is expected to have negative consequences to the species adapted to the colder (= higher) parts of the gradients. Data from a multitaxa (birds, butterflies, carabidae and staphylinids beetles and spiders) inventory from the Italian Alps Vitterbi [79] simulated a moderate temperature increase on species distribution, and spiders were the group that hadthe most negative response, with a decrease in richness. The higher sensitivity of spider communities was attributed to the fact that spider species had more restricted distribution along the sampling plots, being more dependent on microclimatic conditions. The general results of this study also predicted intense changes in composition for all the groups with increasing air temperature, with detrimental consequences for high-elevation species.

High-elevation faunas are threatened everywhere, but the problem is worse in mountains located in tropical places, such as our study site. Environmental gradients are stronger in tropical mountains, due to the narrower climatic tolerance of tropical biotas when compared with temperate ones [5,13,14], and thus effects of climate change are expected to be more pronounced in the tropics.

The situation of Guayana region highlands, or Pantepui, is of particular concern [86], due both to its equatorial location and the richness and endemism levels of its well-studied flora [19]. The singular topography of this region, with several isolated tabletop mountains reaching above 2000 m, stimulated speciation and produced communities with a high proportion of species endemic toa single mountain [87]. If the spiders follow this pattern, we can suppose that there are a lot of high elevation specialists yet to be found in the mountains of the Pantepui, which, at least for the genus *Chrysometa*, is a reasonable expectation [69,80].

However, the flora distribution also shows that the habitat of the high elevation specialists could simply "vanish in the air" [88], if the IPCC predictions for this century are accurate. If spiders also follow this pattern, this implies that several potential endemic new species threaten to disappear before they can even be described, stressing the urgency of this problem.

Author Contributions: A.A.N. led the sampling trip, material identification, data analysis and writing. A.D.B. helped in the identification of the material and provided resources and facilities to help this process. G.P.-N. helped with analysis, writing, submission and reviewing process. E.M.V. conceptualized the study and assisted the first author in all the steps of the study, and provided financial and logistical resources. All authors have read and agreed to the published version of the manuscript.

Funding: A.D.B. acknowledges support by Conselho Nacional de Desenvolvimento Científico e Tecnológico (CNPq, grant PQ 303903/2019-8). E.M.V. was supported by a fellowship from the Conselho Nacional de Desenvolvimento Científico e Tecnológico (CNPq, 308040/2017-1). A.A. Nogueira is supported by a post-doc fellowship from PNPD/CAPES, and the filed expedition was financed by a BECA-IEB/Moore Foundation (B/2007/01/BDP/01) fellowship and by a grant from Wildlife Conservation Society (WCS).

Informed Consent Statement: Not applicable.

Data Availability Statement: The material collected was deposited at the collection of the Instituto Nacional de Pesquisas da Amazônia (INPA), at Manaus (AM), and duplicates are deposited at the Instituto Butantan (IBSP), São Paulo (SP) and at the Museu Paraense Emílio Goeldi (MPEG), Belém (PA). The material sampled in this study can be found in the Arachnological Collections of the following scientific museums: Instituto Nacional de Pesquisas da Amazônia, Instituto Butantan e Museu Paraense Emílio Goeldi.

Acknowledgments: We are grateful to Tomé, Mário, Waldir "Chouriman" Pereira, Nancy Lo-Man-Hung and David Candiani, for their invaluable help in the field. The first author also thanks the PPGEco-INPA, the 5°PEF Maturacá, a frontier squad from the Brazilian army for logistic help, the IBAMA/ICMBio and PARNA Pico da Neblina for the collecting license (Ibama-Sisbio 10560–1), and FUNAI and the Ayrca, a local Yanomami association, for receiving us at the Yanomami Indigenous Land.

Conflicts of Interest: The authors declare no conflict of interest.

References

1. Whittaker, R.H. Vegetation of the Siskiyou Mountains, Oregon and California. *Ecol. Monogr.* **1960**, *30*, 279–338. [CrossRef]
2. McCain, C.M. Elevational gradients in diversity of small mammals. *Ecology* **2005**, *86*, 366–372. [CrossRef]
3. McCain, C.M. Area and mammalian elevational diversity. *Ecology* **2007**, *88*, 76–86. [CrossRef]
4. McCain, C.M. Could temperature and water availability drive elevational species richness patterns? A global case study for bats. *Glob. Ecol. Biogeogr.* **2007**, *16*, 1–13. [CrossRef]
5. Janzen, D.H. Why mountain passes are higher in the tropics? *Am. Nat.* **1967**, *101*, 233–249. [CrossRef]
6. Lomolino, M.K. Elevation gradients of species-density: Historical and prospective views. *Glob. Ecol. Biogeogr.* **2001**, *10*, 3–13. [CrossRef]
7. Sanders, N.J.; Lessard, J.P.; Fitzpatrick, M.C.; Dunn, R.R. Temperature, but not productivity orgeometry, predicts elevational diversitygradients in ants across spatial grains. *Glob. Ecol. Biogeogr.* **2007**, *16*, 640649. [CrossRef]
8. Orme, C.D.L.; Davies, R.G.; Burgess, M.; Eigenbrod, F.; Pickup, N.; Olson, V.A. Global hotspots of species richness are not congruent with endemism or threat. *Nature* **2005**, *436*, 1016–1019. [CrossRef]
9. Rahbek, C. The role of spatial scale and the perception of large-scale species-richness patterns. *Ecol. Lett.* **2005**, *8*, 224–239. [CrossRef]
10. Lieberman, D.; Lieberman, M.; Peralta, R.; Hartshorn, G.S. Tropical forest structure and composition on a largescale elevational gradient in Costa Rica. *J. Ecol.* **1996**, *84*, 137–152. [CrossRef]
11. Vazquez, J.A.; Givnish, T.J. Elevational gradients in tropical forest composition, structure, and diversity in the Sierra de Manantla´n. *J. Ecol.* **1998**, *86*, 999–1020.
12. Brehm, G.; Homeier, J.; Fiedler, K. Beta diversity of geometrid moths (Lepidoptera: Geometridae) in an Andean montane rainforest. *Div. Distrib.* **2003**, *9*, 351–366. [CrossRef]
13. Ghalambor, C.K.; Huey, R.B.; Martin, P.R.; Tewksbury, J.J.; Wang, G. Are mountain passes higher in the tropics? Janzens hypothesis revisited. *Integr. Comp. Biol.* **2006**, *46*, 5–17. [CrossRef]
14. McCain, C.M. Vertebrate range sizes indicate that mountains may be higher in the tropics. *Ecol. Lett.* **2009**, *12*, 550–560. [CrossRef]

15. Patterson, B.D.; Stotz, D.F.; Solari, S.; Fitzpatrick, J.W. Contrasting patterns of elevational zonation for birds and mammals in the Andes of southeastern Peru. *J. Biogeogr.* **1998**, *25*, 593–607. [CrossRef]
16. Jankowski, J.E.; Ciecka, A.L.; Meyer, N.Y.; Rabenold, K.N. Beta diversity along environmental gradients: Implications of habitat specialization in tropical montane landscapes. *J. Anim. Ecol.* **2009**, *78*, 315–327. [CrossRef] [PubMed]
17. Willard, D.E.; Foster, M.S.; Barrowclough, G.F.; Dickerman, R.W.; Cannell, P.F.; Coats, S.L.; Cracraft, J.L.; O'Neill, J.P. The Birds of Cerro de la Neblina. *Fieldiana* **1991**, *65*, 27–51.
18. Huber, O. Geographical and physical features. In *Flora of the Venezuelan Guayana*; Berry, P.E., Holst, B.K., Yatskievych, K., Eds.; Missouri Botanical Garden Press: St. Louis, MO, USA, 1995; Volume 1, pp. 161–191.
19. Berry, P.E.; Riina, R. Insights into the diversity of the Pantepui flora and the biogeographic complexity of the Guayana Shield. *Biol. Skr.* **2005**, *55*, 145–167.
20. Rull, V.; Nogué, S. Potential migration routes and barriers for vascular plants of the Neotropical Guyana Highlands during the Quaternary. *J. Biogeogr.* **2007**, *4*, 1327–1341. [CrossRef]
21. Scharff, N.; Coddington, J.A.; Griswold, C.E.; Hormiga, G.; Bjorn, P.D.P. When to quit? Estimating spider species richness in a northern European deciduos forest. *J. Arachnol.* **2003**, *31*, 246–273. [CrossRef]
22. Cardoso, P.; Arnedo, M.A.; Triantis, K.A.; Borges, P.A.V. Drivers of diversity in Macaronesian spiders and the role of species extinctions. *J. Biogeogr.* **2010**, *37*, 1034–1046. [CrossRef]
23. World Spider Catalog. *World Spider Catalog. Version 22.5*; Natural History Museum Bern: Bern, Switzerland, 2021; Available online: http://wsc.nmbe.ch (accessed on 12 October 2021). [CrossRef]
24. Coddington, J.A.; Griswold, C.E.; Silva, D.; Larcher, L. Designing and testing sampling protocols to estimate biodiversity in tropical ecosystems. In *The Unity of Evolutionary Biology: Proceedings of the Fourth International Congress of Systematic and Evolutionary Biology, College Park, MD, USA, 30 June–7 July 1991*; Dioscorides Press: Portland, OR, USA, 1991; pp. 44–60.
25. Janzen, D.H.; Ataroff, M.; Farinas, M.; Reyes, S.; Rincon, N.; Soler, A.; Soriano, P.; Vera, M. Changes in the arthropod community along an elevational transect in the Venezuelan Andes. *Biotropica* **1976**, *8*, 193–203. [CrossRef]
26. Robinson, J.V. The effect of architectural variation in habitat on a spider community: An experimental field study. *Ecology* **1981**, *62*, 73–80. [CrossRef]
27. Greenstone, M.H. Determinants of web spider species diversity: Vegetation structural diversity vs. prey availability. *Oecologia* **1984**, *62*, 299304. [CrossRef] [PubMed]
28. Ferreira-Ojeda, L.; Florez-Daza, E. Arañas orbitelares (Araneae: Orbiculariae) em tres formaciones vegetales de la Sierra Nevada de Santa Marta (Magdalena, Colombia). *Rev. Iber. Aracnol.* **2007**, *16*, 3–16.
29. Chatzaki, M.; Lymberakis, P.; Markakis, G.; Mylonas, M. The distribution of ground spiders (Araneae, Gnaphosidae) along the elevational gradient of Crete, Greece: Species richness, activity and elevational range. *J. Biogeogr.* **2005**, *32*, 813–831. [CrossRef]
30. Foord, S.H.; Dippenaar-Schoeman, A.S. The effect of elevation and time on mountain spider diversity: A view of two aspects in the Cederberg mountains of South Africa. *J. Biogeogr.* **2016**, *43*, 2354–2365. [CrossRef]
31. Malumbres-Olarte, J.; Crespo, L.; Cardoso, P.; Szűts, T.; Fannes, W.; Pape, T.; Scharff, N. The same but different: Equally megadiverse but taxonomically variant spider communities along an elevational gradient. *Acta Oecolog.* **2018**, *88*, 19–28. [CrossRef]
32. Miller, J.A.; Griswold, C.E.; Yin, C.M. The symphytognathoid spiders of the Gaoligongshan, Yunnan, China (Araneae, Araneoidea): Systematics and diversity of micro-orbweavers. *ZooKeys* **2009**, *11*, 9–195. [CrossRef]
33. Tanasevitch, A.V. A new Palliduphantes Saaristo & Tanasevitch, 1996 from the Elburz Mountains, Iran (Araneae: Linyphiidae). *Zool. Mid. East* **2017**, *63*, 172–175.
34. Dimitrov, D. A review of the linyphiid spider genus *Proislandiana* Tanasevitch, 1985 with description of a new high mountainous species from Turkey and Armenia (Araneae: Linyphiidae). *Zootaxa* **2020**, *4743*, 247–256. [CrossRef] [PubMed]
35. RADAM. *Folha NA19. Pico da Neblina*; Ministério das Minas e Energia: Rio de Janeiro, Brazil, 1978.
36. Pires, J.M.; Prance, T.G. The vegetation types of the Brazilian Amazon. In *Key Environments: Amazonia*; Prance, G.T., Lovejoy, T.E., Eds.; Pergamon Press: Oxford, UK, 1985; pp. 109–145.
37. Berry, P.E.; Huber, O.; Holst, B.K. Introduction. Floristic analysis and phytogeography. In *Flora of the Venezuelan Guayana*; Berry, P.E., Holst, B.K., Yatskievych, K., Eds.; Missouri Botanical Garden Press: St. Louis, MO, USA, 1995; Volume 1, pp. 161–191.
38. Jost, L. Entropy and diversity. *Oikos* **2006**, *113*, 363–375. [CrossRef]
39. Jost, L. Partitioning diversity into independent alpha and beta components. *Ecology* **2007**, *88*, 2427–2439. [CrossRef] [PubMed]
40. Berger, W.H.; Parker, F.L. Diversity of planktonic foraminifera in deep-sea sediments. *Science* **1970**, *168*, 1345–1347. [CrossRef] [PubMed]
41. Kruskal, J.B. Nonmetric multidimensional scaling: A numerical method. *Psychometrika* **1964**, *29*, 115–129. [CrossRef]
42. Fasham, M.J.R. A comparison of nometric multidimentsional scaling, principal components and reciprocal averaging for the ordination of simulated coenoclines, and coenoplanes. *Ecology* **1977**, *58*, 551–561. [CrossRef]
43. Kenkel, N.C.; Orlóci, L. Applying metric and nonmetric multidimensional scaling to ecological studies: Some new results. *Ecology* **1986**, *67*, 919–928. [CrossRef]
44. Brehm, G.; Konrad Fiedler, K. Ordinating tropical moth ensembles from an elevational gradient: A comparison of common methods. *J. Trop. Ecol.* **2004**, *20*, 165–172. [CrossRef]

45. Smouse, P.E.; Long, J.C.; Sokal, R.R. Multiple regression and correlation extensions of the Mantel test of matrix correspondence. *System. Zool.* **1986**, *35*, 627–632. [CrossRef]
46. Legendre, P. Comparison of permutation methods for the partial correlation and partial Mantel tests. *J. Statist. Comput. Simul.* **2000**, *67*, 37–73. [CrossRef]
47. Clarke, K.R. Non-parametric multivariate analyses of changes in community structure. *Austr. J. Ecol.* **1993**, *18*, 117–143. [CrossRef]
48. Dufrêne, M.; Legendre, P. Species assemblages and indicator species: The need for a fl exible asymetrical approach. *Ecol. Monogr.* **1997**, *67*, 345–366.
49. Colwell, R.K. EstimateS: Statistical Estimation of Species Richness and Shared Species from Samples. Version 8.2. User's Guide and Application. 2009. Available online: http://purl.oclc.org/estimates (accessed on 1 November 2010).
50. Hammer, O.; Harper, D.A.T.; Ryan, P.D. PAST: Paleontological Statistics Software Package for Education and Data Analysis. *Palaeontol. Electron.* **2001**, *4*, 9.
51. R Core Team. R: A Language and Environment for Statistical Computing. R Foundation for Statistical Computing, Vienna, Austria. 2020. Available online: https://www.R-project.org/ (accessed on 2 October 2011).
52. Rosenberg, M.S. *PASSAGE. Pattern Analysis, Spatial Statistics, and Geographic Exegesis. Version 1.0.*; Department of Biology, Arizona State University: Tempe, AZ, USA, 2001.
53. McCune, B.; Mefford, M.J. *PC-ORD: Multivariate Analysis of Ecological Data. Version 5·12*; MjM Software: Gleneden Beach, OR, USA, 1999.
54. Nogueira, A.A.; Venticinque, E.M.; Brescovit, A.D.; Lo-Man-Hung, N.F.; Candiani, D.F. List of species of spiders (Arachnida, Araneae) from the Pico da Neblina, state of Amazonas, Brazil. *Check List* **2014**, *10*, 1044–1060. [CrossRef]
55. Choi, S.W.; Jeong-Seop, N.A. Elevational distribution of moths (Lepidoptera) in Mt. Jirisan National Park, South Korea. European. *J. Entom.* **2010**, *107*, 229–245.
56. McGill, B.J.; Etienne, R.S.; Gray, J.S.; Alonso, D.; Anderson, M.J.; Benecha, H.K.; White, E.P. Species abundance distributions: Moving beyond single prediction theories to integration within an ecological framework. *Ecol. Lett.* **2007**, *10*, 995–1015. [CrossRef] [PubMed]
57. Hurlbert, A.H. Species–energy relationships and habitat complexity in bird communities. *Ecol. Lett.* **2004**, *7*, 714–720. [CrossRef]
58. MacArthur, R.H.; Diamond, J.M.; Karr, J.R. Density compensation in island faunas. *Ecology* **1972**, *53*, 330–342. [CrossRef]
59. Longino, J.T.; Colwell, R.K. Density compensation, species composition, and richness of ants on a neotropical elevational gradient. *Ecosphere* **2011**, *2*, 1–20. [CrossRef]
60. Halaj, J.; Ross, D.W.; Moldenke, A.R. Habitat structure and prey availability as predictors of the abundance and community. *J. Arachnol.* **1998**, *26*, 203–220.
61. Toti, D.S.; Coyle, F.A.; Miller, J.A. A structured inventory of Appalachian grass bald and heath bald spider assemblages and a test of species richness estimator performance. *J. Arachnol.* **2000**, *28*, 329–345. [CrossRef]
62. Nogueira, A.A.; Pinto-da-Rocha, R.; Brescovit, A.D. Comunidade de aranhas orbitelas (Arachnida-Araneae) na região da Reserva Florestal do Morro Grande, Cotia, São Paulo, Brasil. *Biota Neotrop.* **2006**, *6*, 1–24. [CrossRef]
63. Lo-Man-Hung, N.F.; Gardner, T.A.; Ribeiro-Júnior, M.A.; Barlow, J.; Bonaldo, A.L. The value of primary, secondary, and plantation forests for Neotropical epigeic arachnids. *J. Arachnol.* **2008**, *36*, 394–401. [CrossRef]
64. Bosmans, R.; Maelfait, J.P.; De Kimpe, A. Analysis of the spider communities in an elevational gradient in the French and Spanish Pyrenees. *Bull. Brit. Arachnol. Soc.* **1986**, *7*, 69–76.
65. Davis, A.L.V.; Scholtz, C.H.; Chown, S.L. Species turnover, community boundaries and biogeographical composition of dung beetle assemblages across an elevational gradient in South Africa. *J. Biogeogr.* **1999**, *26*, 1039–1055. [CrossRef]
66. Bach, K.; Kessler, M.; Gradstein, S.R. A simulation approach to determine statistical significance of species turnover peaks in a species-rich tropical cloud forest. *Div. Distrib.* **2007**, *13*, 863–870. [CrossRef]
67. Wu, F.; Yang, X.J.; Yang, J.X. Additive diversity partitioning as a guide to regional montane reserve design in Asia: An example from Yunnan Province, China. *Div. Distrib.* **2010**, *16*, 1022–1033. [CrossRef]
68. Höfer, H.; Brescovit, A.D. Species and guild structure of a Neotropical spider assemblage (Araneae) from Reserva Ducke, Amazonas, Brazil. *Andrias* **2001**, *15*, 99–119.
69. Nogueira, A.A.; Pena-Barbosa, J.P.P.; Venticinque, E.M.; Brescovit, A.D. The spider genus *Chrysometa* (Araneae, Tetragnathidae) from the Pico da Neblina and Serra do Tapirapecó mountains (Amazonas, Brazil): New species, new records, diversity and distribution along two elevational gradients. *Zootaxa* **2011**, *2772*, 33–51. [CrossRef]
70. Turnbull, A.L. Ecology of the true spiders (Araneomorphae). *Annual. Rev. Entomol.* **1973**, *18*, 305–348. [CrossRef]
71. Jetz, W.; Rahbek, C.; Colwell, R.C. The coincidence of rarity and richness and the potential signature of history in centres of endemism. *Ecol. Lett.* **2004**, *7*, 1180–1191. [CrossRef]
72. Melo, A.S.; Rangel, T.F.L.V.B.; Diniz-Filho, J.A.F. Environmental drivers of beta-diversity patterns in New-World birds and mammals. *Ecography* **2009**, *32*, 226–236. [CrossRef]
73. Polotow, D.; Griswold, C. Chinja, a new genus of spider from the Eastern Arc Mountains of Tanzania (Araneae, Zoropsidae). *Zootaxa* **2018**, *4472*, 545–562. [CrossRef]
74. Steinbauer, M.J.; Field, R.; Grytnes, J.A.; Trigas, P.; Ah-Peng, C.; Attorre, F.; Beierkuhnlein, C. Topography-driven isolation, speciation and a global increase of endemism with elevation. *Glob. Ecol. Biogeogr.* **2016**, *25*, 1097–1107. [CrossRef]

75. Chaladze, G.; Otto, S.; Tramp, S. A spider diversity model for the Caucasus Ecoregion. *J. Insect Conserv.* **2014**, *18*, 407–416. [CrossRef]
76. Deltshev, C.; Komnenov, M.; Blagoev, G.; Georgiev, T.; Lazarov, S.; Stojkoska, E.; Naumova, M. Faunistic Diversity of Spiders (Araneae) in Galichitsa Mountain (FYR Macedonia). *Biodiv. Data J.* **2013**, *1*, e977. [CrossRef] [PubMed]
77. McDiarmid, R.W.; Donnelly, M.A. Herpetofauna of the Guyana Highlands: Amphibians and Reptiles of the Lost World. In *Ecology and Evolution in the TropicsA Herpetological Perspective*; Donnelly, M.A., Crother, B.I., Guyer, C., Wake, M.H., White, M.E., Eds.; The Univestity of Chicago Press: Chicago, IL, USA, 2005; 584p.
78. Aisen, S.; Werenkraut, V.; Márquez, M.E.G.; Ramírez, M.J.; Ruggiero, A. Environmental heterogeneity, not distance, structures montane epigaeic spider assemblages in north-western Patagonia (Argentina). *J. Insect Conserv.* **2017**, *21*, 951–962. [CrossRef]
79. Viterbi, R.; Cerrato, C.; Bionda, R.; Provenzale, A. Effects of temperature rise on multi-taxa distributions in mountain ecosystems. *Diversity* **2020**, *12*, 210. [CrossRef]
80. Levi, H.W. The Neotropical orb-weaver genera Chrysometa and Homalometa (Araneae: Tetragnathidae). *Bull. Mus. Comp. Zool.* **1986**, *151*, 91–215.
81. Henle, K.; Davies, K.F.; Kleyer, M.; Margules, C.; Settele, J. Predictors of Species Sensitivity to Fragmentation. *Biodiv. Conserv.* **2004**, *13*, 207–251. [CrossRef]
82. Dunn, R.R. Modern insect extinctions, the neglected majority. *Conserv. Biol.* **2005**, *19*, 130–136. [CrossRef]
83. Laurance, W.F.; Lovejoy, T.E.; Vasconcelos, H.L.; Bruna, E.M.; Didham, R.K.; Stouffer, P.C.; Gascon, C.; Bierregard, R.O.; Laurance, S.G.; Sampaio, E. Ecosystem decay of amazonian forest fragments: A 22-year investigation. *Conserv. Biol.* **2002**, *16*, 605–618. [CrossRef]
84. Dirnböck, T.; Essel, F.; Rabitsch, W. Disproportional risk for habitat loss of high-altitude endemic species under climate change. *Glob. Chang. Biol.* **2011**, *17*, 990–996. [CrossRef]
85. Kuntner, M. Climate change and spider biodiversity: Redefining araneological research priorities. *Crop Protect.* **2014**, *61*, 109. [CrossRef]
86. Rull, V.; Vegas-Vilarrúbia, T.; Nogué, S.; Huber, O. Conservation of the unique neotropical vascular flora of the Guayana Highlands in the face of global warming. *Conserv. Biol.* **2009**, *23*, 1323–1327. [CrossRef] [PubMed]
87. Nogué, S.; Rull, V.; Vegas-Vilarrúbia, T. Elevational gradients in the neotropical table mountains: Patterns of endemism and implications for conservation. *Diver. Distrib.* **2013**, *19*, 676–687. [CrossRef]
88. Rull, V.; Vegas-Vilarr´ubia, T. Unexpected biodiversity loss under globalwarming in the Neotropical Guayana Highlands. *Glob. Chang. Biol.* **2006**, *12*, 1–9. [CrossRef]

MDPI

St. Alban-Anlage 66

4052 Basel

Switzerland

Tel. +41 61 683 77 34

Fax +41 61 302 89 18

www.mdpi.com

Diversity Editorial Office

E-mail: diversity@mdpi.com

www.mdpi.com/journal/diversity

9 783036 541655